# Tissue Elasticity Imaging

## Volume 2: Clinical Applications

# Tissue Elasticity Imaging

## Volume 2: Clinical Applications


*Edited by*

**S. Kaisar Alam**

*Imagine Consulting LLC*
*Dayton, NJ, United States*

*The Center for Computational Biomedicine Imaging and Modeling (CBIM)*
*Rutgers University*
*Piscataway, NJ, United States*

**Brian S. Garra**

*Division of Imaging, Diagnostics, and Software Reliability*
*Office of Science and Engineering Laboratories, Center for Devices and Radiological Health, FDA, Silver Spring, MD, United States*

Elsevier
Radarweg 29, PO Box 211, 1000 AE Amsterdam, Netherlands
The Boulevard, Langford Lane, Kidlington, Oxford OX5 1GB, United Kingdom
50 Hampshire Street, 5th Floor, Cambridge, MA 02139, United States



**Library of Congress Cataloging-in-Publication Data**
A catalog record for this book is available from the Library of Congress

**British Library Cataloguing-in-Publication Data**
A catalogue record for this book is available from the British Library

ISBN: 978-0-12-809662-8

For information on all Elsevier publications visit our website at
https://www.elsevier.com/books-and-journals

*Publisher:* Susan Dennis
*Acquisition Editor:* Anita Koch
*Editorial Project Manager:* Lindsay Lawrence
*Production Project Manager:* Paul Prasad Chandramohan
*Cover Designer:* Matthew Limbert

Typeset by TNQ Technologies

# Contents

Contributors........................................................................................ xi
About the editors.................................................................................. xiii
Foreword.............................................................................................. xv
Preface ................................................................................................ xvii
Acknowledgments.................................................................................. xix

**CHAPTER 1 Clinical elasticity estimation and imaging: applications and standards ..................................... 1**

*Brian S. Garra*

**1.** Introduction ........................................................................ 1
**2.** Elastography: different methods with different capabilities......... 2
**3.** Clinical applications of strain and shear wave elastography ........ 6
3.1 Strain elastography.......................................................6
3.2 Shear wave elastography ................................................7
3.3 Strain rate imaging.......................................................8
**4.** Elastography applications: translation to clinical use ................ 9
**5.** Future directions ..............................................................15
References.............................................................................16

**CHAPTER 2 Breast elastography ..............................................21**

*Richard G. Barr*

**1.** Introduction/background.....................................................21
**2.** Principles/techniques .........................................................21
2.1 Strain elastography...................................................... 21
2.2 Shear wave elastography .............................................. 27
2.3 Guidelines .................................................................. 29
**3.** Diseases and applications ..................................................29
3.1 Benign lesions............................................................. 29
3.2 Malignant lesions......................................................... 36
3.3 Others........................................................................ 40
**4.** Opportunities....................................................................41
4.1 Quality measures ......................................................... 41
**5.** Artifacts and limitations.....................................................42
5.1 Bull's-eye artifact ........................................................ 42
5.2 Blue, green, and red artifact.......................................... 43
5.3 Worm artifact.............................................................. 43
5.4 Sliding artifact ........................................................... 43
5.5 Bang artifact............................................................... 43

**6.** Summary/conclusions ........................................................43
References........................................................................44

**CHAPTER 3 Clinical applications of elastographic methods to improve prostate cancer evaluation........................47**
*Eduardo Gonzalez, Fanny L. Casado and Benjamin Castaneda*
**1.** Introduction ........................................................................47
**2.** Static deformation by compression........................................49
2.1 Strain elastography................................................... 49
**3.** Dynamic deformation exerted by external mechanical vibrators ........................................................................52
3.1 Sonoelastography........................................................ 52
3.2 Vibroelastography....................................................... 54
3.3 Magnetic resonance elastography................................... 55
**4.** Excitation by acoustic radiation force....................................57
4.1 Acoustic radiation force impulse imaging ........................ 57
4.2 Shear wave elastography.............................................. 58
**5.** Current status and future trends ...........................................60
**6.** Conclusions...........................................................................61
References.................................................................................62

**CHAPTER 4 Cardiovascular elastography .................................67**
*Elisa Konofagou*
**1.** Cardiac imaging.....................................................................67
1.1 Myocardial elastography .............................................. 67
1.2 Electromechanical wave imaging................................... 76
**2.** Vascular imaging...................................................................88
2.1 Stroke......................................................................... 88
2.2 Abdominal aortic aneurysm.......................................... 89
2.3 Pulse-wave velocity (PWV).......................................... 89
2.4 Pulse wave imaging..................................................... 89
2.5 Methods..................................................................... 90
2.6 PWI performance assessment in experimental phantoms..... 92
2.7 Mechanical testing...................................................... 94
2.8 PWI in aortic aneurysms and carotid plaques in human subjects in vivo ................................................ 94
Acknowledgments ........................................................................ 96
References...................................................................................96
Further reading........................................................................... 103

**CHAPTER 5 Ultrasound-based liver elastography** ..................... 109
*Ioan Sporea and Roxana Şirli*
**1.** Introduction to chronic liver disease: etiology, screening, and diagnosis ..................... 109
**2.** Transient elastography ..................... 112
**3.** Point shear wave elastography ..................... 114
3.1 Virtual touch quantification ..................... 115
3.2 ElastPQ technique ..................... 117
3.3 Point shear wave elastography from Hitachi ..................... 118
**4.** Two-dimensional shear wave elastography ..................... 119
4.1 Supersonic shear wave imaging ..................... 119
4.2 2D-SWE.GE ..................... 119
4.3 Two-dimensional shear wave elastography from Toshiba ... 121
**5.** Comparative studies ..................... 122
**6.** Strain elastography ..................... 123
References ..................... 124

**CHAPTER 6 Thermal therapy monitoring using elastography** .... 135
*Kullervo Hynynen*
**1.** Introduction ..................... 135
**2.** Principles and techniques ..................... 136
2.1 Thermal effects on tissues ..................... 136
2.2 Clinical use of thermal exposures for therapy ..................... 137
2.3 Need for exposure monitoring ..................... 139
2.4 Principles of elastography for thermal therapy monitoring ..................... 140
**3.** Elastographic methods for thermal therapy monitoring ..................... 143
**4.** Diseases and applications ..................... 144
4.1 Tumor treatments ..................... 144
4.2 Prostate ..................... 145
4.3 Cardiac ablation ..................... 145
**5.** Future opportunities ..................... 146
**6.** Conclusion ..................... 147
References ..................... 147

**CHAPTER 7 Thyroid elastography** ..................... 157
*Manjiri Dighe*
**1.** Thyroid pathology ..................... 157
**2.** Strain elastography ..................... 157
2.1 Introduction ..................... 157
2.2 Strain histograms ..................... 158

2.3 Strain ratio......160
2.4 Examination technique......160
2.5 Interobserver and intraobserver variabilities......164
2.6 Practical advice, tips, and limitations......164
**3.** Shear wave elastography......166
3.1 Introduction......166
3.2 Different methods of shear wave elastographic imaging of the thyroid......166
3.3 Review of literature......167
3.4 Interobserver and intraobserver variabilities......168
3.5 Examination technique......169
3.6 Practical advice and tips......169
3.7 Interpretation of results......170
**4.** Artifacts in thyroid elastography......172
**5.** Conclusion......174
References......174

**CHAPTER 8 Elastography applications in pregnancy......181**
*Helen Feltovich*
**1.** Introduction......181
**2.** The cervix......181
2.1 Strain elastography......182
2.2 Shear wave elasticity imaging......185
**3.** The placenta......188
3.1 Strain elastography......188
3.2 Shear wave methods......189
**4.** Conclusions......192
References......192

**CHAPTER 9 Musculoskeletal elastography......197**
*M. Abd Ellah, M. Taljanovic and A.S. Klauser*
**1.** Introduction......197
**2.** Compression (strain) elastography......198
**3.** Shear wave elastography......199
**4.** Transient elastography......199
**5.** Applications of sonoelastography in the musculoskeletal system......200
5.1 Tendon disorders......200
5.2 Achilles tendon......200
5.3 Lateral epicondylitis......202
5.4 Medial epicondylitis......204

5.5 Patellar tendinopathy ........205
5.6 Quadriceps tendinopathy ........206
5.7 Rotator cuff tendinopathy ........206
5.8 Finger tendon and trigger fingers ........207
5.9 Joints and ligaments ........207
**6.** Muscles ........210
**7.** Nerves ........212
**8.** Plantar fascia ........213
**9.** Tumor and tumorlike masses ........214
**10.** Future perspectives ........215
**11.** Limitations and conditions of good practice ........215
11.1 Strain elastography/sonoelastography ........215
11.2 Shear wave elastography ........216
**12.** Conclusion ........217
References ........217

Index ........225

# Contributors

**M. Abd Ellah**
Department of Radiology, Medical University of Innsbruck, Innsbruck, Tyrol, Austria; Department of Diagnostic Radiology, South Egypt Cancer Institute, Assiut University, Assiut, Egypt; Radiology/Neuroradiology Department Rehabilitationskliniken Ulm, Germany

**Richard G. Barr**
Radiology, Northeast Ohio Medical University, Radiology Consultants Inc., Youngstown, OH, United States

**Fanny L. Casado**
Instituto de Ciencias Ómicas y Biotecnología Aplicada, Pontificia Universidad Católica del Perú, Lima, Perú

**Benjamin Castaneda**
Laboratorio de Imágenes Médicas, Pontificia Universidad Católica del Perú, Lima, Peru

**Manjiri Dighe**
Department of Radiology, Abdominal imaging section, University of Washington, Seattle, WA, United States

**Helen Feltovich**
Maternal-Fetal Medicine, Intermountain Healthcare, Provo, UT, United States; Quantitative Ultrasound Laboratory, Department of Medical Physics, University of Wisconsin, Madison, WI, United States

**Brian S. Garra**
Division of Imaging, Diagnostics and Software Reliability, Office of Science and Engineering Laboratories, CDRH, FDA, Silver Spring, Maryland, MD, United States

**Eduardo Gonzalez**
Laboratorio de Imágenes Médicas, Pontificia Universidad Católica del Perú, Lima, Peru; Department of Biomedical Engineering, Johns Hopkins University, Baltimore, MD, United State

**Kullervo Hynynen**
Physical Sciences Platform, Sunnybrook Research Institute; Department of Medical Biophysics and Institute of Biomaterials and Biomedical Engineering, University of Toronto, Toronto, ON, Canada

**A.S. Klauser**
Department of Radiology, Medical University of Innsbruck, Innsbruck, Tyrol, Austria

**Elisa Konofagou**
Columbia University, New York, NY, United States

**Roxana Şirli**
Department of Gastroenterology and Hepatology, Victor Babeş University of Medicine and Pharmacy, Timişoara, Romania

**Ioan Sporea**
Department of Gastroenterology and Hepatology, Victor Babeş University of Medicine and Pharmacy, Timişoara, Romania

**M. Taljanovic**
Department of Medical Imaging, University of Arizona, College of Medicine, Banner- University Medical Center, Tucson, AZ, United States

# About the editors


**S. Kaisar Alam, Ph.D.**

President and Chief Engineer, Imagine Consulting LLC, Dayton, NJ, United States

Visiting Research Faculty, Center for Computational Biomedicine Imaging and Modeling (CBIM), Rutgers University, Piscataway, NJ, United States

Adjunct Faculty, Electrical & Computer Engineering, The College of New Jersey (TCNJ), Ewing, NJ, United States


Dr. S. Kaisar Alam received his B.Tech (Honors) from IIT, Kharagpur, India. Following a 3-year stint as a Lecturer at RUET, Bangladesh, he came to the University of Rochester, Rochester, New York, for graduate studies and received his M.S. and Ph.D. degrees in electrical engineering in 1991 and 1996, respectively. After spending 3 years (1995–1998) as a postdoctoral fellow at the University of Texas Health Science Center, Houston, Dr. Alam was a Principal Investigator at Riverside Research, New York, from 1998 to 2013, working on a variety of research topics in biomedical imaging. He was the Chief Research Officer at Improlabs Pte Ltd, an upcoming tech startup in Singapore until 2017. Then he founded his own consulting company for biomedical image analysis, signal processing, and medical imaging. He has also been involved in training and mentoring high school students. He has been a visiting research professor at CBIM, Rutgers University, Piscataway, New Jersey (since 2013), a visiting professor at IUT, Gazipur, Bangladesh (2010 and 2012), and an adjunct faculty at The College of New Jersey (TCNJ), Ewing, New Jersey (since 2017).

Dr. Alam has been active in research for more than 30 years. His research interests include diagnostic and therapeutic applications of ultrasound and optics, and signal/image processing with applications to medical imaging. The areas of his most active research include elasticity imaging and quantitative ultrasound; he is among a few researchers with experience in both quasistatic and dynamic elasticity imaging. Dr. Alam has written over 40 papers in international journals and holds several patents. He is a coauthor of the textbook *Computational Health Informatics* (to be published late 2019 or early 2020 by *CRC Press*). He is a Fellow of AIUM, a Senior Member of IEEE, and a Member of Sigma Xi, AAPM, ASA, and SPIE. Dr. Alam has served in the AIUM Technical Standards Committee and the Ultrasound Coordinating Committee of the RSNA Quantitative Imaging Biomarker Alliance (QIBA). He is an Associate Editor of *Ultrasonics* (Elsevier) and *Ultrasonic Imaging* (Sage). Dr. Alam was a recipient of the prestigious Fulbright Scholar Award in 2011–2012.

**Brian S. Garra, M.D.**

Division of Imaging, Diagnostics, and Software Reliability, Office of Science and Engineering Laboratories, Center for Devices and Radiological Health, FDA, Silver Spring, MD, United States

Dr. Brian S. Garra completed his residency training at the University of Utah and spent 3 years as an Army radiologist in Germany before returning to Washington DC and the National Institutes of Health in the mid 1980s. After 4 years at the NIH, he joined the faculty of Georgetown University as Director of Ultrasound. In 1998, he left Georgetown to become Professor & Vice Chairman of Radiology at the University of Vermont/Fletcher Allen Healthcare. In 2009, Dr. Garra returned to the Washington DC area as Chief of Imaging Systems & Research in Radiology at the Washington DC Veterans Affairs Medical Center. In April 2010, he also joined the FDA as an Associate Director in the Division of Imaging and Applied Mathematics/OSEL. In 2018, he left the VA and currently splits his time between the FDA and private practice radiology in Florida.

Dr. Garra's clinical activities include spinal MRI and general ultrasound. His research interests include PACS, digital signal processing, and quantitative ultrasound including Doppler, ultrasound elastography, and photoacoustic tomography. He was chair of the FDA radiological Devices Panel from 1999 to 2002 and has been involved in the approval of several new technologies including high resolution breast ultrasound, the first digital mammographic system, the first computer-aided detection system for mammography, and the first computer-aided nodule detection system for chest radiographs as well as the ultrasound contrast agent albunex. He also led the team that developed the AIUM breast ultrasound accreditation program, and helped develop the ARDMS registry in breast ultrasound. He is currently also Vice Chairman of the Ultrasound Coordinating Committee of the RSNA Quantitative Imaging Biomarker Alliance (QIBA) and is the Principal Author of the forthcoming QIBA Ultrasound Shear Wave Speed Profile which will provide a standard approach to acquisition of shear wave speed data for research, clinical application, and regulatory testing.

# Foreword

Given the heavy relatively successful use of manual palpation over the past few thousand years, the ultrasound community, and medicine in general, was very excited to understand and realize the possibility of measuring and imaging the stiffness of tissues. This included tissues too deep for manual palpation. Improving the spatial and quantitative fidelity of elasticity images was addressed aggressively. Also pursued were many extensions related to elastic properties, such as the anisotropy of elasticity, the complex elastic modulus (viscous and elastic components), and elasticity as a function of time under compression.

This two-volume book *Tissue Elasticity Imaging* extensively covers the principles, implementation, and applications of all these approaches to image the biomechanical properties of tissues. The achieved and future biomedical applications of these many capabilities are also well explained, as are important optical and magnetic resonance imaging techniques that followed, and that sometimes leaped ahead of the many ultrasound developments.

These rapid advances are brought to life for the reader of these books by physicians and other imaging scientists and engineers who made leading advances in each of the covered areas. I initially wished to list key lead authors with a summary of their contributions, but that would essentially be repeating most of the table of contents. The editors of these books, Drs. Brian Garra and S. Kaisar Alam, excelled in recruiting the many luminaries to author the various chapters, defining the topics, and editing the work for readability by the target audience of imaging scientists, engineers, entrepreneurs, clinicians, and operators of the systems. The work should serve as a definitive reference for those teaching and those writing shorter explanations for various groups. This is a much-needed work in the field. Luckily, it will not be the last, as advances are and will continue to be made.

**Paul L. Carson, Ph.D.**
University of Michigan
Ann Arbor, Michigan
United States
July 14, 2019

# Preface

Volume 2 of Tissue Elasticity Imaging reviews the state of clinical elastographic applications as of late 2018. It begins with an introductory chapter that contains a brief explanation of elastography basics, discusses attempts to standardize quantitative elastography for applications such as detection and imaging of liver fibrosis, and discusses some of the reasons why elastography is not used more widely (at least in the United States) 20 years after the technique was first introduced. Potential avenues for future work in elastography are also discussed. The later chapters discuss various specific applications and their potential for widespread clinical use from the perspective of leaders in clinical elastography who author those chapters.

As clinical elastography in all of its forms is progressing rapidly and evolving as new clinical applications are explored, some of the specifics in each chapter may quickly become somewhat outdated. The reader can however use the authors of the chapters and the references given in each chapter as resources to obtain guidance on the latest in each area of clinical elastography. The reader may contact the chapter and reference authors by email or other means and may search via google or other search engines on the type of elastography and the author name(s) to obtain new information on each clinical elastographic application and information on new applications coming online.

I hope that clinicians new to elastography will find this volume useful as a way to get introduced to the use of elastography in your area of clinical interest and expertise. If the brief explanations of the basic science and methods by which elastograms are created are not sufficient, I suggest that the reader refer to the more detailed discussions in Volume 1. The readers can also consult the companion website for this book at https://www.elsevier.com/books-and-journals/book-companion/978-0-12-809662-8.

For those clinicians already using elastography, I hope you will find in this volume some tips on how to refine your elastographic techniques and insight into new ways to use elastography that you may not have already considered. For basic scientists interested in elastography, this volume will introduce to you the wide variety of clinical applications and problems with using the current technology for those applications. It should help you to explore ways in which elastography technology can be improved to provide easier to use elastographic systems that provide higher quality imaging and quantification for clinical use.

I look forward to participate with you all in helping elastography realize its vast potential for improving the health of patients. The future will be exciting!

**Brian S. Garra**
May 31, 2019

**Chapter reviewers:**
*Volume 2: Clinical applications*
Arrigo Fruscalzo
~~Caroline Malecke~~
David Cosgrove (now deceased)
Eleni E Drakonaki
Giovanna Ferraioli
Jonathan Langdon
Manjiri D. Dighe
Matthew Urban
Mohammad Mehrmohammadi
Ogonna K. Nwawka
Qian Li
Remi Souchon
Richard Barr
Richard Lopata
Siddhartha Sikdar

# Acknowledgments

Editing this important reference book was much harder and at the same time, much more fulfilling than I could have ever imagined. First and foremost, I want to thank the Almighty. He gave me the power to pursue my dreams and this book. I could never have done this without my faith in Him. This book happened because He wished it to be.

I am ever grateful to my deceased parents who always encouraged me to pursue my dreams. Thank you my dear wife, daughter, and son for your constant patience and support, especially during difficult times. My younger brother and sister have been my source of strength since they were born. Their spouses and children have been a source of inspiration and joy for me. I have a large number of uncles, aunts, cousins, nephews, and nieces, who have always supported me. I am lucky to have all of you as my family.

I also want to thank many individuals whom I regard as mentors and friends. They include my childhood mentor Dr. Kazi Khairul Islam, my doctoral advisor Dr. Kevin J. Parker, my postdoc supervisor late Dr. Jonathan Ophir, my former supervisors Dr. Ernie Feleppa and late Dr. Fred Lizzi, and my coeditor Dr. Brian Garra. (Brian also provided the artwork used to design the cover.).

I am also indebted to many family members, friends, and colleagues, and it would be impossible to thank them all individually. I am lucky to have been your family, friend, and colleague. Thank you all!

Last but not the least, thanks to everyone in the Elsevier team. Special thanks to our Acquisition Editor (Dr. Anita Koch), Editorial Project Managers (Lindsay Lawrence, Jennifer Horigan, and Amy Clark), Project Manager (Paul Prasad Chandramohan), Cover Designer (Matthew Limbert), and many other individuals who worked behind the scenes to make this book a reality.

**S. Kaisar Alam**
Dayton, New Jersey, USA
October 1, 2019

CHAPTER

# Clinical elasticity estimation and imaging: applications and standards

1

**Brian S. Garra**

*Division of Imaging, Diagnostics and Software Reliability, Office of Science and Engineering Laboratories, CDRH, FDA, Silver Spring, Maryland, MD, United States*

## 1. Introduction

Since the initial development of elastographic imaging in the early 1990s [1], the method has been tested on a wide variety of disorders including cancers and noncancerous diseases both focal and diffuse. Although sonoelasticity imaging [2] was developed slightly earlier and was maturing at the same time [3], elastography in the form of strain elastography of the breast was the first method to be adopted clinically because it requires no external vibration source, it can produce high-quality images when performed correctly, and it is able to accurately classify breast tumors as benign or malignant [4–7]. With the subsequent development of shear wave elastography (SWE), evaluation of breast lesions using SWE was generally successful [8], although reports of significant numbers of false-negative cases have more recently appeared [9,10]. As SWE is capable of providing quantitative estimates of stiffness, the technique is being applied successfully to the evaluation of liver fibrosis [11–13], demonstrating its usefulness in the evaluation of diffuse liver disease.

As is often the case with new technologies, enthusiastic initial reports on the breast gave way to more sobering studies with uneven results, and currently breast elastography use is considerable, but not as widespread as it could be. Elastographic evaluation of most other disorders remains in the experimental stage. An exception is in liver evaluation where SWE use for the estimation of liver fibrosis has quickly assumed a dominant role in evaluating diffuse liver disease. It has become one of the most important tools for management of patients with chronic hepatitis [14], replacing liver biopsy in many if not most centers for evaluation and management of chronic hepatitis C.

Although elasticity imaging and quantification in all its forms has produced variable clinical results so far, the technology is maturing, is evolving, and shows great promise for future success in several areas of clinical practice. The term

Tissue Elasticity Imaging. https://doi.org/10.1016/B978-0-12-809662-8.00001-2

elastography has become the most common name for all elasticity imaging and quantification methods and so will be used throughout the remainder of this chapter.

## 2. Elastography: different methods with different capabilities

All elastography is based on tissue response to the applied pressure or pressure waves. However, different methods of creating elastograms exist, which produce elastograms displaying tissue stiffness differently. Elastographic methods are currently classified in two main ways (Fig. 1.1). The first scheme classifies the major methods into three types of mechanical deformation applied to tissue.

The first method, termed "static" or "quasi-static" elastography, is performed by compressing tissue or other material slowly and measuring the movement of the material in response to the compression at different distances from the compressor. Each compression therefore produces a "pre"dataset and a "post"dataset that are compared to one another to measure tissue movement as a function of depth.

In the second method, termed "dynamic" elastography, a vibration is applied to the tissue (amounting to multiple small compressions at some frequency). The resulting waves of tissue displacement are displayed or their speed of travel through the tissue is measured and reported or imaged. Sonoelasticity imaging is one form of dynamic elastography, but the far more widely used method is SWE that uses either a mechanical vibrator (as in magnetic resonance elastography [15]) or an acoustic radiation force impulse (ARFI) to stimulate the production of shear waves in tissues [16].

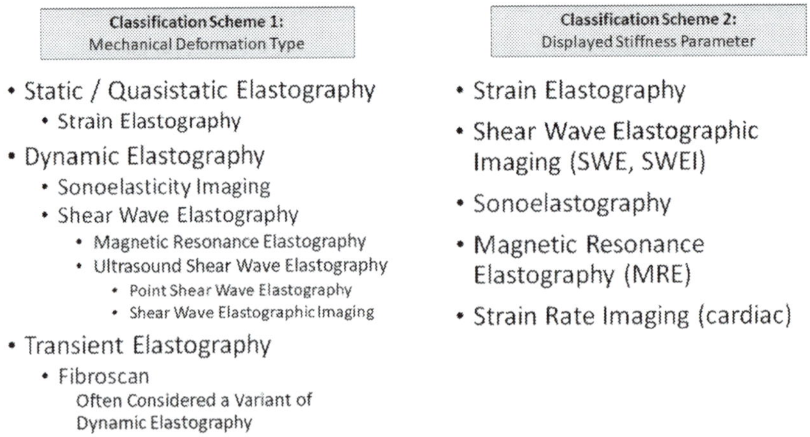

**Classification Scheme 1:**
Mechanical Deformation Type

- Static / Quasistatic Elastography
  - Strain Elastography
- Dynamic Elastography
  - Sonoelasticity Imaging
  - Shear Wave Elastography
    - Magnetic Resonance Elastography
    - Ultrasound Shear Wave Elastography
      - Point Shear Wave Elastography
      - Shear Wave Elastographic Imaging
- Transient Elastography
  - Fibroscan
    Often Considered a Variant of Dynamic Elastography

**Classification Scheme 2:**
Displayed Stiffness Parameter

- Strain Elastography
- Shear Wave Elastographic Imaging (SWE, SWEI)
- Sonoelastography
- Magnetic Resonance Elastography (MRE)
- Strain Rate Imaging (cardiac)

**FIGURE 1.1**

Elastography classification.

In the third method, termed "transient" elastography [17], successive rapid compressions are applied and after each compression, the resulting wave of tissue displacement (shear wave) is observed using a-mode ultrasound and the speed of the wave is measured.

The second classification scheme is based on the parameter that is imaged or computed as a surrogate for the Young's (elastic) modulus. In the quasi-static method, tissue displacement is measured by comparison of the pre- and postradiofrequency ultrasound data, and from the displacement data, an image of the strain values is created. For this reason the method is known as strain elastography.

In the most common dynamic elastography method, the speed of travel of a shear wave is measured and an image of the shear wave speed (SWS) values is created. This method is therefore termed SWE.

In another dynamic approach well suited to rapidly moving tissues, Doppler ultrasound or other methods are used to track the change in tissue velocity over short-distance ranges giving the strain rate (the change in strain over short-distance intervals). This method is therefore termed strain rate imaging [18].

The second classification scheme is currently the most popular in medical imaging because the parameter being imaged is explicitly described.

The transient elastography approach previously mentioned does not produce an image at all, only a single numerical value for tissue stiffness. Strain rate imaging is primarily performed in the heart where Doppler ultrasound is already widely used.

With respect to strain elastography and SWE, there are advantages and disadvantages to each method.

For strain elastography the advantages are that it is easier to implement at a lower cost on current ultrasound systems because it involves only different processing of received radiofrequency signals without changes to hardware. In addition, the signal processing is relatively simple and not processor-intensive. It is a relatively mature technique having been initially implemented over 25 years ago and produces high-quality images when performed properly.

Disadvantages of the method include the following:

1. The amount of tissue displacement and strain depend on the amount of pressure applied (the "stress" in Hooke's law [19] that states that "stress is proportional to the strain" or $F = kx$, where F is the stress in units of pressure, k is the proportionality constant, and $x =$ strain). Strain is the change in length or thickness of a material in response to the force. For freehand real-time elastography the force applied cannot be reliably reproduced from one image to another. The strain values therefore cannot be used to estimate absolute tissue stiffness because the strain depends on the tissue stiffness and the applied pressure (which is unknown). For this reason, strain images are not mapped to a physical property of tissue but are mapped relative to the median value of the specific image they belong to, much like MRI images. These images of "relative" strain are excellent for the detection and characterization of focal lesions

but cannot be used to estimate overall stiffness or diffuse stiffening of tissue filling an entire image.

2. High-quality strain elastograms require meticulous attention to the acquisition technique and even a minor variation in the compression technique can result in very poor elastograms with many artifacts and much noise (Fig. 1.2). The good thing about strain elastography is that images can be generated in real time, making large numbers of images available for review. However, of the 20–30 images generated during a single compression, only about 3–4 may be of high enough quality to be interpretable.
3. Manufacturers typically compute and display a quality index for each image to help inform the user of the images that should not be used for diagnosis due to quality issues (Fig. 1.3). But many times a given lesion may appear quite different on each of the several "good" images, making it hard for a reader to determine which findings are real and which are not.

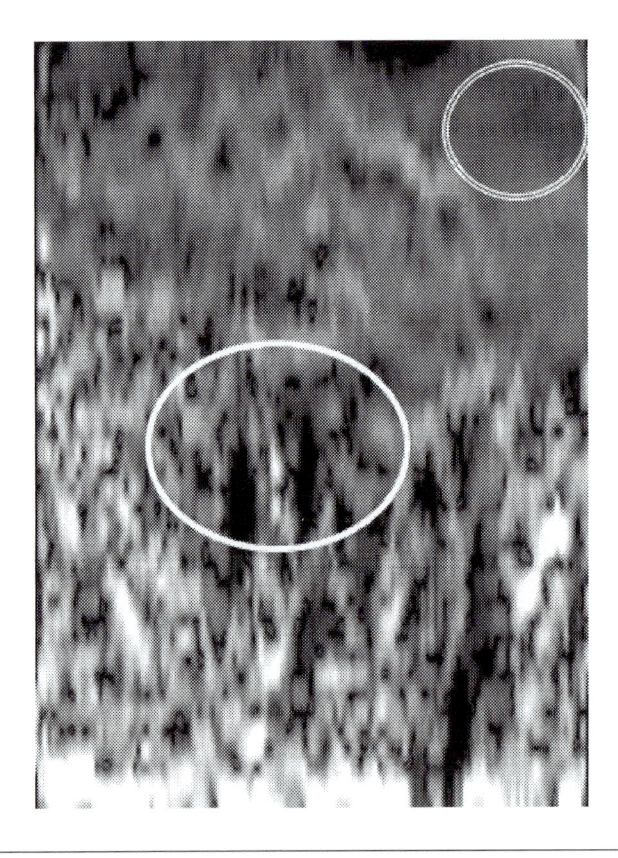

**FIGURE 1.2**

Poor-quality strain elastogram in which most of the image is dominated by noise. A noisy section is shown in the white ellipse and a low-noise area is outlined by the small double-walled ellipse.

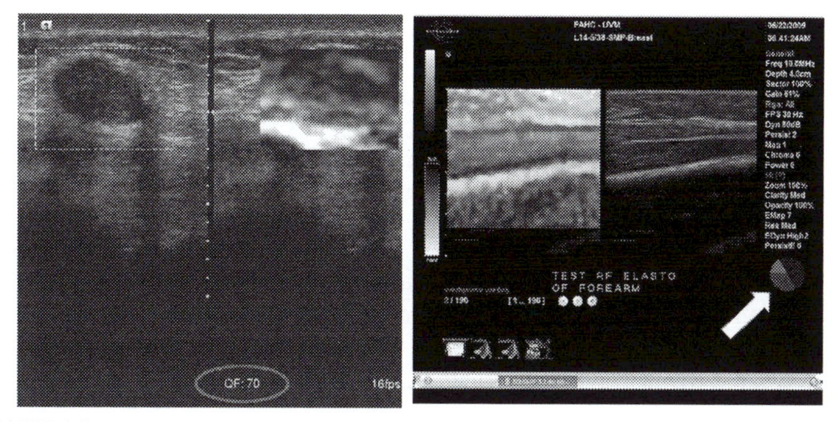


**FIGURE 1.3**

Two examples of quality indicators. (A) A quality factor (QF). If the QF is below 70 the elastogram image is not displayed. (B) The arrow points to the pie chart quality indicator. The more the sections of the pie, the higher the quality.

SWE uses SWS as a surrogate for Young's (elastic) modulus because, like ultrasound compressional waves, shear wave propagation speed is dependent on the stiffness of the medium through which the waves are traveling. Shear wave speed can be converted to shear modulus by the equation $V_S = \sqrt{\frac{G}{\rho}}$ (where $V_s$ is the shear wave velocity, $\rho$ is the density of the material, and $G$ is the shear modulus) and to the elastic or Young's modulus ($E$) by the equation $E = 3\rho V_S^2$ where $\rho$ is the tissue density and $V_s$ is the shear wave velocity when the material is homogeneous and isotropic. Therefore, SWS images and region of interest quantification may be displayed as SWS, shear modulus, or Young's modulus and can be used to estimate the overall stiffness in an image or region rather than the relative stiffness of one region of an image relative to the overall image value. This inherent quantification capability allows SWE to be used for the evaluation of diffuse liver diseases, such as liver fibrosis and cirrhosis, and diffuse diseases of other organs if they are composed of isotropic tissue (not the kidney; for example see [20]).

Additionally SWE requires no skilled compression by the operator because the compression is most often applied by an acoustic radiation force pulse rather than manually using the ultrasound probe. This leads to more consistent and reproducible results even when different operators are acquiring the SWS data.

SWE may also be used for the evaluation of focal diseases, but in those cases (as for tissue anisotropy–see above), SWS or shear modulus should be reported because the simple equation for conversion to Young's modulus no longer holds. However, FDA allows the conversion, as reporting Young's modulus in kilopascals was already widespread in the literature when the first devices were submitted for FDA review and clearance. The thinking was that only allowing reporting of SWS would require operators and physicians to convert the tables in units of kilopascals in the literature back to SWS, potentially leading to confusion and diagnostic errors.

Disadvantages of SWE are that it is more expensive to implement requiring new hardware because the higher intensity ARFI pulse that creates the shear waves must be generated. As noted earlier, Young's modulus estimation currently makes the assumption of homogeneous and isotropic material, which is violated in most tissues. For focal diseases, SWE may yield erroneous Young's modulus results due to reflections at the lesion boundary interface, which result in wave cancellation and erroneous SWS estimates with current software. Unfortunately, the erroneous SWS values lead to incorrectly low stiffness values in the breast. This is troubling because low SWS values could result in an incorrect diagnosis of a benign lesion when in fact a cancer is present. Incorrect SWS values are rarely seen in mass lesions outside the breast. This may be due to the smaller difference in stiffness between the lesion and the surrounding tissue, leading to fewer boundary reflections. The incorrect SWS problem has not been systematically explored for lesions in organs other than the breast.

## 3. Clinical applications of strain and shear wave elastography

Despite the disadvantages of SWE noted earlier, the method has become the dominant form of elastography owing to its quantification capability and somewhat greater ease of acquisition. Still, both types of imaging continue to be applied clinically.

### 3.1 Strain elastography

Although strain elastography has been tested in almost every organ and disease, breast mass evaluation to diagnose breast cancer was the first major application and remains the most popular application of strain imaging. Breast cancers, especially the most common, invasive ductal carcinoma, are very often markedly stiffer than the surrounding glandular and fatty tissues. This plus the fact that microscopic invasion into the surrounding tissues is common with resulting fibrotic (desmoplastic) reaction makes cancers very easy to identify in most cases using a stiffness estimate and the difference between the sonographically identified lesion size and the elastographic lesion size. Some workers have reported nearly perfect performance of strain elastography in the correct classification of focal breast masses [6] but most studies report lower accuracy with receiver operating characteristic areas being in the 0.90–0.92 range. The most widespread lesion characterization criteria are from the Tsukuba scale originating in Japan using the Hitachi equipment. Unfortunately the reported results of strain elastography have been uneven [21] and are probably even more so in normal clinical practice.

Other approaches to strain elastogram quantification besides the measurement of pressure applied during creation of the strain image have been tested. One can use the ratio of two strain values at different parts of the image as an indicator of stiffness. This is the most common approach and may provide higher accuracy than use of the Tsukuba scale.

It is also possible to take the ratio of the axial strain and lateral strain. Lateral strain is related to the sideways movement of a material when axial compression is applied. For homogeneous, isotropic, elastic materials, the axial dimension decreases during compression, so the lateral dimension must increase. This results in lateral movement of the material during compression, which can be converted into strain values. For a three-dimensional object, compressing in the axial direction produces lateral movement in the other two dimensions with half of the movement in each direction. Thus in a purely elastic material, the ratio of axial to lateral strain, known as the Poisson's ratio, in the plane of the ultrasound transducer is 0.5. If the material is not isotropic but stiffer in one direction, or if the material is not purely elastic, the Poisson's ratio will be either higher or lower than 0.5 depending on the material properties. As most tissues and materials are not purely elastic nor isotropic, it is reasonable to expect that the Poisson's ratios of different tissues and tissue states will vary and this property could be used to detect and classify abnormalities. Because the lateral strain is harder to measure accurately, this technique has not yet been tested clinically.

Another approach for the quantification of strain elastography is to analyze the strain histogram of a lesion normalized to the maximum strain value in the entire strain image. The analysis can be performed at a single compression level or at multiple levels [22]. Various parameters related to the distribution of relative strain values can then be computed. Some of these include the mean relative strain, standard deviation, percentage of values within a range of strain values relative to the mean, and skewness and kurtosis of the distribution [23]. This approach has been shown to be as effective as strain ratios without having to place reference regions of interest [24]. In preliminary work, it also performs well in the evaluation of liver fibrosis [22] and is equivalent to or better than SWE [25] for the quantification of elasticity in phantoms.

Multiple other lesions in other organs have been studied using strain elastography, with promising results in the evaluation of prostate lesions, lymph nodes, thyroid nodules, and even pancreatic lesions (via endoscopic elastography) [26]. Abnormalities in these organs typically appear as areas of increased stiffness but without the size discrepancy between B-mode ultrasound and elastography that is seen in breast cancer. But for most of these structures, the technical difficulty of obtaining high-quality strain images has limited the adoption of the method by the medical community. The advent of SWE, with its promise of quantification and easier data acquisition, has caused interest in strain elastography to wane.

## 3.2 Shear wave elastography

Because of its ability to quantify liver stiffness objectively, SWE has become the dominant method for noninvasively evaluating liver fibrosis and cirrhosis as the causes of liver parenchyma stiffening. Much of this dominance is related to widespread adoption of the FibroScan® nonimaging elastography device, which can be used in a clinician's office and whose performance has been documented in over a 1000 clinical studies [27]. Although the FibroScan® is well known, relatively few clinicians are aware of the fact that ultrasound maging systems can also generate quantitative stiffness estimates based on SWS. In fact, clinicians are more aware of magnetic resonance elastography than they are of shear wave imaging on high-end clinical ultrasound scanners. Several publications have suggested that ultrasonic shear wave imaging can obtain SWS estimates more reliably than can the nonimaging FibroScan® [28,29] in part because one can see the liver and be sure the data are being acquired from the liver and from approximately the same location in the liver. This reduces variability in the stiffness estimates.

The other applications of shear-wave-based elasticity imaging are much less widespread. The method has been used for evaluation of breast masses, even though the underlying assumptions of homogeneous isotropic tissue are violated. There are concerns about this application because of the falsely low stiffness values often obtained from within very stiff breast cancers [30] that could mislead clinicians into incorrectly classifying a cancer as benign. In many cases, the interpreter must look for a small region of high stiffness at the periphery of a breast lesion to make a correct diagnosis of cancer because the remainder of the high-stiffness cancer is masquerading as a low-stiffness lesion. As noted previously, this may be due to reflections and wave cancellation at the boundary of the breast mass.

Studies on the use of SWE in the prostate have been conducted and results have been promising [31] but this application has not yet become popular. There is also interest in using SWE to monitor the changing stiffness of organs such as the uterine cervix during pregnancy to determine whether labor and delivery of a fetus are imminent [32]. Many other organs and pathologic processes have been studied, including lymph nodes, musculoskeletal, cardiac, fat induration, arterial wall, parathyroid glands thyroid nodules, various types of inflammation, and even erectile dysfunction. These applications are still far away from widespread use.

## 3.3 Strain rate imaging

The use of strain rate imaging and quantification is limited to evaluation of myocardial function during the echocardiographic exam using data from either tissue Doppler imaging or two-dimensional (2D) speckle tracking. The global strain parameter has been shown to be more useful than ejection fraction for the evaluation of left ventricular myocardial abnormality [33]. Although the technique is well established, there are many implementations and there has so far been no standardization of the method or of results across different ultrasound scanners [34]. It is not a part of the standard echocardiographic examination and is currently recommended as only a supplemental diagnostic method until further clinical studies can provide data on whether use of the method in decision making will result in better patient outcomes.

## 4. Elastography applications: translation to clinical use

New imaging technologies such as elastography often take an extended period before clinical applications gain widespread popularity in the clinic. For example, color Doppler technology was first developed 10–15 years before its use became widespread. Strain elastography was developed in 1991 and was not widespread till well after 2001. Initial reports covering the use of elastography (both strain and shear wave) in a wide assortment of organs and disease processes [35] showed variable but generally positive results. As the technology matured and become more widespread, users have gotten more widely variable results and for this reason, acceptance of the technology for most of the early applications has stalled. For example, in the breast, initial work showing improved characterization of breast tumors using strain elastography has generally been confirmed, but it is still far from being widely used in the United States, even though it is used much more commonly in other countries. In fact, usage has declined in the United States often in favor of SWE despite the presence of clear evidence that SWE can give misleading results when used for breast mass characterization.

The problems with strain elastography primarily relate to the continuing immaturity of the technology and insufficient user training. Problems related to system performance include

a. lack of 2D arrays to allow tracking of out-of-plane motion;
b. beam width and element spacing in arrays, leading to inaccurate correction for lateral motion;
c. limited correction for lateral motion resulting from suboptimal algorithms and lack of processing speed to provide corrections in real time;
d. inadequate monitoring software to inform the operator of poor-quality strain images.

Typical problems related to training include the following:

a. *Application of precompression*, which leads to increased lateral and out-of-plane motion and decreased stiffness difference between lesions and surrounding tissue. The first occurs because the tissue may be maximally compressed and therefore moves laterally to an area beyond the face of the transducer where the compression pressure is less. The second occurs because softer tissues initially compress more and as they are compressed, they get progressively stiffer. With enough precompression, initially soft tissues become nearly as stiff as initially hard tissues.
   This problem of precompression is exacerbated by the previous experience with grayscale gained by sonographers and sonologists where increased pressure usually produces better images. Special techniques are required to ensure minimal precompression.
b. The opposite problem may also occur where *insufficient compression* is applied resulting in low strain values. This often occurs in the situation where too much

precompression has been applied, the combination producing very poor quality elastograms.

**c.** *Excessive lateral motion during acquisition.* This results from failure to compensate for lateral and out-of-plane motion during acquisition and is usually caused by poor positioning of the patient and transducer. Pressure applied by the transducer should be such that the tissue below the transducer compresses (Fig. 1.4) rather than moving sideways (Fig. 1.5). Lateral motion can be caused by the transducer not being perpendicular to the surface deep to the tissue being compressed—in the case of breast elastography, the curving bony and muscular part of the chest wall is the surface against which the breast tissue is being compressed and which promotes sideways movement (Fig. 1.6). Sideways movement during strain elastography can be seen well as compression on the real-time B-mode image (potential video). To compensate, the transducer can be tilted so more pressure is applied to one end forcing the breast tissue in the opposite direction from the direction of lateral displacement (Fig. 1.7). Out-of-plane motion is compensated for by starting with the chest wall as parallel to the transducer face as possible and then tilting the transducer in the elevational plane to prevent a nodule or other structure from disappearing as it moves out of the imaging plane. This must be done by trial and error.

**d.** *Inadequate instruction.* Inadequate training in elastography was and is inevitable when the sonographer/applications personnel primarily responsible for providing manufacturer training and support only received minimal training themselves. Few other training materials are available and no inexpensive courses are being given. Without proper training, users quickly become discouraged and stop using the technology. Thus strain elastography quickly gained a reputation for being inconsistent and inaccurate for distinguishing benign from cancerous tumors.

Similar problems with accuracy and variability have more recently befallen SWE of both the liver and breast. The main clinical variables affecting the accuracy of SWS estimation are as follows:

**a.** Tissue heterogeneity and anisotropy. The equations used to estimate SWS and calculate Young's modulus assume that the tissue is homogeneous and isotropic [36]. Heterogeneity will affect the measured SWS by altering both the direction of travel of the wave and the smoothness of the wave front. In the breast, a tumor mass clearly creates marked heterogeneity and it is thought that shear wave reflection at the boundary of tumors is the cause of the falsely low SWS values often recorded from breast cancers. To avoid erroneously diagnosing a cancer as benign, use of grayscale features and strain elastography (if available) is critical. In the liver, the actual disease process can be patchy, thus not only degrading the measurements but also introducing sampling error that may result in under- or overestimation of the overall severity of liver fibrosis. Anisotropy not only affects the stiffness of tissue when measured in different directions but also requires a more complex equation to transform shear wave velocity to Young's

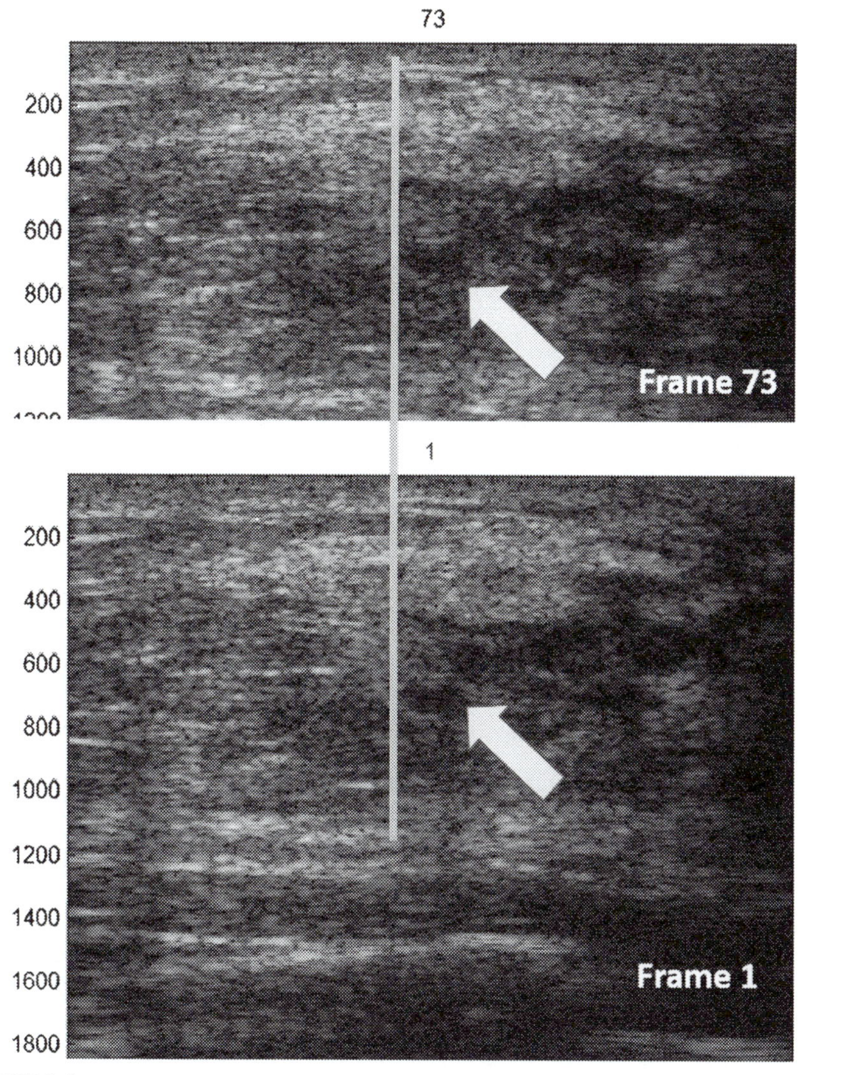


**FIGURE 1.4**

A small hypoechoic lesion (*arrows*) shows no sideways shift on precompression (frame 1) image compared with the postcompression image (frame 73) relative to the vertical line.

modulus. This equation requires additional measurements that cannot be assumed and making assumptions introduces random errors into the stiffness estimation.

**b.** Precompression and tissue movement. As for strain elastography, precompression increases the stiffness of tissues, making the stiffness measurement

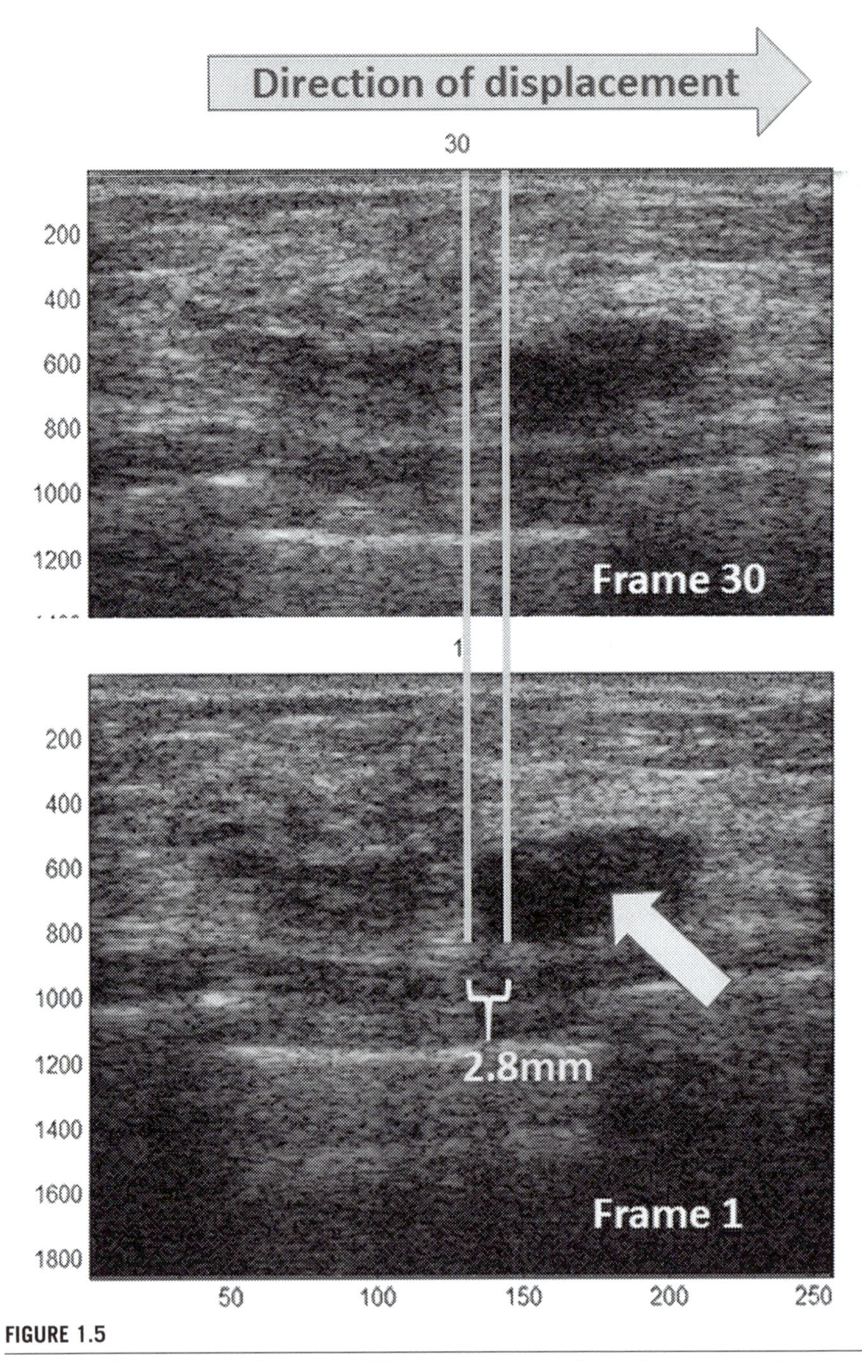


**FIGURE 1.5**

A hypoechoic lesion (*arrow*) displaces 2.8 mm sideways on the postcompression image (frame 30) compared with the precompression image (frame 1).

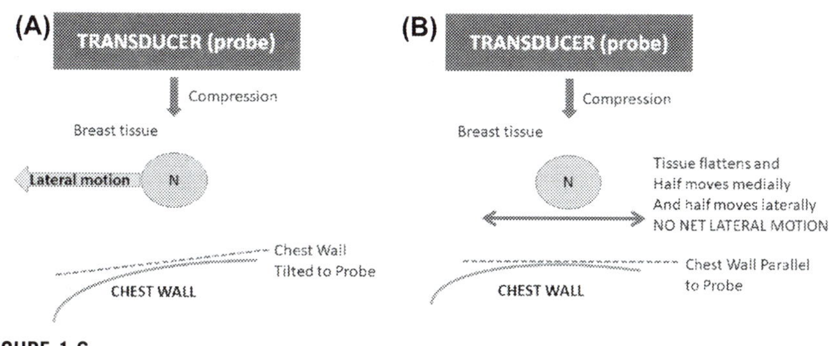


**FIGURE 1.6**

Tissue lateral motion. (A) When the surface against which the transducer is pressing (chest wall in this case) is tilted with respect to the transducer face, the breast tissue and nodule (N) tend to move laterally toward the region of greater tissue thickness and less pressure. (B) If the transducer is parallel to the chest wall the breast tissue and nodule (N) flatten out, with the compressed tissue moving sideways in both directions with no net lateral motion.

operator dependent. Tissue movement introduces errors in the SWS estimate and thus in the estimated stiffness values.

**c.** Depth dependence and limited depth of acquisition. Experimental work both on phantoms and clinical have shown that ultrasound imaging systems give different stiffness values at different depths. This may be due to depth-dependent differences in lateral resolution, interpolation, focusing, or other

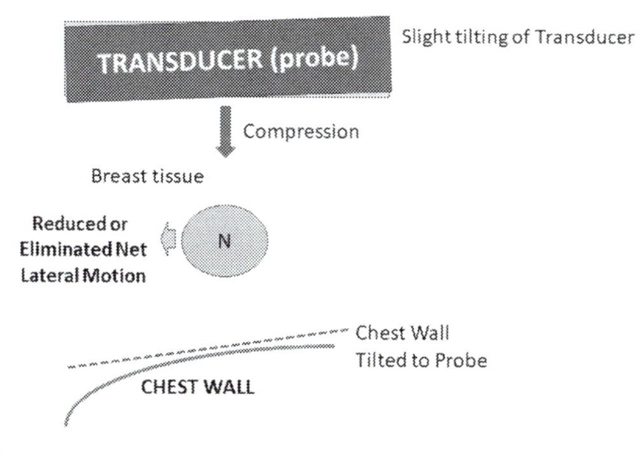


**FIGURE 1.7**

Correcting lateral motion. Tilting the transducer downward slightly on the side toward which the tissue is moving laterally increases pressure on that side slightly tending to force the tissue back, thus reducing lateral motion.

factors not accounted for in the computation software. Also, the depth of acquisition is limited to about 6 cm or less due to attenuation of the ARFI beam, making it impossible to examine many parts of the liver or even the liver itself in large patients.

These and possibly other factors have resulted in some level of mistrust of ultrasound shear wave imaging and quantification, even though use of the technology in the form of FibroScan® is widespread for evaluating diffuse liver disease. Differences in the SWS values given by different machines and differing cutoff values for various levels of fibrosis have also contributed to slow adoption of shear wave imaging using ultrasound imaging systems.

To address the concerns about elastography variation and inaccuracy, various ultrasound organizations have attempted to standardize acquisition and interpretation. The Japan Society of Ultrasonics in Medicine (JSUM) issued practice guidelines for performance of SWE of the liver in 2013 [37], and in the same year, the European Federation of Societies for Ultrasound in Medicine and Biology (EFSUMB) issued recommendations for all forms of elastography including SWE [38,39]. In 2017, EFSUMB issued an update on the clinical use of liver ultrasound elastography dealing specifically with SWS estimation of liver fibrosis [40]; this document provides a comprehensive discussion of SWE technology, diffuse liver disease, elastographic procedures, expected results in liver disease, and even reimbursement issues for Europe.

In the United States, efforts to standardize the acquisition and interpretation of SWE data are being headed by the Radiological Society of North America (RSNA) Quantitative Imaging Biomarker Alliance (QIBA) and by the Society of Radiologists in Ultrasound (SRU). In 2015 the SRU published the report of a consensus panel on the use of SWE for the evaluation of liver fibrosis [29]. The consensus panel developed a suggested acquisition protocol and also provided tables for the clinical interpretation of SWS values for various ultrasound systems because different manufacturers' ultrasound systems provide somewhat different results. In the SRU grading grid, liver fibrosis is divided into three categories: no clinically significant fibrosis risk, moderate risk of clinically significant fibrosis, and high risk of clinically significant fibrosis. This was selected based on the literature showing what levels of fibrosis SWE can distinguish and the criteria used for clinical action such as initiation of antiviral therapy. Although pathologic scoring of liver fibrosis using the Metavir or Batts-Ludwig system is useful, clinical decision making in patients does not always align well with the pathology-based fibrosis scores. In fact, it is possible that the nearly continuous scale of SWS values will be more useful for staging and clinical decision making than the relatively crude five- to nine-point scales employed by the pathology grading systems.

The second major effort at standardization involves the development of an acquisition protocol by the SWS subcommittee of QIBA [41]. The SWS biomarker committee of QIBA was established in 2012 and consists of representatives from clinical practice, academia, industry, and the FDA. The goal of the committee is to produce a well-documented "profile" that describes a standard method of acquisition and analysis of SWS data that will yield SWS results with a known (and relatively small) bias and variance. The SWS biomarker committee first studied the existing literature to identify known sources of variance and then proceeded to conduct phantom tests using both elastic and viscoelastic phantoms [42,43]. Those tests demonstrated that scanners from different manufacturers give similar results in a controlled environment, suggesting that clinical reports of SWS variability may be the result of differences in the acquisition method. The controlled acquisition methods in the profile should help reduce those problems.

A draft profile is undergoing the QIBA approval process for public release. In the profile, performance claims are made for three ranges of fibrosis corresponding to the ranges suggested in the SRU consensus document rather than a single blanket claim. This approach couples performance claims to ranges currently considered clinically relevant. The goal of the profile is to achieve within subject coefficients of variation of 5% in the 1.2−2.2 m/s range, as this corresponds to the range in which treatment decisions will often be made with the help of liver fibrosis estimates. Higher coefficients of variation (approximately 10%) are claimed for SWS values above 2.2 m/s. Variance at high SWS is known to exist, and in this stiffness range (liver cirrhosis), management decisions are usually made based on clinical factors rather than on the degree of cirrhosis.

Although the profile makes claims for bias and variance based on phantom tests, it is expected that clinical testing of the profile will yield similar results or will demonstrate new problems that will be addressed in future versions of the profile. After the public comment stage, a confirmation study to demonstrate that the profile can be executed in a clinical environment is conducted (the "technical confirmation" stage). Then a study is conducted to confirm that the profile used in a clinical environment meets the profile claims (the "claim confirmation" stage). Finally, a large multicenter trial may be conducted to confirm that the profile claims can be achieved in a large number of patients and clinical sites. This final "clinically confirmed" stage is beyond the scope of the QIBA and may never actually be conducted. The profile may however become very useful in research, treatment monitoring, and the clinical environment, even without reaching the clinically confirmed stage.

The QIBA process can be a template for improving the usefulness of other elastographic applications to help ensure that maximum clinical benefit is derived from both strain elastography and SWE.

## 5. Future directions

For SWE, further refinements in acquiring shear wave tracking and computation of SWS can be expected. These include

1. correction of depth dependence of SWS measurements,
2. better agreement in SWS values between manufacturers,

3. better correction for shear wave frequency dependence in viscoelastic in vivo tissues.

It is likely that the possibility of estimating tissue viscosity using SWS acquisition at multiple frequencies will be further explored for clinical usefulness. Also, SWS estimation in heterogeneous tissues and anisotropic tissues, as well as focal lesions, will be refined to give more correct values and lower variance. This should enhance the usefulness of SWE in the evaluation of breast cancer and other lesions.

Also, improved SWS imaging is likely with reduction in the number of imaging artifacts, increase in frame rates, and increase in the maximum depth of acquisition. This is now limited to 6–7 cm but could be increased greatly if acoustic power limits are increased by the FDA.

For strain elastography, application of advanced models to compute relative elastic modulus from strain values (the inverse problem) is expected. The development of pressure sensors integrated with ultrasound probes will allow these relative elastic modulus values to be converted to true Young's modulus, making strain elastographic quantification competitive with SWE once again. With improved training in strain acquisition, the method may replace SWE for many applications.

Other applications of existing elastographic techniques, including the analysis of stiffness anisotropy for musculoskeletal and wound healing applications, may be developed and may become clinical useful. Techniques exploiting the Poisson's ratio imaging may be refined for effective application in lymphedema and may even become a useful tool for the characterization of other lesions and tissues similar to the use of diffusion-weighted imaging in MRI. Finally, shear strain imaging may be further developed to explore tissue boundaries and lesion margins as useful diagnostic tools for lesion characterization and monitoring of slip and nonslip boundaries when tissue adhesions are suspected.

The future value and application of elastographic techniques is vast and further work is already underway to refine techniques and training beyond the first-generation methods of today. This important ongoing work will help ensure that elastography achieves its full potential.

## References

[1] J. Ophir, I. Céspedes, H. Ponnekanti, Y. Yazdi, X. Li, Elastography: a quantitative method for imaging the elasticity of biological tissues, Ultrason. Imaging 13 (2) (1991) 111–134.

[2] R.M. Lerner, K.J. Parker, Sono-elasticity imaging in ultrasonic tissue characterization and echographic imaging, in: J.M. Thyssen (Ed.), Proceedings of the 7th European Communities Workshop, European Communities, Luxembourg, 1987. Nijmegen, The Netherlands.

[3] R.M. Lerner, S.R. Huang, K.J. Parker, "Sonoelasticity" images derived from ultrasound signals in mechanically vibrated tissues, Ultrasound Med. Biol. 16 (1990) 231–239.

[4] B.S. Garra, E.I. Cespedes, J. Ophir, S.R. Spratt, R.A. Zuurbier, C.M. Magnant, M.F. Pennanen, Elastography of breast lesions: initial clinical results, Radiology 202 (1) (1997) 79–86.

[5] E.S. Burnside, T.J. Hall, A.M. Sommer, et al., Differentiating benign from malignant solid breast masses with US strain imaging, Radiology 245 (2007) 401–410.

[6] R.G. Barr, Real-Time ultrasound elasticity of the breast: initial clinical results, Ultrasound Q. 26 (2010) 61–66.

[7] X.1 Gong, Q. Xu, Z. Xu, P. Xiong, W. Yan, Y. Chen, Real-time elastography for the differentiation of benign and malignant breast lesions: a meta-analysis, Breast Cancer Res. Treat. 130 (1) (2011) 11–18, https://doi.org/10.1007/s10549-011-1745-2. Epub 2011 Aug 26.

[8] W.A.1 Berg, D.O. Cosgrove, C.J. Doré, F.K. Schäfer, W.E. Svensson, R.J. Hooley, R. Ohlinger, E.B. Mendelson, C. Balu-Maestro, M. Locatelli, C. Tourasse, B.C. Cavanaugh, V. Juhan, A.T. Stavros, A. Tardivon, J. Gay, J.P. Henry, C. Cohen-Bacrie, BE1 Investigators. Radiology. Shear-wave elastography improves the specificity of breast US: the BE1 multinational study of 939 masses, Radiology 262 (2012) 435–449, https://doi.org/10.1148/radiol.11110640. Epub.

[9] R.G. Barr, Shear wave imaging of the breast: still on the learning curve, J. Ultrasound Med. 31 (3) (2012) 347–350.

[10] Y.K. Shu, J. So, K.H. Ko, J. So, H.K. Jung, H. Kim, Shear wave elastography: is it a valuable additive method to conventional ultrasound for the diagnosis of small (=2? cm) breast cancer? Medicine 94 (42) (2015) e1540.

[11] M. Friedrich-Rust, M.-F. Ong, S. Martens, et al., Performance of transient elastography for the staging of liver fibrosis: a meta-analysis, Gastroenterology 134 (2008), 960-974.e968.

[12] D. Kirk Gregory, J. Astemborski, H. Mehta Shruti, et al., Assessment of liver fibrosis by transient elastography in persons with hepatitis C virus infection or HIV – hepatitis C virus coinfection, Clin. Infect. Dis. 48 (2009) 963–972.

[13] G. Ferraioli, P. Parekh, A. Levitov, C. Filice, Shear wave elastography for evaluation of liver fibrosis, J. Ultrasound Med. 33 (2014) 197–203.

[14] Guidelines for the Screening, Care and Treatment of Persons with Chronic Hepatitis C Infection, World Health Organization, Geneva Switzerland, 2016, pp. 85–93.

[15] Y.K. Mariappan, K.J. Glaser, R.L. Ehman, Magnetic resonance elastography: a review, Clin. Anat. 23 (5) (2010) 497–511.

[16] R.M. Sigrist, J. Liau, A.E. Kaffas, M.C. Chjammas, J.K. Willman, Ultrasound elastography: review of techniques and clinical applications, Theranostics 7 (5) (2017) 1303–1329.

[17] L. Sandrin, B. Fourquet, J. Hasquenoph, et al., Transient elastography: a new noninvasive method for assessment of hepatic fibrosis, Ultrasound Med. Biol. 29 (12) (2003) 1705–1713.

[18] M. Dandel, H. Lehmkuhl, C. Knosalla, N. Suramelashvili, R. Hetzer, Strain and strain rate imaging by echocardiography – basic concepts and clinical applicability, Curr. Cardiol. Rev. 5 (2009) 133–148.

[19] Hooke's Law Was Originally Described (in 1660) for Small Displacements in Springs. For Small Displacements (Compressions or Stretching of a Material) the Constant Value Is Truly Constant. For Materials in General, the Constant Is Known as Young's Modulus and the Value of Young's Modulus May Not Be Constant when Large Displacements of

the Elastic Material Occur. For Springs, K Is the Shear Modulus rather than the Elastic (Young's) Modulus.

[20] X. Liu, N. Li, T. Xu, et al., Effect of renal perfusion and structural heterogeneity on shear wave elastography of the kidney: an in vivo and ex vivo study, BMC Nephrol. 18 (2017) 265, https://doi.org/10.1186/s12882-017-0679-2.

[21] J.F. Carlsen, C. Ewertsen, L. Lonn, M.B. Nielsen, Strain elastography ultrasound: an overview with emphasis on breast cancer diagnosis, Diagnostics 3 (2013) 117–125, https://doi.org/10.3390/diagnostics3010117.

[22] J. Carlsen, C. Ewertsen, A. Saftoiu, L. Lönn, M.B. Nielsen, Accuracy of visual scoring and semi-quantification of ultrasound strain elastography – a phantom study, PLoS One 9 (2014) e88699, https://doi.org/10.1371/journal.pone.0088699.

[23] L. Ge, B. Shi, Y. Song, Y. Li, S. Wang, X. Wang, Clinical value of real-time elastography quantitative parameters in evaluating the stage of liver fibrosis and cirrhosis, Exp. Ther. Med. 10 (3) (2015) 983–990, https://doi.org/10.3892/etm.2015.2628.

[24] J.F. Carlsen, C. Ewertsen, S. Sletting, M.-L. Talman, I. Vejborg, M. Bachmann Nielsen, Strain histograms are equal to strain ratios in predicting malignancy in breast tumors, in: R. Rota (Ed.), PLoS One 12 (10) (2017) e0186230, https://doi.org/10.1371/journal.pone.0186230.

[25] J.F. Carlsen, M.R. Pedersen, C. Ewertsen, S. Rafaelsen, A. Saftoiu, L.B. Lonn, M.B. Nielsen, Diagnostic Performance of Strain- and Shear-Wave Elastography in an Elasticity Phantom: A Comparative Study, Exhibit ECR, 2014, https://doi.org/10.1594/ecr2014/C-2135.

[26] C.F. Dietrich, R.G. Barr, A. Farrokh, M. Dighe, M. Hocke, C. Jenssen, Y. Dong, A. Saftoiu, R. Havre, Strain elastography – how to do it? Ultrasound Int. Open 3 (4) (2017) E137–E149.

[27] http://www.fibroscan.com/en/products last visited 12'22'2017.

[28] T. Poynard, T. Pham, H. Perazzo, M. Munteanu, E. Luckina, et al., Real-time shear wave versus transient elastography for predicting fibrosis: applicability, and impact of inflammation and steatosis. A non-invasive comparison, PLoS One 11 (10) (2016) e0163276. https://doi.org/10.1371/journal.pone.0163276.

[29] R.G. Barr, G. Ferraioli, M.L. Palmeri, Z.D. Goodman, G. Garcia Tsao, J. Rubin, B. Garra, R. Myers, S. Wilson, D. Rubens, D. Levine, Elastography assessment of liver fibrosis: society of Radiologists in ultrasound consensus conference statement, Radiology 276 (3) (2015) 845–861.

[30] R.G. Barr, Z. Zhang, Shear-wave elastography of the breast: value of a quality measure and comparison with strain elastography, Radiology 275 (1) (2015) 45–53.

[31] J. Correas, A. Tissler, A. Khairoune, V. Vassiliou, A. Mejean, O. Helenon, R. Memo, R. Barr, Prostate cancer: diagnostic performance of shear wave elastography, Radiology 275 (2015) 280–289.

[32] M. Muller, D. Ait-Belkacem, M. Hessabi, J. Gennisson, G. Grange, F. Goffinet, E. Lecarpentier, D. Cabrol, M. Tanter, V. Tsatsaris, Assessment of the cervix in pregnant women using shear wave elastography: a feasibility study, Ultrasound Med. Biol. 41 (11) (2015) 2789–2797.

[33] D.H. MacIver, I. Adeniran, H. Zhang, Left ventricular ejection fraction is determined by both global myocardial strain and wall thickness, Int. J. Cardiol. Heart Vasc. 7 (2015) 113–118, https://doi.org/10.1016/j.ijcha.2015.03.007.

[34] O.A. Smiseth, H. Torp, A. Opdahl, K.H. Haugaa, S. Urheim, Myocardial strain imaging: how useful is it in clinical decision making? Eur. Heart J. 37 (15) (2016) 1196–1207, https://doi.org/10.1093/eurheartj/ehv529.

[35] A. Sarvazyan, T.J. Hall, M.W. Urban, M. Fatemi, S.R. Aglyamov, B.S. Garra, An overview of elastography-an emerging branch of medical imaging, Curr. Med. Imag. Rev. 7 (2011) 255–282.

[36] An Isotropic Material Whose Physical Properties Do Not Depend on the Direction of Measurement i.e. The Properties Are the Same in All Directions. Materials That Are Not Isotropic Are Termed Anisotropic.

[37] M. Kudo, T. Shiina, F. Moriyasu, et al., JSUM ultrasound elastography practice guidelines: liver, J. Med. Ultrason. 40 (2013) 325–357.

[38] J. Bamber, D. Cosgrove, C.F. Dietrich, et al., EFSUMB guidelines and recommendations on the clinical use of ultrasound elastography. Part 1: basic principles and technology, Ultraschall der Med. 34 (2) (2013) 169–184.

[39] D. Cosgrove, F. Piscaglia, J. Bamber, et al., EFSUMB guidelines and recommendations on the clinical use of ultrasound elastography. Part 2: clinical Applications, Ultraschall der Med. 34 (3) (2013) 238–253.

[40] C. Dietrich, J. Bamber, A. Berzigotti, et al. EFSUMB Guidelines and Recommendations on the Clinical Use of Liver Ultrasound Elastography, Update 2017 (Long Version), doi https://doi.org/10.1055/s-0043-103952.

[41] https://qibawiki.rsna.org/index.php/Ultrasound_SWS_Biomarker_Ctte.

[42] T.J. Hall, A. Milkowski, B. Garra, et al., RSNA/QIBA: shear wave speed as a biomarker for liver fibrosis staging, in: 2013 IEEE International Ultrasonics Symposium (IUS), Prague, 2013, pp. 397–400, https://doi.org/10.1109/ULTSYM.2013.0103.

[43] M. Palmeri, K. Nightingale, S. Fielding, et al., RSNA QIBA ultrasound shear wave speed Phase II phantom study in viscoelastic media, in: 2015 IEEE International Ultrasonics Symposium (IUS), Taipei, 2015, pp. 1–4, https://doi.org/10.1109/ULTSYM.2015.0283.

CHAPTER

# Breast elastography

2

**Richard G. Barr**

*Radiology, Northeast Ohio Medical University, Radiology Consultants Inc., Youngstown, OH, United States*

## 1. Introduction/background

Stiffness has been used for thousands of years to evaluate breast masses [1]. It is well know that stiff, nonmobile lesions in the breast have a high probability of being malignant. Krouskop [2] found in in vitro studies that there is significant elastographic difference between benign and malignant lesions, with little overlap between them. This suggests that elastography of the breast should have high sensitivity and specificity for characterizing breast lesions.

Both strain elastography (SE) and shear wave elastography (SWE) have been used to evaluate breast pathology with high sensitivity and specificity [3,4]. The breast has unique elastographic properties that are not seen in other organs; for reasons not fully understood, breast malignant lesions appear larger on elastography than on B-mode, whereas benign lesions appear smaller [4]. Therefore a semiquantitative measure of the length ratio of a lesion on elastography to B-mode can be used to characterize breast lesions [4]. On SWE, many cancers have properties that are not well understood, do not allow for good shear wave propagation, and are often not color-coded (a shear wave speed [SWS] could not be calculated) or can be color-coded as soft. The addition of a quality measure (QM) to SWS estimates can be used to confirm that adequate shear waves are generated for an appropriate stiffness value [5,6].

## 2. Principles/techniques

### 2.1 Strain elastography

#### *2.1.1 Techniques*

A detailed description of SE is discussed in detail in Volume 1 of this book. Here we highlight the features specific for breast imaging. In SE a compression/release force is applied to the breast using the transducer. There is an optimum amount and frequency of compression and release for each vendor. There is a learning curve on how to perform compression/release for SE, which varies depending on the ultrasound system being used. Every system has a bar, graph, or number that provides

Tissue Elasticity Imaging. https://doi.org/10.1016/B978-0-12-809662-8.00002-4

real-time feedback on the appropriateness of the compression/release being applied. Some systems require minimal displacement often provided with patient breathing or heart motion, while others require compression and release of the transducer of a few millimeters. The elastographic quality monitor is optimized (higher bar or number) when the appropriate displacement/release is applied. Too much or too little displacement will cause the monitor to decrease.

For accurate SE images the displacement must be applied in the same location of the lesion throughout the displacement/release cycle. If the lesion is moving in and out of the plane while obtaining the elastogram, the resultant elastogram will be inaccurate. Often a ring of softness around the lesion is identified when the lesion is moving relative to the imaging plane. This "sliding sign" indicates that the lesion is moving relative to the adjacent tissue suggesting it is benign. Malignant lesions do not move relative to the surrounding tissue, as they have invaded the adjacent tissue [7]. The patient should be positioned so that the transducer is perpendicular to the floor and the patient rotated so that the patient's breathing moves the lesion only within the imaging plane.

When performing elastography, both SE and SWE, of the breast, it is critical not to apply any pressure with the transducer other than the displacement/release. When the breast is compressed, the stiffness of all tissue types increases and this increase varies with the tissue being compressed. This is called precompression. A small amount of precompression is usually applied when performing B-mode imaging, as it displaces Cooper ligaments so that they become parallel to the transducer, reducing refractive artifacts. To obtain optimal elastograms, only very light contact of the transducer with the skin is required. A reproducible method for obtaining elastograms with minimal precompression has been described [8]. Using ample amounts of coupling gel is helpful. After obtaining the B-mode image choose a structure such as a rib in the far field. As the transducer is lifted the structure will move deeper in the image. When the structure is as deep in the far field as possible and while adequate transducer contact is maintained, the elastogram is obtained.

SE imaging is qualitative; the exact stiffness of the tissue is not obtained. Only the relative stiffness of one tissue to another is obtained with SE. There are several maps, from gray scale to multiple color maps, used to display the relative stiffness in the SE elastogram. There is no standard, and some systems display results with blue as stiff and red as soft, whereas others display with red as stiff and blue as soft. It is therefore critical to also include the color map used with the elastogram. The dynamic range of the display map varies based on the stiffness values present in the elastogram. Therefore the "color" of a lesion will vary depending on what other tissues are present in the field of view (FOV). To have a more constant dynamic range of display, it is important to include various tissue types in the FOV. The FOV should be as large as possible and should include the lesion, fat, glandular tissue, and pectoralis muscle. By doing this, if the lesion is benign the fat will be the softest tissue, while the pectoralis muscle will be the stiffest tissue and will be similar to other elastograms in the same patient as well as in other patients. If, however, the lesion is malignant, it will be the stiffest tissue; therefore, other tissues will display as significantly softer.

In addition to the manual compression method (using the transducer or patient motion to cause the displacement/release), acoustic radiation force impulse (ARFI) can be used to obtain an SE image [4,7]. With this technique the transducer is held still on the patient with minimal precompression and the ARFI pulse is used to cause the stress on the tissue. The displacement of the tissues is monitored and a strain elastogram calculated. Similar color maps used in the displacement/release technique are used to display the results. The ARFI pulse is attenuated as it traverses the tissues but usually an adequate elastogram can be obtained up to 4-cm depth.

### 2.1.2 Interpretation

Breast cancer lesions appear larger on the SE elastogram than on the B-mode image, whereas benign lesions appear smaller on the elastogram than on the B-mode image (Fig. 2.1A and B). This appears to be unique to breast imaging. As SE is relative, evaluation of only the "color" on the elastogram does not provide adequate information for high sensitivity and specificity of characterizing breast masses. However, the size changes that occur are extremely sensitive and specific for characterizing breast masses.

There are three methods that have been proposed to interpret SE images of the breast (Fig. 2.1):

1. elastographic to B-mode length ratio (E/B ratio),
2. 5-point color scale,
3. comparing the stiffness of the lesion to the stiffness of fat (strain ratio).

### 2.1.3 Elastographic to B-mode length ratio

The breast has an unique elastic property that malignant lesions appear larger on elastography than on B-mode imaging, while benign lesions appear smaller [3,4,9,10]. Therefore we can use the length of a lesion on elastography divided by the length of the lesion on B-mode, i.e., the E/B ratio, as a method to characterize breast lesions [4]. The E/B ratio has been shown to be related to the aggressiveness of the malignancy with less aggressive lesions such as mucinous cancers or ductal carcinoma in situ (DCIS) with ratios of 1.0 or slightly larger and more malignant lesions such as invasive ductal cancer (IDC) having ratios up to 3. A cutoff value of $<1$ for benign lesions and $\geq 1.0$ for malignant lesions has been used with high sensitivity and specificity for lesion characterization. Sensitivities of 98%–99% and specificities of 85%–90% have been reported with this technique [6,9–11]. False-negative results have been lymphomas, both primary and metastatic, which are softer lesions and have a ratio of $<1.0$ [3,6].

One confounding factor using the E/B ratio is fibroadenomas and fibrocystic changes have similar elastic properties to those of glandular tissue. When measuring the lesion on SE the measurement may also include glandular tissue resulting in a false-positive finding. Another confounding factor is the presence of two lesions adjacent to each other. These may appear as one lesion on the B-mode image. Close inspection of the elastogram can distinguish the two lesions. In these cases, care

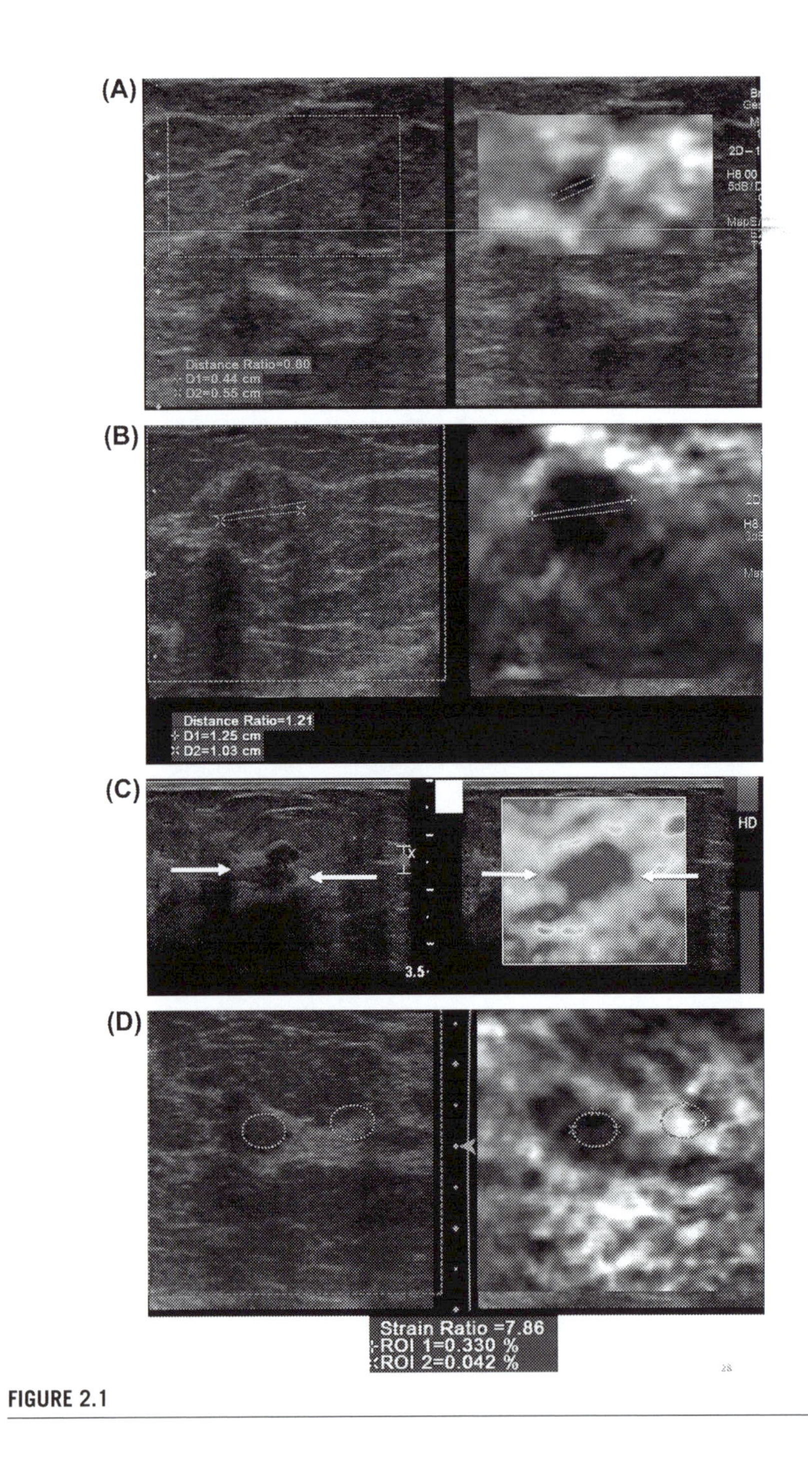
(A)
Distance Ratio=0.80
D1=0.44 cm
D2=0.55 cm
(B)
Distance Ratio=1.21
D1=1.25 cm
D2=1.03 cm
(C)
HD
3.5
(D)
Strain Ratio =7.86
ROI 1=0.330 %
ROI 2=0.042 %

FIGURE 2.1

must be taken in performing measurements. If different results are obtained in different lesion positions, the possibility of the "lesion" being two adjacent lesions should be considered. Always use the results of the larger E/B ratio. If such a lesion is biopsied, always try to biopsy the portion of the lesion that has the larger E/B ratio.

### 2.1.4 5-Point color scale

A 5-point color scale can be used to characterize breast masses [12,13]. In this method a score of 1 is given when the lesion is completely soft based on the color scale used, a score of 2 is given if the lesion has mixed soft and stiff components, a score of 3 if the lesion is stiff and smaller than that on B-mode, a score of 4 is given if the lesion is stiff and the same size as that in B-mode, and a score of 5 is given when the lesion is stiff and larger than that on B-mode (Fig. 2.1C). A cutoff value between 3 and 4 is used to characterize lesions as benign or malignant. Sensitivities of 87%–93% and specificities of 83%–90% have been obtained with this technique [7,13].

### 2.1.5 Strain ratio

SE is a qualitative technique. The exact stiffness of a tissue is not obtained in SE, but just its relative stiffness to other tissues in the FOV. However, we can obtain a semi-quantitative measure of stiffness by comparing the stiffness of the lesion to a

There are three strain elastography (SE) methods that have been used to characterize breast lesions as benign or malignant. The size change of breast lesions, which occurs in elastography, can be used to characterize lesions. Benign lesions appear smaller on elastography. (A) A 45-year-old patient presented with a new mammographic abnormality. On ultrasonography the lesion is well circumscribed, is isoechoic, and measures 5.5 mm (left image). On elastography the lesion (black, right image) is stiffer than the surrounding fat and measures 4.4 mm, giving an elastographic to B-mode length ratio (E/B ratio) of 0.8. On biopsy the lesion was a benign fibroadenoma. (B) A similar well-circumscribed isoechoic lesion measures 10 mm (left image). On the elastogram (right image) the lesion is stiffer (black) than the surrounding fat and measures 12 mm. The E/B ratio is 1.2. The lesion was found to be an invasive ductal cancer on core biopsy. The 5-point color-scale method evaluates lesions based on a 5-point scale, in which the stiffness and size are evaluated. For a score of 1 the lesion is soft (similar to fat), a score of 2 has both soft and hard components, for a score of 3 the lesion is stiff and smaller than the lesion on B-mode, for a score of 4 the lesion is stiff and the same size as that in B-mode, and for a score of 5 the lesion is stiff and larger than that on B-mode. An example is provided in (C). The scale used in elastography is blue (dark gray in print version) is stiff and red (gray in print version) is soft. This invasive ductal carcinoma is stiff and larger than that on B-mode and is therefore classified as a score of 5. The third method of SE lesion characterization is a semiquantitative method in which the stiffness of the lesion is compared to the stiffness of fat, the strain ratio. (D) A region of interest (ROI) is placed in the lesion and an ROI is placed in the adjacent fat. In this case the Strain Ratio (SR) was 7.9 suggestive of a malignant lesion. On core biopsy the lesion was found to be an invasive ductal cancer.

reference tissue [14]. For breasts, we use fat as the reference tissue. Most ultrasound systems can calculate this by placing a region of interest (ROI) in the lesion and an ROI in fat. The system will then provide a ratio of the stiffness values (Fig. 2.1D). Ultrasound systems use different methods to determine the relative strain of the tissues, and ratios obtained on one vendor's equipment may not be the same as that from another vendor's equipment. This measurement also requires that the same amount of stress be applied to both the lesion and the reference tissue. Therefore the ROIs should be placed at the same depth from the transducer, if possible. In some patients, especially in Asian women, there may not be reference fat tissue so normal glandular tissue can be used but the cutoff values to characterize the lesion will be different than when fat is used as the reference. When using fat as the reference tissues, cutoff values of 2.3–4.8 have been reported to characterize breast lesions as benign or malignant [14–16].

### 2.1.6 Review of the literature

Hall used an E/B ratio of >1.2 for a lesion to be malignant based on the receiver operating characteristic (ROC) curve of a small dataset. With these criteria, a sensitivity of 100% and a specificity of 75.4% was obtained; however, their series did not include low-grade cancers such as DCIS and mucinous cancer. Barr [11] in a single center trial of 123 biopsy-proven breast lesions observed a sensitivity of 100% and a specificity of 95% in characterizing benign from malignant breast lesions by using an E/B ratio of <1.0 as benign and ≥1.0 as malignant. A large multicenter, unblinded trial evaluating 635 biopsy-proven cases using Barr's criteria had a sensitivity of 99% and a specificity of 87% [9]. There were 3 cancers out of 222 that had a ratio of less than 1. In retrospect, one lesion was measured incorrectly on B-mode. The second "lesion" may have been two adjacent lesions, one benign and one malignant, confounding the measurements. The third increased in size in anterior to posterior dimension but got smaller in width. In cases of DCIS and invasive lobular carcinoma, which are poorly visualized on B-mode imaging, this technique cannot be used. In a study using E/B ratio, results demonstrate that a lesion with a pretest probability of 50% (Breast Imaging Reporting and Data System [BI-RADS] 4B) can be downgraded to a 2% posttest probability (BI-RADS 3) [6].

A single center trial of 230 lesions showed 99% sensitivity, 91.5% specificity, positive predictive value of 90%, and negative predictive value of 99.2% using the E/B ratio [17].

The E/B ratio has been shown to correlate with tumor grade [18]. For less aggressive tumors such as DCIS, mucinous, or colloid cancer the ratio is close to 1.0. For IDCs the ratio increases with the grade. The clinical utility of this finding is unclear at this time. Some reports suggest a greater specificity with a cutoff value of 1.2; however, with this cutoff, low-grade malignancies such as DCIS or mucinous cancers can be misclassified as benign, thus trading higher specificity at the expense of a lower sensitivity.

Using the 5-point color scale, breast SE has been shown to objectively evaluate tumor or tissue stiffness, in addition to morphology and vascularity [13]. Itoh [13] found a sensitivity of 86.5% and a specificity of 89.8% using this technique with a cutoff point between 3 and 4. SE has also been shown to visualize nonmass lesions or peritumoral ductal lesions [19].

Raza [20] reported a prospective clinical trial with a sensitivity of 92.7% and a specificity of 85.8%. Chang et al. [21] analyzed factors that affect the accuracy of elasticity scores in a prospective study, and they reported their findings on the accuracy control of SE. They reported that the breast being thin at the location of the lesion (the target lesion being located in a shallow area) was the biggest factor affecting elastographic image quality, and they mentioned that the accuracy of elastography differed depending on the depth of the lesion and that accuracy control was necessary.

Initial studies [14–16] have found strain ratio to be valuable in determining if a lesion was benign or malignant. Thomas [15] compared B-mode BI-RADS category score, the 5-point color scale, and the strain ratio in 227 breast lesions. Based on the ROC curve, they selected a cutoff of 2.46 to distinguish benign from malignant lesions using the strain ratio. The mean ratio for malignant lesions was $5.1 \pm 4.2$, while for benign $1.6 \pm 1.0$ ($P < .001$). They found a sensitivity and specificity of 96% and 56%, respectively, for B-mode imaging, 81% and 89% for 5-point color scale, and 90% and 89% for the lesion-to-fat ratio.

Zhi et al. [22] in a similar study compared the strain ratio and the 5-point color scale in 559 breast lesions. They found the strain ratio of benign lesions was $1.83 \pm 1.22$, while for malignant lesions, $8.38 \pm 7.65$. These were significantly different ($P < .00,001$). Bases on their ROC curves, they selected a cutoff point of 3.05. The area under the curve for the 5-point color system was 0.89, whereas that of the strain ratio was 0.94 ($P < .05$). In a different study of 408 lesions [14] a strain ratio cutoff of 4.8 had a sensitivity of 76.6% and specificity of 76.8%.

Farrokh [16] reported a sensitivity of 94.4% and specificity of 87.3% with a cutoff above 2.9 in a prospective study using strain ratio. Alhabshi [23] reported that E/B ratio and SR were the most useful methods of lesion characterization, with a cutoff value of 1.1 for E/B and a cutoff value of 5.6 for SR. In this study, B-mode imaging and the 5-point color scale were less accurate. Stachs [24] demonstrated the strain ratio utility in 224 breast masses, with 215 reporting that the strain ratio was predominantly higher in malignant tumors, i.e., $3.04 \pm 0.9$ (mean $\pm$ standard deviation) for malignant tumors versus $1.91 \pm 0.75$ for benign tumors.

The appropriate cutoff value for this technique varies greatly between studies. When using this technique the appropriate cutoff value should be determined with your technique and the equipment in your laboratory.

## 2.2 Shear wave elastography

### *2.2.1 Techniques*

As opposed to SE, SWE provides a quantitative stiffness value based on the SWS. For breast SWE the transducer is placed on the breast with light pressure. No movement of the transducer is required, in fact, that is contraindicated. Precompression is a major concern as it is in SE, and the discussion and methods to standardize light compression in SE apply also to SWE [3].

Both point shear wave elastography (p-SWE) and two-dimensional shear wave elastography (2D-SWE) have been used to evaluate breast lesions [3]. Breast masses, especially malignancies, tend to be very heterogeneous in stiffness. Therefore the 2D-SWE technique is preferred, as the larger FOV can depict the differences in stiffness and the area of highest stiffness can be identified.

### *2.2.2 Interpretation*

Most research suggests that the maximum stiffness value (Emax) of the lesion or surrounding tissue (approximately 3 mm) should be used to characterize a lesion [6,25–28].

Breast cancers are often very heterogeneous and disorganized. Shear waves may not generate or propagate in some breast cancers. In these cancers the lesion may not color-code and thus accurate SWS estimates cannot be calculated. Often there is a rim of high stiffness around the lesion that can be used to characterize the lesion. In some cancers the lesion may color-code soft and no rim of high stiffness is identified. These lesions are always positive on SE. The addition of a QM that assesses the accuracy of the SWS estimate is helpful in identifying cases that would otherwise show false-negative results [6].

To distinguish benign from malignant lesions, SWS cutoff values of 4.5 m/s (60 kPa) to 5.2 m/s (80 kPa) have been proposed [6,28,29]. It has been suggested that most clinical value of SWE is to upgrade or downgrade BI-RADS category 3 or 4a lesions. If a BI-RADS category 3 lesion is negative on elastography, it can be downgraded to a BI-RADS category 2 lesion and followed up in 1 year. If a BI-RADS category 3 lesion is positive on SWE the lesion should be upgraded to BI-RADS category 4a and biopsied. If a BI-RADS category 4a lesion is negative on elastography the lesion can be downgraded to BI-RADS category 3.

### *2.2.3 Review of the literature*

Chang [29] in a study of 158 consecutive patients found that the mean elasticity values were significantly higher in malignant masses (153 kPa ±58) than in benign masses (46 kPa ±43) ($P < .0001$). With a cutoff value of 80 kPa, they had a sensitivity and specificity of 88.8% and 84.9%, respectively. The area under the ROC curve was 0.90 for conventional ultrasonography, 0.93 for SWE, and 0.98 for the combined data. In a study of 48 breast lesions Athanasiou et al. [26] found similar results with similar stiffness values for benign lesions (45 ± 41 kPa) and malignant lesions (147 kPa ±40) ($P < .001$). They suggested that the addition of SWE to conventional ultrasonography could decrease the number of biopsies performed in benign lesions. In Tozaki's series of 161 masses including 43 malignancies, using an SWS cutoff of 3.6 m/s (38 kPa), a sensitivity of 91% and a specificity of 80.6% was obtained [30]. In a small series, Evans et al. [27] found the sensitivity and specificity for SWE was 97% and 83%, respectively, which was greater than

B-mode alone. In their series, they used a cutoff value of 50 kPa. They also confirmed that the technique is highly reproducible.

Based on a large multicenter study (BE1), a cutoff value of 80 kPa (5.2 m/s) was determined to distinguish benign from malignant lesions [28]. This large multicenter trial demonstrated that when added to BI-RADS classification in B-mode imaging, SWE increased the diagnostic accuracy. They found that the evaluation of SWE signal homogeneity and lesion-to-fat ratios was the best way to differentiate between benign and malignant lesions. The addition of SWE increased the characterization of lesion over BI-RADS alone, with a sensitivity and specificity of 93.1% and 59.4%, respectively, in BI-RADS and 92.1% and 7.4%, respectively, with the addition of SWE. The authors comment that the major value of the addition of SWE is in BI-RADS 3 and 4a lesions, where the SWE results are used to upgrade or downgrade the lesion [28]. Lesion size measurements were highly reproducible (>0.9) and interobserver reliability for maximum and mean elasticity values was also highly reproducible (>0.8) [31].

### 2.3 Guidelines

Guidelines recommending the use of elastography for characterizing breast lesions have been published by the European Federation of Societies for Ultrasound in Medicine and Biology (EFSUMB) [32] and the World Federation for Ultrasound in Medicine and Biology (WFUMB) [33]. Both guidelines recommend the addition of elastography to conventional ultrasonography to improve the characterizations of breast lesions as benign or malignant. The WFUMB guidelines provide a detailed description of how to perform breast elastography, how to interpret the results, and how to correlate findings obtained on SE and SWE. WFUMB also has guidelines and recommendations for the basic science of elastography [34]. BI-RADS version 5 allows for upgrading BI-RADS 3 lesions to BI-RADS 4A if elastography is suggestive of a malignant lesion. It also allows for downgrading a BI-RADS 4A lesion to BI-RADS 3 if the elastography is suggestive of a benign lesion [35].

## 3. Diseases and applications

### 3.1 Benign lesions

Benign lesions have an E/B ratio <1.0 on SE, a 5-point color-scale value of 3 or less on SE, and a strain ratio of <4.5 (may vary by vendor or laboratory). The stiffness values of benign lesions on SWE are less than 4.5 m/s (60 m/s) in some series and 5.2 m/s (80 kPa) in other series [33].

#### *3.1.1 Cysts*

Simple cysts are easily diagnosed with B-mode imaging. Simple cysts have a thin wall, through transmission, and are anechoic without internal echoes. Elastography does not add any additional diagnostic information for simple cysts. Complicated cysts have internal echoes that are mobile, whereas complex cystic masses have internal echoes that are not mobile. However, for complicated cysts and complex cystic masses, elastography significantly adds to the lesion characterization. With some SE systems, both simple and complex cysts have a "bull's-eye" artifact [36] that has a 100% sensitivity and specificity for characterizing a lesion as a benign simple or complicated cyst (Fig. 2.2). Other SE systems have a three-layered colored artifact with a blue, green, and red pattern (BGR; blue for stiff). This "BGR" pattern has not been studied for its sensitivity and specificity for characterization of a lesion as a benign simple or complicated cyst [3].

The bull's-eye artifact can help distinguish a complicated cyst from a complicated cystic mass. This artifact occurs because there is decorrelation with moving material. Therefore complicated cysts where the internal echoes are moving do not affect the bull's-eye pattern. If there is a solid component within the cyst, there is a deformity in the bull's-eye pattern confirming the presence of a "solid" component [36] (Fig. 2.3). This does not mean that the cystic lesion is a malignancy but does require further workup of the lesion. Both benign and malignant lesions can result in the deformity of the bull's-eye artifact. Benign lesions such as an acorn cyst where the internal debris is fixed will lead to the defect in the bull's-eye pattern.

Simple cysts in SWE do not support shear wave propagation and therefore may be color-coded black in 2D-SWE and will give x.xx or 0.00 in p-SWE. Complicated cysts will color-code soft in 2D-SWE and will have the elastographic appearance similar to a fibroadenoma. In p-SWE, low SWSs (<2.5 m/s) will be obtained.

### 3.1.2 Fibroadenomas

The elastographic appearance of fibroadenomas can be from very soft to stiff. The majority of fibroadenomas will have stiffness values from 1.9 m/s (12 kPa) to 3.8 m/s (42 kPa) [3] (Fig. 2.4). However, some fibroadenomas can be very soft, especially lipofibroadenoma, whereas others can be stiff and have a stiffness value above the cutoff value. In our experience, approximately 5% of fibroadenomas will have stiffness values suggestive of a malignant lesion. If the fibroadenoma has calcifications the stiffness value will be increased.

In SE, if the stiffness of the fibroadenoma is similar to glandular tissue and the fibroadenoma is located in glandular tissue, it will blend in with the adjacent glandular tissue. This could lead to erroneous E/B ratio, as the elastographic measurement may include both the fibroadenoma and the adjacent glandular tissue.

### 3.1.3 Fibrocystic change and related pathology

In general, fibrocystic change is hypoechoic on B-mode and usually very conspicuous. However, on elastography the stiffness of fibrocystic change is similar to glandular tissue. Therefore fibrocystic change usually blends in with adjacent glandular tissue on elastography. On SE, it is has similar stiffness to glandular tissue, and on SWE, it has stiffness values ranging from 1.2 m/s (7 kPa) to 2.2 m/s (15 kPa) [3].

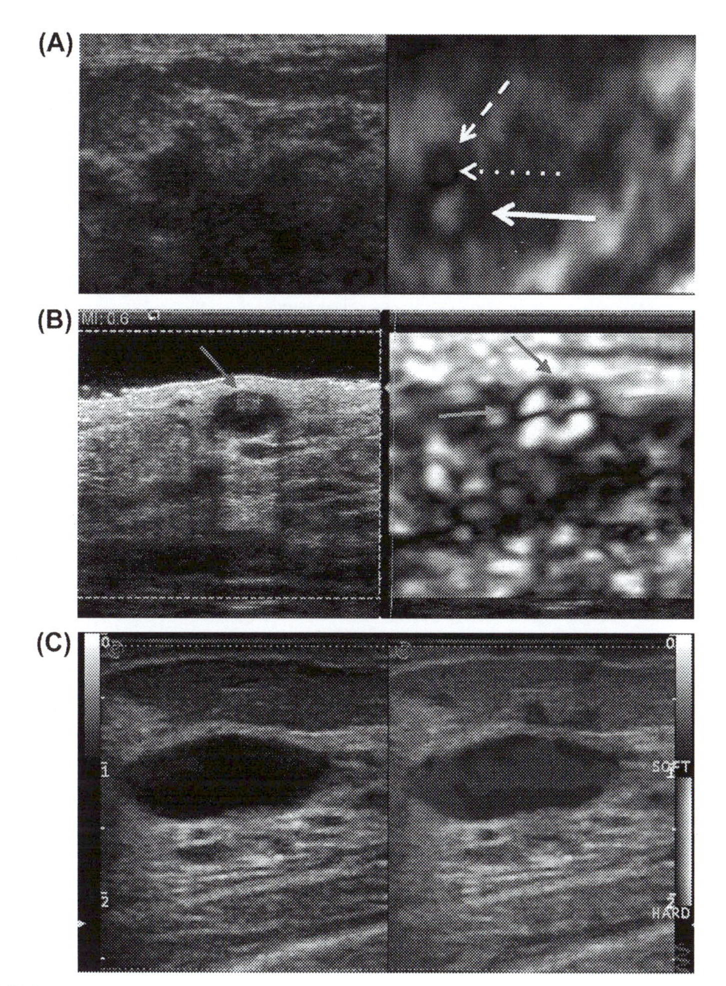


**FIGURE 2.2**

With some ultrasound systems (Siemens ultrasound and Philips ultrasound), both simple and benign complicated cysts have a bull's-eye artifact (A). This artifact has a black outer rim (*dashed arrow*), a central white spot (*dotted arrow*), and a distal white spot (*solid arrow*). All three components are needed to classify the lesion as a benign simple or complicated cyst. (B) If there is a "solid" component in the cyst (i.e., a complicated cystic mass) the solid component will be a defect in the bull's-eye pattern. The red (black in print version) arrows point to the bull's-eye artifact and the blue (gray in print version) arrow points to a 2-mm biopsy-proven benign papilloma. Note the black defect in the bull's-eye artifact. (C) For other ultrasound systems, cysts have a tricolor pattern of blue, green (light gray in print version), and red.

*(B) Reproduced with permission from Barr RG and Lackey AE "The Bull's Eye Artifact on Breast Elastography Imaging in Reducing Breast Lesion Biopsy Rate" Ultrasound Quarterly 2011; 27:151–155.*

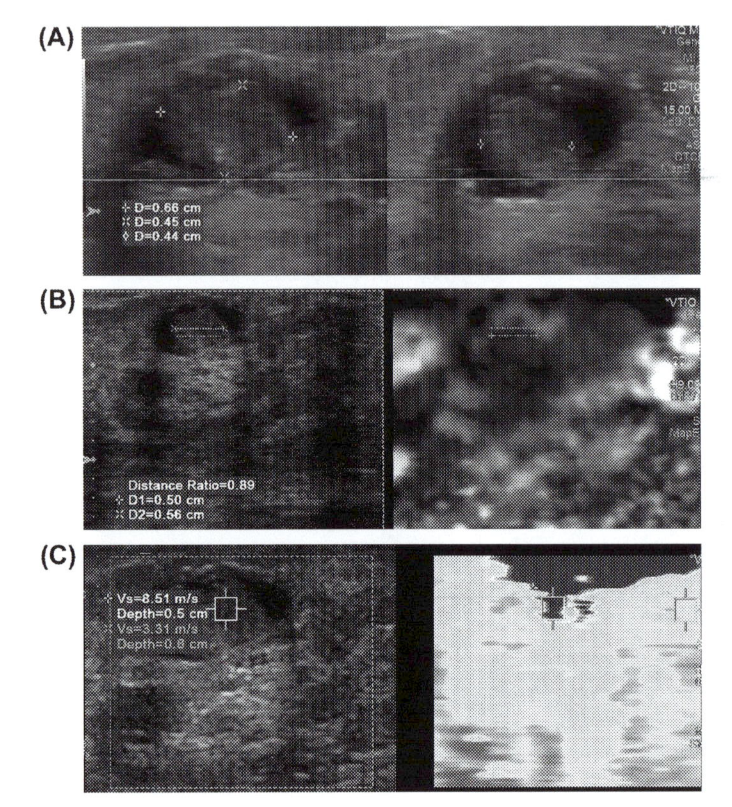


**FIGURE 2.3**

A 65-year-old female presented with a palpable mass near the nipple. (A) The lesion is cystic with a "solid component" in the cyst measuring 7 mm. Both the transverse and longitudinal images of the lesion are presented. On strain elastography the fluid component is soft (*white*) and the solid component is stiff (*black*). There is no bull's-eye artifact. The distance ratio is 0.9 suggestive of a benign lesion. (B) The elastogram is presented on the right and the B-mode image is on the left. (C) On two-dimensional shear wave elastography the solid component has a stiffness value of 8.5 m/s (right image) suggestive of a malignant lesion. On biopsy the lesion was found to be fat necrosis.

### *3.1.4 Papillary lesions*

Papillomas of the breast can be divided into solitary papillomas, multiple papillomas, and juvenile papillomatosis. Solitary or central papillomas arise in the large retroareolar ducts. Pathologically, a papilloma is a masslike projection that consists of papillary fronds attached to the inner mammary duct wall by a fibrovascular core that is covered with ductal epithelial and myoepithelial cells. Ductal epithelial cells may undergo apocrine metaplasia, hyperplasia, or atypia.

Central papillomas are usually solitary, but multiple central papillomas have been reported [37]. Central papillomas are typically small and are often

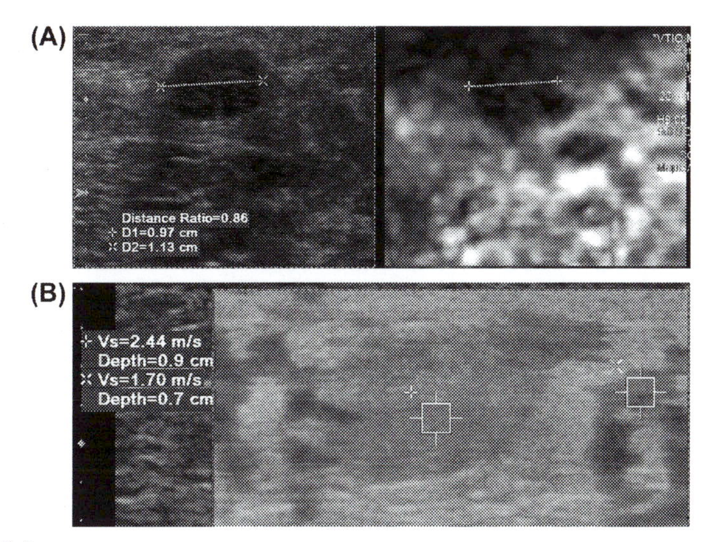


**FIGURE 2.4**

A 48-year-old female presented with a well-circumscribed hypoechoic mass. (A) On strain elastography the elastographic to B-mode length ratio is 0.9 suggestive of a benign lesion. (B) On two-dimensional shear wave elastography the lesion has a stiffness value of 2.44 m/s consistent with a benign lesion. On core biopsy the lesion was found to be a fibroadenoma.

mammographically occult. Sonography or ductography is usually necessary for visualization of the lesion. On sonography, a papilloma is seen as an intraductal mass in a dilated duct, an intracystic mass, or a solid mass with a well-defined border [38]. In ductography a cannula is inserted into the duct that has a fluid discharge and iodinated contrast is injected into the duct. A mammogram is then obtained to determine if there is a filling defect corresponding to a mass lesion.

The characteristic ultrasonographic finding of a papilloma is a solid mural nodule within a dilated duct. Other features include an intracystic mass or a well-circumscribed hypoechoic solid mass. The vascular pedicle within the mural nodule can often be identified on color Doppler imaging.

Differentiating benign from malignant papillary lesions can be difficult. A nonparallel orientation, echogenic halo, posterior acoustic enhancement, and associated microcalcifications are reported to be more frequent in malignant lesions [39].

Four types of papillary lesions have been described: type 1, intraluminal mass; type 2, extraductal mass; type 3, purely solid mass; and type 4, mixed. Type 1 lesions can be further subdivided based on the degree of expansion and filling of the duct by the mass into (1) intraductal type, (2) intracystic type, and (3) solid type with anechoic rim [40].

On conventional ultrasonography, this lesion has the appearance of a complex cystic lesion. The anechoic portion of the lesion demonstrates a bull's-eye appearance suggesting that portion is a low viscosity fluid. The solid component has a benign appearance based on the E/B distance ratio and lesion-to-fat ratio. The lesion SE findings can be seen in a complicated cyst with adherent debris or high-viscosity fluid. The "solid" component has elevated SWS, which is not concordant with the SE findings.

### 3.1.5 Mastitis

The edema of tissues in mastitis is often poorly visualized on B-mode imaging. The inflammation and edema of tissues shows a significant increase in stiffness values. On SE, mastitis appears as a stiff lesion, and on SWE, it has stiffness values often suggestive of a malignancy. The stiffness values range from 3.5 m/s (35 kPa) to 6.4 m/s (120 kPa) [41] (Fig. 2.5).

If the patient has mastitis with abscess formation, the abscess will appear soft. Therefore mastitis with abscess formation has a "donut" appearance with a soft center (abscess) and a thick stiff rind (edema and inflammation).

In general, breast cancers have stiff avascular central areas and not soft necrotic areas; therefore, a soft center in the lesion should always raise the possibility that the lesion is not a primary breast cancer.

### 3.1.6 Fat necrosis

Fat necrosis can be seen after surgery, radiation therapy, or trauma. Patients with fat necrosis are usually asymptomatic but can present with a palpable mass that can be tender. These lesions can often be diagnosed by mammography by the classic appearance of oil cysts and dystrophic calcifications.

On sonography the appearance of fat necrosis varies according to the chronicity of the process. The appearance can range from a solid mass, complex mass with mural nodules or echogenic bands, to an isoechoic or anechoic mass with or without shadowing or posterior acoustic enhancement.

Fat necrosis is one lesion that can be a false-positive finding on both SE and SWE [3] (Fig. 2.6). A detailed study of fat necrosis appearance on elastography has not been performed. The appearance on elastography is variable, with some cases being very soft and others being very stiff as in this case. The elastic properties of fat necrosis may vary depending on the chronicity of the lesion. When there is acute inflammation associated with fat necrosis, the stiffness should increase similar to that seen in mastitis.

In this case the strain findings (E/B ratio and strain ratio) are suggestive of a malignant lesion. On the 5-point color scale the lesion would be classified with a score of 4 or 5. On both ultrasound shear wave systems used, the lesion is very stiff, coding well within the malignant range. The QM of the shear wave image was good (green).

Whenever SE and SWE give discordant results, an explanation based on the principles of those techniques should be sought. If one cannot be identified, the more suspicious findings should be used in characterizing the lesion.

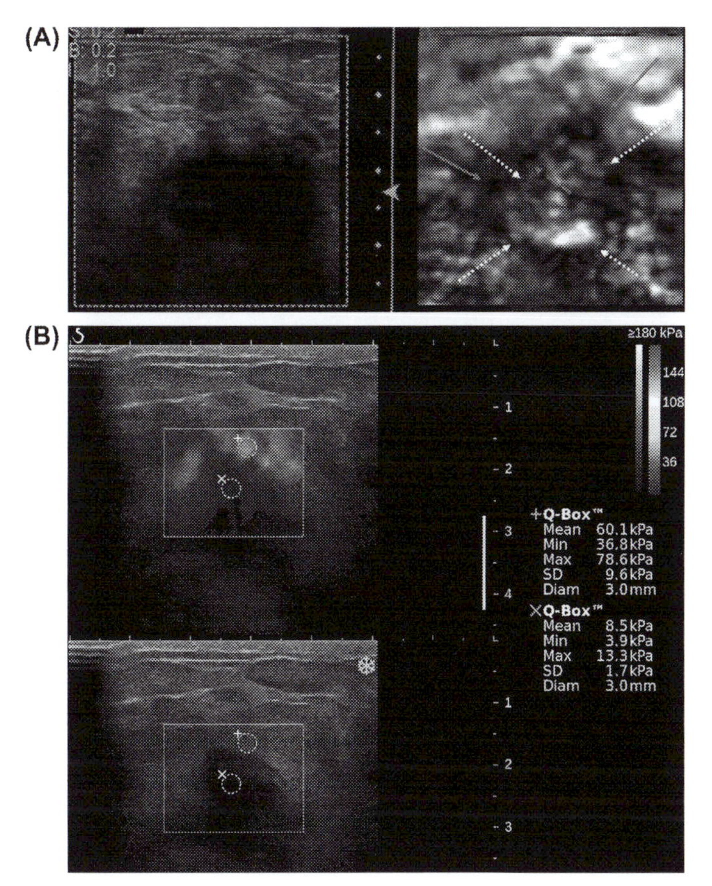


**FIGURE 2.5**

A 43-year-old female presented with a painful mass in her right breast. (A) On strain elastography the lesion has a soft (*dotted yellow arrows* [white in print version]) center and a stiff outer rim (*red arrows* [black in print version]). (B) On two-dimensional shear wave elastography the lesion has a stiffness value of 0.8 m/s (8.5 kPa) in the center and a stiffness value of 4.5 m/s (65 kPa) in the rim.

*Reproduced with permission from Sousaris N, Barr RG. Sonoelastography of Mastitis. J Ultrasound Med 2016; 35:1791–1797.*

### 3.1.7 Hematoma

Hematomas are typical after surgery, image-guided biopsy, or trauma, but they can occur spontaneously, particularly if the patient is on anticoagulant therapy. The age of the blood products determines the imaging appearance of the lesion. They are avascular and any vascularity in the lesion should raise the question of an associated lesion and a malignancy should be excluded.

In general the bull's-eye artifact or BGR appearance is not seen with hematomas [3]. Even when the hematoma is anechoic, there is enough viscosity in the lesion to

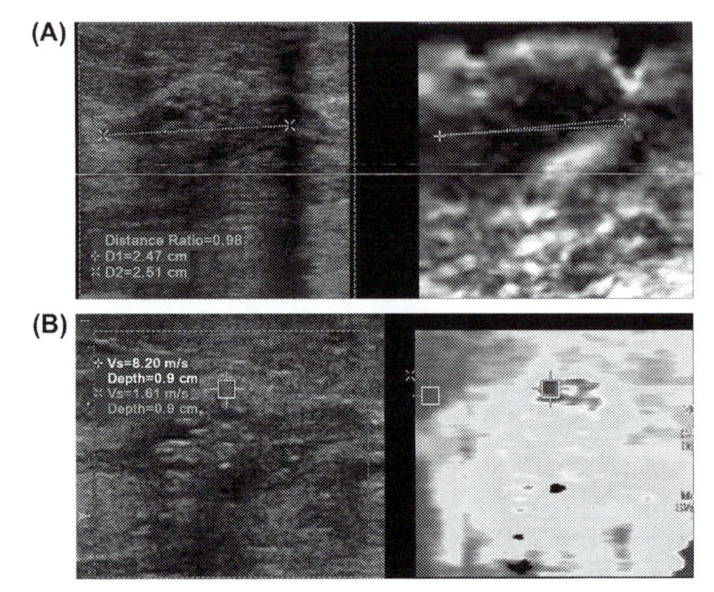


**FIGURE 2.6**

A 69-year-old female presented with a new palpable mass. (A) The lesion is at the site of a prior biopsy for benign papilloma. The lesion measures 2.51 cm and has an elastographic to B-mode length ratio of 0.98. (B) On two-dimensional shear wave elastography the lesion has a maximum stiffness value of 8.20 m/s. On biopsy the lesion was found to be fat necrosis.

prevent the occurrence of these strain artifacts. If the hematoma is chronic and the fluid within the lesion is mostly serous with or without some free-floating debris, the lesion may have the cyst artifact appearance. As in this case the usual appearance of SE is a soft (white) lesion with a stiff (black) fine wall.

On shear wave imaging, the hematoma will usually allow for generation of shear waves regardless of chronicity. The lesion will usually color-code as soft.

If there is a stiff area within or adjacent to the soft hematoma, the possibility of a second lesion should be considered. In this situation a malignancy should be excluded.

## 3.2 Malignant lesions

Malignant lesions have an E/B ratio 1 on SE, a score of 4 or 5 on the 5-point color scale, and a strain ratio of 4.5 (may vary based on equipment and laboratory). On SWE, malignant lesions have stiffness values greater than 4.5 m/s (60 kPa) in some series, whereas other series have a cutoff value of 5.2 m/s (80 kPa) [33].

### *3.2.1 Ductal carcinoma in situ*

DCIS is the most common type of noninvasive breast cancer.

DCIS rarely produces symptoms or a palpable mass, and it is mostly found on screening mammography as microcalcifications [42,43]. DCIS has become one of the most commonly diagnosed breast conditions; now accounting for 20% of breast cancers and precancers that are detected through screening mammography [43].

On SE, DCIS has an E/B ratio of almost 1. It is not uncommon to not be able to obtain accurate measurements on B-mode, making the E/B ratio difficult to use in some cases. On SWE, DCIS will have stiffness values of a malignancy; however, a well-designed study has not been performed.

### 3.2.2 Mucinous cancer

Mucinous carcinoma accounts for about 2% of breast cancers [44]. These cancers tend to manifest as soft masses because of the large amount of mucin within the lesion. Because of the predominance of mucin, this carcinoma typically manifests as a low-density, relatively well-defined, or microlobulated oval or lobular mass at mammography [45].

As mucinous cancers are softer than other breast cancers, there is a concern that their elastographic features may be different than those of other malignancies. In our experience, mucinous cancers are stiff enough to produce elastographic features similar to those of other breast malignancies. The E/B ratio is often 1.0 or close to 1.0. This is one reason why we prefer to use an E/B ratio $\geq$1.0 as our cutoff value for distinguishing between benign and malignant lesions. Some studies have found that an E/B ratio of 1.2 provides increased specificity; however, these studies did not include mucinous cancers or many cases of DCIS [3]. The mucin is very viscous and does not produce a bull's-eye artifact [36].

### 3.2.3 Invasive ductal cancer

The morphology of breast cancer is heterogeneous, as breast cancer is not one disease.

On SE the E/B ratio is greater than 1, the 5-point color scale score is 4 or 5, and the strain ratio is >4.5 (may vary by vendor). On SWE the mass stiffness is often very heterogeneous (Fig. 2.7). The stiffness value of the mass or the surrounding 3 mm has a stiffness value $\geq$4.5 m/s (60 kPa). There is a problem with shear wave generation in IDC [5,6]. Some IDCs will not color-code (are black) but there may be a rim of high stiffness value surrounding the mass (Fig. 2.8). In a small number of cases the IDC may color-code blue or green (Fig. 2.9). Evaluation of the displacement curves used to calculate the SWS in these cases has found noise that is interpreted as a slow SWS [6]. The addition of a QM can help determine if these cases are either false-negative results [6] or positive results on SE.

### 3.2.4 Invasive lobular cancer

Invasive lobular cancer (ILC) accounts for 6%–9% of breast cancers.

At ultrasonography, the numerous sheets of tumor cells will frequently cause architectural distortion and posterior acoustic shadowing and, as at mammography, often without a discrete mass. It may occasionally be difficult to distinguish the mild posterior acoustic shadowing that may be seen with fibrocystic changes from that of ILC.

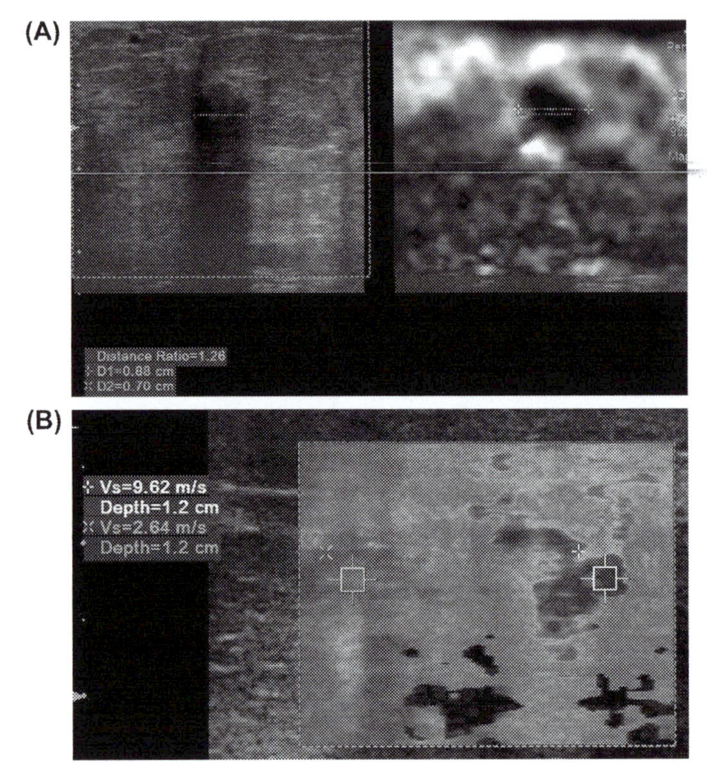


**FIGURE 2.7**

A 66-year-old female presented with a new breast lesion on screening mammography. On ultrasonography a 7-mm well-circumscribed lesion with marked shadowing corresponds to the mammographic lesion. (A) On strain elastography the lesion has a strain ratio of 1.3 suggestive of a malignant lesion. (B) On two-dimensional shear wave elastography the lesion has high stiffness through out with a stiffness value of 9.6 m/s consistent with a malignancy. On biopsy the lesion was found to be an invasive ductal carcinoma.

Usually, SE and SWE findings are concordant and highly suggestive of a malignancy. On SE, as a "mass" is not often identified using the 5-point color scale, comparing the stiffness of the area of concern to the glandular tissue is helpful.

### *3.2.5 Lymphoma*

Breast lymphomas account for approximately 0.15% of malignant breast carcinomas. Approximately half are primary lymphomas and the other half are metastatic to the breast. Lymphomas appear drastically different on elastography than primary breast cancers, with the lesion mostly being soft. These lesions tend to have markedly increased blood flow on color Doppler imaging. In patients with a history of lymphoma and a very vascular soft breast mass, lymphoma should be considered. Diagnosing primary breast lymphomas is more difficult, as these usually present

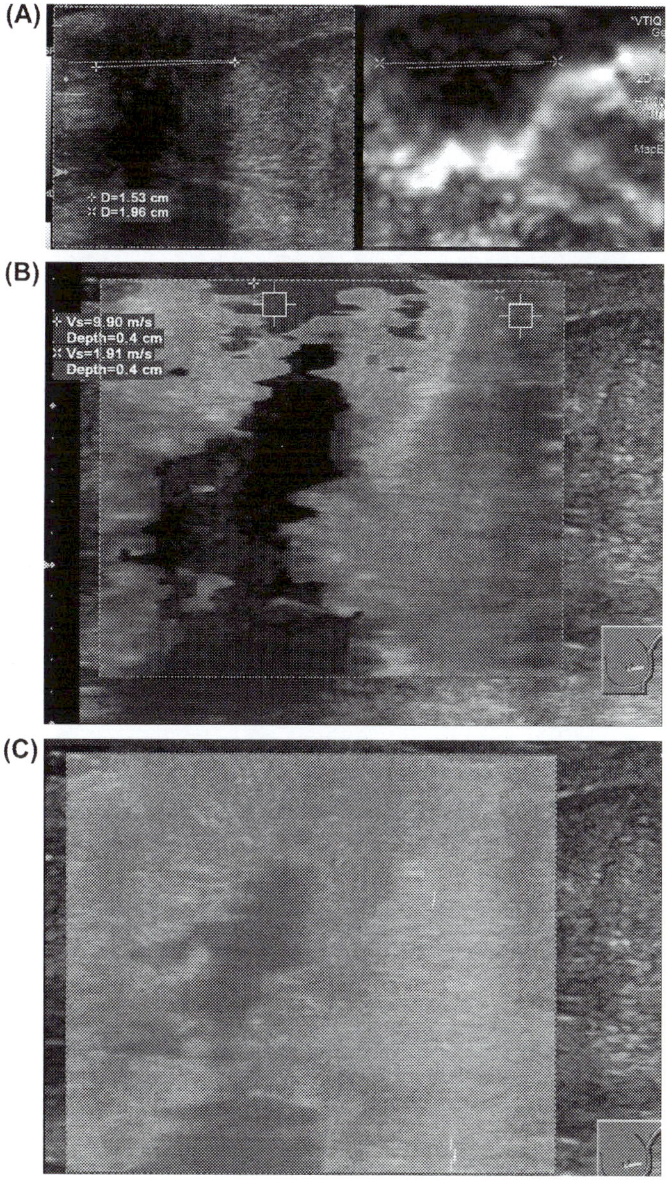

(A)

(B)

(C)

**FIGURE 2.8**

An 84-year-old female presented with a palpable mass. (A) On strain elastography the mass is stiff with an elastographic to B-mode length ratio of 1.3. (B) On two-dimensional shear wave elastography the majority of the mass does not color-code. However, there is a stiff rim surrounding the lesion, with a maximum stiffness value of 9.9 m/s. (C) The quality map confirms that estimates of shear wave speed in areas not color-coded are poor.

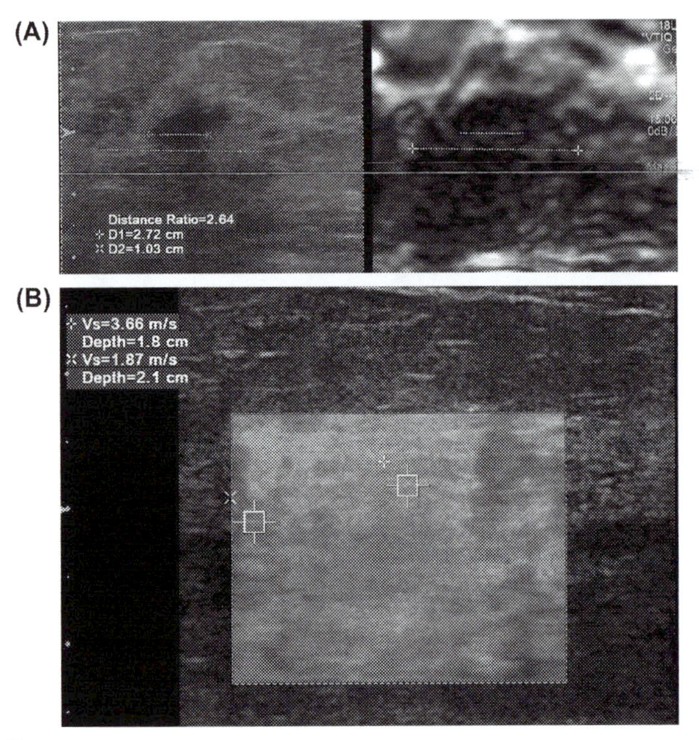


**FIGURE 2.9**

A 64-year-old female presented with a new lesion on screening mammography. (A) On strain elastography (SE) the lesion has an SR of 2.6 suggestive of a malignant lesion. (B) On two-dimensional shear wave elastography (SWE) the lesion has a low stiffness value of 3.7 m/s suggestive of a benign lesion. In these "soft" cancers, on SWE, the SE images are always suggestive of a malignant lesion. This is due to poor shear wave propagation and can be identified with a quality map of the shear waves.

with a well-circumscribed mass that is soft on elastography; however, markedly increased blood flow should raise the possibility of primary breast lymphoma [46].

## 3.3 Others

### *3.3.1 Metastatic diseases of the breast*

The incidence of metastases to the breast from nonbreast primary cancers is less than 1%. On elastography, this lesion has a unique characteristic of a soft center, representing necrosis, with a stiff outer rim suggestive of malignancy. In general, primary

(*orange-amber* [dark gray in print version]). On biopsy the lesion was found to be an invasive ductal carcinoma.

breast cancers do not demonstrate central necrosis but have a stiff avascular center. This appearance of a central soft area with a very stiff rim can also be seen with mastitis. This can be differentiated from a cyst using the bull's-eye sign, as the distal white spot is not present. The appearance of a "donut" lesion with a soft center and stiff rim should raise the question of mastitis or a nonbreast malignancy [47].

#### *3.3.2 Recurrence at surgical scar*

Surgical scars are relatively soft opposed to the finding on clinical palpation where they "feel" very stiff. However, residual or recurrent tumors are very stiff and can be identified attached to or adjacent to the surgical scar by their relative stiffness. When evaluating a surgical scar, imaging of the entire surgical scar and surrounding tissue should be performed. In this case the patient was able to locate the recurrence by the change in the surgical scar targeting elastographic evaluation. As most scars have significant shadowing, recurrences within the areas of shadowing, including distal to the scar, cannot be evaluated with elastography. No studies have been published on this topic, so the accuracy of elastography in detecting recurrence is not known.

For areas of residual or recurrent tumor, shear wave velocities values greater than 4.5 m/s (60 kPa) would be expected. The use of three-dimensional shear wave is helps screen the scar and the surrounding tissue for areas of increased stiffness suspicious for recurrent tumor.

#### *3.3.3 Gynecomastia*

Gynecomastia is one of the most common diseases of the male breast. It is a benign proliferation (hypertrophy) of the ductal and glandular elements in the male breast.

No studies have been published on the accuracy of elastography in the diagnosis of gynecomastia versus male breast carcinoma. In our experience, gynecomastia has a similar appearance to benign glandular tissue in female patients. On SE the lesion has a similar stiffness to the surroundings, whereas on SWE, gynecomastia has a Vs value similar to the glandular breast tissue in female patients.

## 4. Opportunities

Breast elastography has very high sensitivities and specificities for the characterization of breast lesions. However, the learning curve of SE has limited its clinical acceptance. With SWE the learning curve is less steep but the problem of poor or no shear wave propagation in breast cancers has led to the concern of possible false-negative results.

### 4.1 Quality measures

In very hard lesions such as invasive cancers, the shear wave may not propagate in an orderly fashion. No results are therefore obtained and the area with no results is not color-coded (Fig. 2.8). In these areas, interpretation is not possible. However, in general the periphery of the tumor will be hard and appear as a hard halo surrounding the lesion. Even if the entire mass is not coded as hard, heterogeneity of the SWE is part of the criteria for a suspicious lesion. Care must be taken with precompression, as this can also create the same appearance in a benign lesion [8].

In a large number of malignant lesions, the area identified on B-mode as the hypoechoic mass often does not code on SWE because a shear wave is not identified or may code with a low SWS. Bai [48] found that 63% of breast malignancies have this finding. Preliminary work in the evaluation of this phenomenon suggests that shear waves may not propagate as expected in some malignant lesions [5]. Evaluation of the shear waves in these malignant lesions demonstrates significant noise that may be incorrectly interpreted as a low shear wave velocity by the algorithm.

The addition of a QM that evaluates the shear waves generated and determines if they are adequate for an accurate Vs (or kPa) measurement will help in eliminating possible false-negative cases [5,6]. On the Siemens system a Quality Map has been added. The "traffic light" display aids in faster assessment of the shear wave quality and therefore in the degree of confidence in the accuracy of the velocity results, especially in cases where the velocity map may not correlate with the clinical assessment. On the Supersonic Imagine (SSI) system the shear wave quality is incorporated into the velocity map by not color-coding the areas with poor shear wave quality. The modification of SWE algorithms to account for the noisy shear waves in breast cancers is needed to limit the possibility of false-negative breast cancer results. A study [6] reported that the addition of a QM to shear wave imaging of the breast can limit false-negative findings (sensitivity without QM 22/46 (48%, 95% confidence interval [CI], 33%–63%) and with QM 42/46 (91%, 95% CI, 79%–98%, $P < .0001$)).

## 5. Artifacts and limitations

There are several artifacts that occur in both SE and SWE. Some of these are due to poor technique, while others contain diagnostic information.

### 5.1 Bull's-eye artifact

A unique artifact is identified within cystic lesions in some systems, called the bull's-eye artifact [36]. This artifact is characterized by a white central signal within a black outer signal and a bright spot posterior to the lesion (Fig. 2.2). This artifact is caused because the fluid is moving and there is decorrelation between images. This artifact has been described in detail. This artifact has a high predictive value for the lesion being a benign simple or complicated cyst. If there is a solid component in the cyst, it will appear as a solid lesion within the pattern. Although limited cases have been reported, this artifact is not seen in mucinous or colloid cancers.

The bull's-eye cyst pattern can be seen with lesions that appear solid and suspicious on B-mode imaging. These lesions have been shown to be complicated cysts. These lesions can be aspirated to confirm if the lesion resolves after aspiration. Note that this artifact is seen with Siemens and Philips equipment and may not be seen in other manufacturers' equipment. With other systems, cysts may give a layering color pattern (BRG artifact discussed later).

This artifact can be used to decrease the number of biopsies performed. In one series, 10% of complicated cysts appeared solid on B-mode are complicated cysts that can be identified with this technique [36]. If core biopsies are performed, notifying the pathologist that the lesion is a complicated cyst as opposed to a solid mass will lead to better pathology/imaging correlation.

### 5.2 Blue, green, and red artifact

Some ultrasound systems have a different artifact that occurs in cysts (Fig. 2.2C). There is a 3-color-layering pattern of BGR (with blue for stiff) identified in cystic lesions. A detail study evaluating the sensitivity and specificity of this artifact has not been performed [3].

### 5.3 Worm artifact

If there is very little variability in the elastic properties of the tissues in the FOV a pattern of varying noise is seen. This can occur when only one tissue is in the FOV or when significant precompression is applied. This artifact can be eliminated by the use of minimal precompression and by including various tissue types in the FOV [3].

### 5.4 Sliding artifact

A soft ring or group of wavelike soft rinds around a lesion in SE indicates the lesion is moving in and out of the imaging plane while the elastogram is being obtained. Maintaining the same imaging plane in the lesion during elastogram acquisition can eliminate this artifact. This artifact can be seen with fibroadenomas or lipomas. This artifact is usually only seen with benign lesions, as it requires the lesion to move separately from the surrounding tissues. In malignant lesions the lesion is usually attached to the surrounding tissue [3].

### 5.5 Bang artifact

If precompression is applied, there will be a stiff area in the near field. This can be minimized by not applying precompression. The use of ample amounts of coupling gel is helpful in eliminating this effect [3].

## 6. Summary/conclusions

Elastography, both SE and SWE, has shown high sensitivity and specificity for characterizing breast lesions as benign or malignant. Each technique has its advantages and disadvantages. The advantage of SE is that it can be used as long as a B-mode image can be obtained regardless of the lesion depth. The disadvantages of SE are that it is a relative technique and does not provide the actual stiffness measurement. SE has a steeper learning curve than SWE. The advantage of SWE is that it provides the actual stiffness measurement and has a less steep learning curve. The disadvantage of SWE is that it is limited in deep lesions and some malignancies may give a false-negative result. The use of both techniques can lead to increased confidence in the findings [7]. The main disadvantage of SWE is that many cancers do not allow for adequate shear wave propagation, which can lead to no SWS calculated or a slow SWS resulting in a false-negative finding. As there is some variability between systems and techniques, the appropriate interpretation should be based on the literature of the system used.

Version 5 of BI-RADS allows for downgrading a BI-RADS 4A lesion to BI-RADS 3 if the elastography is suggestive of a benign lesion. It also allows for upgrading a BI-RADS 3 lesion to a BI-RADS 4a lesion if the elastography is suggestive of a malignant lesion. In a meta-analysis of the E/B ratio method, the overall negative likelihood ratio suggests that a lesion with a pretest probability of 50% (all BI-RADS 4B lesions) can be downgraded to a 2% probability (BI-RADS 3).

Another possible application of elastography that has not been investigated is to perform biopsy in the area of highest stiffness for the best radiology-pathology correlation.

In some systems the bull's-eye artifact can be used to characterize cystic lesions as benign complicated cysts or complex cystic masses. This has been shown to substantially decrease the number of benign biopsies [36].

Elastographic systems continue to evolve and new tools and new evidence will likely emerge in the near future.

## References

[1] M. Tanter, et al., Quantitative assessment of breast lesion viscoelasticity: initial clinical results using supersonic shear imaging, Ultrasound Med. Biol. 34 (9) (2008) 1373–1386.

[2] T.A. Krouskop, et al., Elastic moduli of breast and prostate tissues under compression, Ultrason. Imaging 20 (4) (1998) 260–274.

[3] R.G. Barr, Breast Elastography, Thieme Publishers, New York, NY, 2015.

[4] R.G. Barr, Sonographic breast elastography: a primer, J. Ultrasound Med. 31 (5) (2012) 773–783.

[5] R.G. Barr, Shear wave imaging of the breast: still on the learning curve, J. Ultrasound Med. 31 (3) (2012) 347–350.

[6] R.G.A.Z Barr, Z, Shear Wave Elastography of the Breast: Value of a Quality Measure and Comparison to Strain Elastography 275 (2015) 45–53.

[7] R.G. Barr, Breast Elastography, Thieme Publishers, New York, NY, 2014.

[8] R.G. Barr, Z. Zhang, Effects of precompression on elasticity imaging of the breast: development of a clinically useful semiquantitative method of precompression assessment, J. Ultrasound Med. 31 (6) (2012) 895–902.

[9] R.G. Barr, et al., Evaluation of breast lesions using sonographic elasticity imaging: a multicenter trial, J. Ultrasound Med. 31 (2) (2012) 281−287.
[10] T.J. Hall, Y. Zhu, C.S. Spalding, In vivo real-time freehand palpation imaging, Ultrasound Med. Biol. 29 (3) (2003) 427−435.
[11] R.G. Barr, Real-time ultrasound elasticity of the breast: initial clinical results, Ultrasound Q. 26 (2) (2010) 61−66.
[12] E.I.A. Ueno, Diagnosis of breast cancer by elasticity imaging, Eizo Joho Med. 36 (2004) 2−6.
[13] A. Itoh, et al., Breast disease: clinical application of US elastography for diagnosis, Radiology 239 (2) (2006) 341−350.
[14] E. Ueno, T. Umemoto, H. Bando, E. Tohno, K. Waki, T. Matsumura, New quantitative method in breast elastography: fat lesion ratio (FLR), in: Paper Presented at: Radiological Society of North America 93rd Scientific Assembly and Annual Meeting; November 25−30, 2007, 2007. Chicago, IL.
[15] A. Thomas, et al., Significant differentiation of focal breast lesions: calculation of strain ratio in breast sonoelastography, Acad. Radiol. 17 (5) (2010) 558−563.
[16] A. Farrokh, S. Wojcinski, F. Degenhardt, Diagnostic value of strain ratio measurement in the differentiation of malignant and benign breast lesions, Ultraschall der Med. 32 (4) (2011) 400−405.
[17] S. Destounis, et al., Clinical experience with elasticity imaging in a community-based breast center, J. Ultrasound Med. 32 (2) (2013) 297−302.
[18] J.R. Grajo, R.G. Barr, Strain elastography in the prediction of breast cancer tumor grade, J. Ultrasound Med. 33 (2014) 129−134.
[19] K. Nakashima, T. Moriya, Comprehensive ultrasound diagnosis for intraductal spread of primary breast cancer, Breast Cancer 20 (1) (2013) 3−12.
[20] S. Raza, et al., Using real-time tissue elastography for breast lesion evaluation: our initial experience, J. Ultrasound Med. 29 (4) (2010) 551−563.
[21] J.M. Chang, et al., Breast mass evaluation: factors influencing the quality of US elastography, Radiology 259 (1) (2011) 59−64.
[22] H. Zhi, et al., Ultrasonic elastography in breast cancer diagnosis: strain ratio vs 5-point scale, Acad. Radiol. 17 (10) (2010) 1227−1233.
[23] S.M. Alhabshi, et al., Semi-quantitative and qualitative assessment of breast ultrasound elastography in differentiating between malignant and benign lesions, Ultrasound Med. Biol. 39 (4) (2013) 568−578.
[24] A. Stachs, et al., Differentiating between malignant and benign breast masses: factors limiting sonoelastographic strain ratio, Ultraschall der Med. 34 (2) (2013) 131−136.
[25] S. Chang, et al., Inflammatory breast carcinoma incidence and survival: the surveillance, epidemiology, and end results program of the National Cancer Institute, 1975−992, Cancer 82 (12) (1998) 2366−2372.
[26] A. Athanasiou, et al., Breast lesions: quantitative elastography with supersonic shear imaging–preliminary results, Radiology 256 (1) (2010) 297−303.
[27] A. Evans, et al., Quantitative shear wave ultrasound elastography: initial experience in solid breast masses, Breast Cancer Res. 12 (6) (2010) R104.
[28] W.A. Berg, et al., Shear-wave elastography improves the specificity of breast US: the BE1 multinational study of 939 masses, Radiology 262 (2) (2012) 435−449.
[29] J.M. Chang, et al., Clinical application of shear wave elastography (SWE) in the diagnosis of benign and malignant breast diseases, Breast Cancer Res. Treat. 129 (1) (2011) 89−97.

[30] M. Tozaki, S. Isobe, M. Sakamoto, Combination of elastography and tissue quantification using the acoustic radiation force impulse (ARFI) technology for differential diagnosis of breast masses, Jpn. J. Radiol. 30 (8) (2012) 659–670.
[31] D.O. Cosgrove, et al., Shear wave elastography for breast masses is highly reproducible, Eur. Radiol. 22 (5) (2012) 1023–1032.
[32] D. Cosgrove, et al., EFSUMB guidelines and recommendations on the clinical use of ultrasound elastography. Part 2: clinical applications, Ultraschall der Med. 34 (3) (2013) 238–253.
[33] R.G. Barr, et al., WFUMB guidelines and recommendations for clinical use of ultrasound elastography: part 2: breast, Ultrasound Med. Biol. 41 (5) (2015) 1148–1160.
[34] T. Shiina, et al., WFUMB guidelines and recommendations for clinical use of ultrasound elastography: part 1: basic principles and terminology, Ultrasound Med. Biol. 41 (5) (2015) 1126–1147.
[35] American College of Radiology, American College of Radiology Breast Imaging Reporting and Data System (BIRADS) Ultrasound. Fifth edition, first ed., American College of Radiology, Reston VA, 2013.
[36] R.G. Barr, A.E. Lackey, The utility of the "bull's-eye" artifact on breast elasticity imaging in reducing breast lesion biopsy rate, Ultrasound Q. 27 (3) (2011) 151–155.
[37] G. Cardenosa, G.W. Eklund, Benign papillary neoplasms of the breast: mammographic findings, Radiology 181 (3) (1991) 751–755.
[38] W.T. Yang, M. Suen, C. Metreweli, Sonographic features of benign papillary neoplasms of the breast: review of 22 patients, J. Ultrasound Med. 16 (3) (1997) 161–168.
[39] T.H. Kim, et al., Sonographic differentiation of benign and malignant papillary lesions of the breast, J. Ultrasound Med. 27 (1) (2008) 75–82.
[40] B.K. Han, et al., Benign papillary lesions of the breast: sonographic-pathologic correlation, J. Ultrasound Med. 18 (3) (1999) 217–223.
[41] N. Sousaris, R.G. Barr, Sonographic elastography of mastitis, J. Ultrasound Med. 35 (2016) 1791–1797.
[42] H.G. Welch, S. Woloshin, L.M. Schwartz, The sea of uncertainty surrounding ductal carcinoma in situ–the price of screening mammography, J. Natl. Cancer Inst. 100 (4) (2008) 228–229.
[43] V.L. Ernster, et al., Detection of ductal carcinoma in situ in women undergoing screening mammography, J. Natl. Cancer Inst. 94 (20) (2002) 1546–1554.
[44] J.W. Berg, R.V. Hutter, Breast cancer, Cancer 75 (Suppl. l) (1995) 257–269.
[45] J.A. Harvey, Unusual breast cancers: useful clues to expanding the differential diagnosis, Radiology 242 (3) (2007) 683–694.
[46] N. Sousaris, R.G. Barr, Sonoelastography of Breast Lymphoma, Ultrasound Q, 2016.
[47] N. Sousaris, G. Mendelsohn, R.G. Barr, Lung cancer metastatic to breast: case report and review of the literature, Ultrasound Q. 29 (3) (2013) 205–209.
[48] M. Bai, et al., Virtual touch tissue quantification using acoustic radiation force impulse technology: initial clinical experience with solid breast masses, J. Ultrasound Med. 31 (2) (2012) 289–294.

CHAPTER

# Clinical applications of elastographic methods to improve prostate cancer evaluation

3

**Eduardo Gonzalez[1,3], Fanny L. Casado[2], Benjamin Castaneda[1]**
*[1]Laboratorio de Imágenes Médicas, Pontificia Universidad Católica del Perú, Lima, Peru; [2]Instituto de Ciencias Ómicas y Biotecnología Aplicada, Pontificia Universidad Católica del Perú, Lima, Perú; [3]Department of Biomedical Engineering, Johns Hopkins University, Baltimore, MD, United State*

## 1. Introduction

Worldwide, prostate cancer (PCa) is the second most common cancer in men. Its incidence varies considerably among different countries very likely due to different screening programs, and there is a 10-fold difference in mortality outcomes. The highest incidence as age-standardized rate (ASR) per 1000 people has been reported in 2012 in Norway (129.7 ASR) and the lowest in Bhutan (1.2 ASR). PCa has the highest incidence (30.4 ASR) and mortality (14.9 ASR) for cancers in Peruvian men [1]. Among the several factors that improve life expectancy, such as risk stratification and proper treatment, fast and reliable diagnosis of cancerous lesions plays an important role.

For the past three decades, PCa has been diagnosed after a lengthy process of elimination and invasive tests. First, prostate abnormalities are screened by a combination of digital rectal examination (DRE), which palpates the posterior region of the prostate, and prostate-specific antigen (PSA) measurements. Abnormal results in these tests will lead to a 12-core sextant biopsy guided by B-mode transrectal ultrasonography (TRUS). TRUS aids in the visualization of the prostatic gland and placement of the needle during biopsy. However, it features a sensitivity and specificity for PCa detection that is affected by the stage of the lesion [2]. For instance, Cornud et al. [3] reported a sensitivity and specificity of 84% and 46%, respectively, when detecting lesions of PSA $<10$ ng/mL ($n = 131$), whereas for higher stages (PSA $>10$ ng/mL, $n = 141$) they were 76% and 62%, respectively. The major challenge of the current systematic use of biopsies is the high false-negative result rate [4], which can be improved [5] by increasing the predictive power of a larger number of biopsies (up to 40) but at a higher cost, morbidity, and overdiagnosis [6].

According to our current models of tumor initiation in PCa, a series of events are necessary to occur from a physiologic healthy state to a pathologic state. Our

Tissue Elasticity Imaging. https://doi.org/10.1016/B978-0-12-809662-8.00003-6

understanding of the cellular processes during tumor progression described 30 years ago [7] suggests that healthy tissue stroma reacts to damage caused by highly proliferative cells [8]. A similar process occurs during wound healing by collagen deposition orchestrated with immune mediators that lead to changes in mechanical properties such as its elastic modulus. The paradigm shift regarding the critical contribution of tumor microenvironment to sustain cancer has been particularly useful to avoid studying PCa as a single cell disease [9]. Stromal cells from Gleason 3 PCa showed an increased myofibroblast phenotype and exhibited elevated collagen I synthesis and expression of tenascin and fibroblast activation protein, suggesting features of stroma wounded and repaired [10]. In addition, the collagen accumulation in PCa has been shown to be exacerbated by molecular inhibition of signaling pathways responsible for cellular matrix metabolism [11]. Chung and colleagues [12] proposed the need for interaction between the PCa cells and their microenvironment for progression to androgen independence and metastasis. Their interpretation was that newly evolved PCa cell clones dominate cancer metastasis after cell-cell and cell-matrix interactions with the host microenvironment, rather than selected or expanded preexisting PCa cell clone(s). Other properties such as microvessel density and changes in stromal to glandular ratios may also be considered as biomarkers for PCa. Nevertheless, elasticity has endured as a marker for PCa despite the challenges of developing robust methods for its measurement.

Elastographic imaging methods propose to noninvasively assess the mechanical properties of soft tissues, such as their elastic modulus or stiffness. According to Hook's law, this elastic modulus is directly correlated with the strain (a measure of relative deformation) and stress (force per unit area acting on the body), with the latter applied from external sources or the imaging system [13]. As local strain differs between nonpathologic and pathologic tissues, elastography detects changes in physical tissue characteristics (elastic modulus) for use as a biomarker to discriminate between healthy and diseased states. Clinical diagnosis-treatment protocols designed to increase the usefulness of elastography in the management of PCa have been developed [14]. Seminal work proposing that increases in the elastic moduli of prostate tissue are due to collagen deposition provide a sound rationale for using elastography during diagnosis and localization of masses [15]. Support for this initial proposal comes from the work by Zhang et al. [16], in which mechanical measurements were estimated using a Kelvin-Voigt fractional derivative viscoelastic model. The mechanical stress relaxation data from ex vivo human prostatic tissue indicated an elastic modulus of $15.9 \pm 5.9$ kPa and $40.4 \pm 15.7$ kPa at 150 Hz in healthy and cancerous tissues, respectively.

Quantitative imaging elastographic techniques have also provided evidence supporting elasticity as a biomarker. Elastic ranges to characterize healthy and cancerous tissues are summarized in Table 3.1. Although all the different techniques show increases of elasticity in cancerous tissue, caution should be exercised when comparing the quantitative values among techniques. In fact, there is a current initiative whose goal is to standardize the estimations among different imaging methods [17].

**Table 3.1** Experimental ranges of elastic moduli characterizing healthy and cancerous tissues under various elastographic imaging methods.

| Imaging technique | Observations | Healthy tissue (kPa) | Cancerous tissue (kPa) | References |
|---|---|---|---|---|
| CrW sonoelastography | Ex vivo; posterior regions | $15.9 \pm 5.9$ | $40.4 \pm 15.7$ | [16] |
| CrW sonoelastography | Ex vivo; apex, middle gland, and close to the base | $31.9 \pm 14.8$ | $67.9 \pm 24.8$ | [18] |
| ARFI + SWE | Ex vivo; peripheral, central, and transition zones | $4.8 \pm 0.6$ | $10.0 \pm 1.0$ | [19] |
| SWE | In vivo | $21.2 \pm 11.8$ | $58.0 \pm 20.7$ | [20] |
| SWE | In vivo; peripheral zone | $21 \pm 12$ | $53 \pm 43$ | [21] |
| MRE | In vivo; peripheral, central, and transition zones | $9.83 \pm 2.37$ | $8.13 \pm 3.10$ | [22] |
| MRE (healthy volunteer) | In vivo; peripheral, central, and transition zones | $13.2 \pm 5$ | — | [22] |

ARFI, *acoustic radiation force impulse;* CrW, *crawling wave;* MRE, *magnetic resonance elastography;* SWE, *shear wave elastography.*

Even though the mechanical properties of prostate tumors may not always agree with the assumption that cancerous tissue has higher stiffness, this notion is highly accepted; perhaps because it is concordant with palpation, the time-tested approach of physicians to screen for abnormal growth in regions close to the skin's surface. Overall, the empirical data available present elastography as a promising method for reducing or circumventing altogether the need for a large number of invasive biopsies during diagnosis and therapy planning by taking advantage of the benefits of multiple elastographic modalities that are currently used or ready for clinical assessment in PCa. The following discussion groups the different elastographic methods based on the strategies used to apply pressure on the prostatic gland and promote tissue displacement to locate tumors.

## 2. Static deformation by compression

### 2.1 Strain elastography

Currently, this is the most popular prostate elastographic technique used clinically because it has been commercially available for over a decade and it is already part of clinical protocols worldwide [23,24]. In strain elastography (SE), a B-mode image of the prostate is acquired [25], next the organ is compressed under specific conditions for a certain amount of time, and a second image is obtained. The tissue displacement is estimated using correlation-based techniques to compute the strain difference of the pre- and postbeam formed radiofrequency data between the two states [26]. Then a strain map is generated where softer regions are represented as high strain values relative to its vicinity and vice versa. Given that the color map displayed is commonly scaled according to the highest and lowest strain values in the examined region, implementation of algorithms for automatic differentiation from lesions and healthy tissue are subject to severe limitations.

The main assumptions for SE are that the operator applies a homogeneous compression on the prostate and that there is no out-of-plane motion. This is a challenging task, as elastographic procedures involve freehand examination, with the radiologist controlling the pressure applied to the prostate to meet the aforementioned requirements. Therefore a well-trained technician is usually needed to maintain a constant compression during the examination and lessen the estimation error subject to the operator expertise. In order to decrease this dependency and increase the detection of stiffer masses, automatic balloon inflation devices have been developed and installed around the endorectal probe, balancing the induced pressure and generating a uniform pressure considerably wider than the conventional method (83.4 vs. 39.6 degree angle of visualization in the transrectal probe) [27]. For instance, Sumura et al. [28] conducted this method in 87 patients with PCa and reported an improved sensitivity and specificity of 72.5% and 97.7%, respectively, whereas the conventional strain method showed a sensitivity and specificity of 71.9% and 85.8%, respectively.

Quantitative approaches have been widely studied in SE using inversion schemes for computing the mechanical properties within soft tissues [29]. However, to date, this approach has not been demonstrated in a real-time clinical environment due to the limited preliminary information available about the tissue and the computational complexity of the algorithms. Although direct inversion methods have been validated in tissue-mimicking phantoms and excised tissues, most of them assume the plain-stress condition, which is not relevant for most clinical cases [30]. Furthermore, all these methods require accurate assessment of the pressure applied, which, contrary to the calibrated piston used in phantom testing, cannot be controlled objectively in real practice because it is the technician who must ensure a constant pressure during a clinical examination. Nevertheless, iterative solutions of the inverse problem for Young's modulus assessments have been successfully implemented in simulation phantoms as well as ex vivo and in vitro data using the well-known Hessian-based [31,32] and gradient-free [33,34] optimization methods.

Being that histopathology is the gold standard for tumor diagnosis, systematic prostate biopsy is normally used in conjunction with imaging modalities such as TRUS and magnetic resonance imaging (MRI). Although TRUS is considered a faster and cheaper tool for visualization of the prostate in comparison with MRI, it often cannot detect lesions. Sensitivity and specificity of B-mode TRUS images for PCa detection reported in the literature ranged from 44% to 90% and 30% to 74%, respectively [35]. Thus SE is used principally for biopsy guidance. The comparison of performance between TRUS and TRUS + SE has been thoroughly studied in the literature.

König et al. first implemented SE with TRUS for prostate biopsy. In their study, strain images and B-mode images from 404 men with suspicious DRE were compared with systematic sextant biopsies. From the 151 patients diagnosed with PCa, SE reported a sensitivity of 84.1% in comparison to the 64.2% using conventional B-mode images from TRUS [36]. Later, Zhang et al. [37] conducted a similar study with 83 patients in the prostate peripheral zone, reporting a sensitivity and specificity of 74.5% and 83.3%, respectively.

In the framework of Pallwein et al., comparison of SE with systematic biopsy findings was conducted on 492 PSA screening volunteers, where the peripheral zone of the prostate was divided into 3 regions: base, mid-gland, and apex. Average sensitivity and specificity of the entire prostate was 86% and 72%, respectively, while the highest sensitivity was found in the apex (79%). However, at the basal prostate areas, the authors associated the false-positive findings with chronic inflammation and atrophy [38]. A study involving 109 patients who underwent radical prostatectomy (RP) was later conducted by Salomon et al., reporting 75% sensitivity and 77% specificity for SE in the entire gland. Furthermore, the sensitivity and specificity improved when evaluating lesions of higher Gleason score sum (74% and 78%, respectively, for $\leq 6$; 93% and 93%, respectively, for $>7$) [39]. Finally, in 2014, Zhang et al. conducted an extensive meta-analysis of several frameworks that assessed the performance of SE using RP as validation. From a total of 508 patients with PCa, the sensitivity and specificity were 72% and 76%, respectively [40].

In addition to TRUS, SE performance for target biopsies has been compared with color Doppler ultrasonography (US) in target biopsies. For instance, Kamoi et al. evaluated the diagnostic performance of SE in comparison with power Doppler ultrasonography (PDUS) and TRUS on 107 patients, reporting 68% sensitivity and 81% specificity for SE, similar to those of PDUS (70% and 75%, respectively) and higher than the sensitivity (50%) of TRUS. In the detection of lesions with a high Gleason score, SE outperformed TRUS while presenting sensitivity similar to that of PDUS (SE: 13/13, PDUS: 8/13, TRUS: 5/13) [41]. Similarly, Nelson et al. compared the performance detection of PCa and distribution of Gleason scores with TRUS, PDUS, and SE. After correlating pathologic results with imaging findings, PCa was detected in 60 of 137 patients (43.8%) and in 241 (14%) of 1703 biopsy cores, including 90 (20%) of 448 targeted cores, 106 (13%) of 818 sextant cores, and 45 (10%) of 437 transition zone cores. Although targeted cores were more likely to detect cancer than sextant cores, no sonographic abnormality was found in 53.8% of positive sextant sites. These results showed that PDUS was strongly associated with Gleason score lesions between 8 and 10, whereas SE showed association starting at a Gleason score of 5. While the SE detection accuracy was still below the standard sextant biopsy accuracy, PCa detection from all the aforementioned methods had correlation with the Gleason scores [23].

Likewise, SE has been widely compared with multiparametric MRI (mpMRI)-guided biopsies [42,43]. In a 33-patient study, Aigner et al. [44] reported sensitivity of 84.6% and a negative predictive value (NPV) of 86.7%, which closely followed T2-weighted MRI performance (84.3% and 83.3%, respectively). Later, Pelzer et al. presented a higher sensitivity of SE (92%) in comparison to mpMRI (82%) in 50 examined patients scheduled for RP. Although SE showed advantages in apical and middle regions for PCa detection, both had limitations in the gland's base and ventral parts [45]. In a similar study, Junker et al. compared SE findings with those of mpMRI in a whole-mount step section analysis, obtaining an average sensitivity and specificity of 67.2% and 86.9%, respectively. The detailed report showed a significant decrease in SE accuracy in the transitional zone (18.2%) and anterior part (18.2%) of the prostate as well as detection lesions in prostates with volume greater than 40 $cm^3$ (42.1%) [46].

In other studies, Nygård et al. evaluated the performance of SE combined with PCa antigen 3 (PCA3), frequently used as a biomarker, in 124 patients. Seventy patients with PCa were diagnosed and divided in the low-risk (30%), intermediate-risk (46%), and high-risk (24%) categories. The results showed an enhanced sensitivity of 96% and a NPV of 90% for the intermediate- and high-risk PCa groups [47].

In conclusion, a large variety of studies have demonstrated the feasibility and validation of SE for guidance of target biopsies. Despite the limitations in the application of the technique, compensation procedures obtained by the automatic balloon inflation cope with the irregular distribution of pressure by the user. Moreover, SE has showed higher sensitivity than B-mode TRUS images while presenting enhanced accuracy when mixed with other imaging modalities such as color Doppler imaging or biomarkers such as PCA3. However, according to the analysis of sextant cores for PCa detection, the current studies suggest that standard sextant or 10-core biopsies are required in conjunction with SE [36].

## 3. Dynamic deformation exerted by external mechanical vibrators

### 3.1 Sonoelastography

Sonoelastography, also known as sonoelasticity imaging, was first proposed by Lerner et al. as an imaging technique measuring the amplitude response of soft tissue under harmonic excitation using ultrasonic Doppler methods. First, the tissue is excited with an external shear wave source produced by an external piston. Then, a spectral frequency variance estimator computes a sonoelasticity map from the acquired pulsed-repeated radiofrequency data [48]. In this map, stiffer regions produce disturbance in the vibration pattern and are usually seen as low tissue response, allowing one to distinguish it from healthy tissue [26]. Due to the low frequency (20–100 Hz) and low amplitude (20–100 μm) of the excitation signals, sonoelastography enables deep penetration in the tissue without increasing the risk of overheating the tissue [18].

Another advantage of this technique is the relatively low complexity of the frequency variance estimator used to compute the qualitative sonoelasticity map, which is comparable to the color Doppler algorithm used in commercial US scanners. Therefore real-time implementations have been conducted and validated for PCa detection [49]. For instance, Taylor et al. performed sonelasticity imaging on 19 specimens with G1 lesions (average volume of 3.0 ± 2.1 $cm^3$). The reported sonoelastographic performance was considerably better than the B-mode US in accuracy (55% vs. 17%) and sensitivity (71% vs. 29%) [50]. A similar study was conducted later by Castaneda et al. [51] on 10 patients who underwent RP, obtaining a sensitivity and specificity of 85% and 84%, respectively (over 80% accuracy for tumors larger than 4 mm in diameter), which were higher than the US results (30% and 100%) when compared with pathologic findings. Fig. 3.1 depicts an example of qualitative sonoelastography on an ex vivo prostate gland, in which the lesion is clearly differentiated in elasticity map in comparison with the B-mode image.

Quantitative characterization of the prostate in sonoelastography is likewise possible with the application of an additional vibration source for harmonic excitation. By setting a slight offset frequency between two opposing sources, an interference moving pattern is generated in the sonoelasticity map, where wider stripes are related to stiffer regions. Later, a shear wave speed (SWS) estimator calculates the local elasticity modulus, given the vibration frequency applied to the tissue [26]. In vitro studies conducted by Castaneda et al. [18] in 15 prostate glands after RP reported 80% accuracy in cancer detection for quantitative sonoelastography in comparison with qualitative sonoelastography (63%). The elasticity moduli reported for cancerous and healthy tissues were 67.9 ± 24.8 kPa and 31.9 ± 14.8 kPa, respectively. These results correlate with the framework of Zhang et al. [16], in which they reported a Young's elasticity modulus of 40.4 ± 15.7 kPa and 15.9 ± 5.9 kPa for cancerous and normal tissues, respectively, from 17 in vitro samples. Fig. 3.2 shows a SWS image in meters per second of two ex vivo prostate

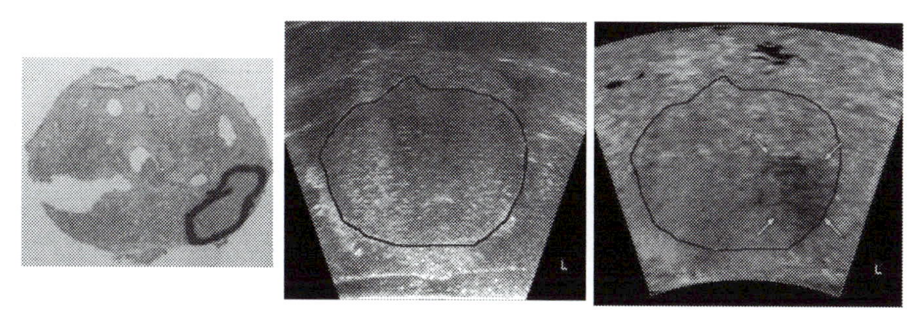

**FIGURE 3.1**

A case of qualitative sonoelastography on an ex vivo prostate gland visualizing a cross section close to the apex region. (A) Histology slice. (B) B-mode image. (C) Qualitative sonoelastography image depicting a mass at the bottom-right corner.

*Modified from B. Castaneda, K. Hoyt, K. Westesson, L. An, J. Yao, L. Baxter, J. Joseph, J. Strang, D. Rubens, K. Parker, Performance of three-dimensional sonoelastography in prostate cancer detection: a comparison between ex vivo and in vivo experiments, Int. Ultrasonics Ultrason. Symp. (2009) 519–522.*

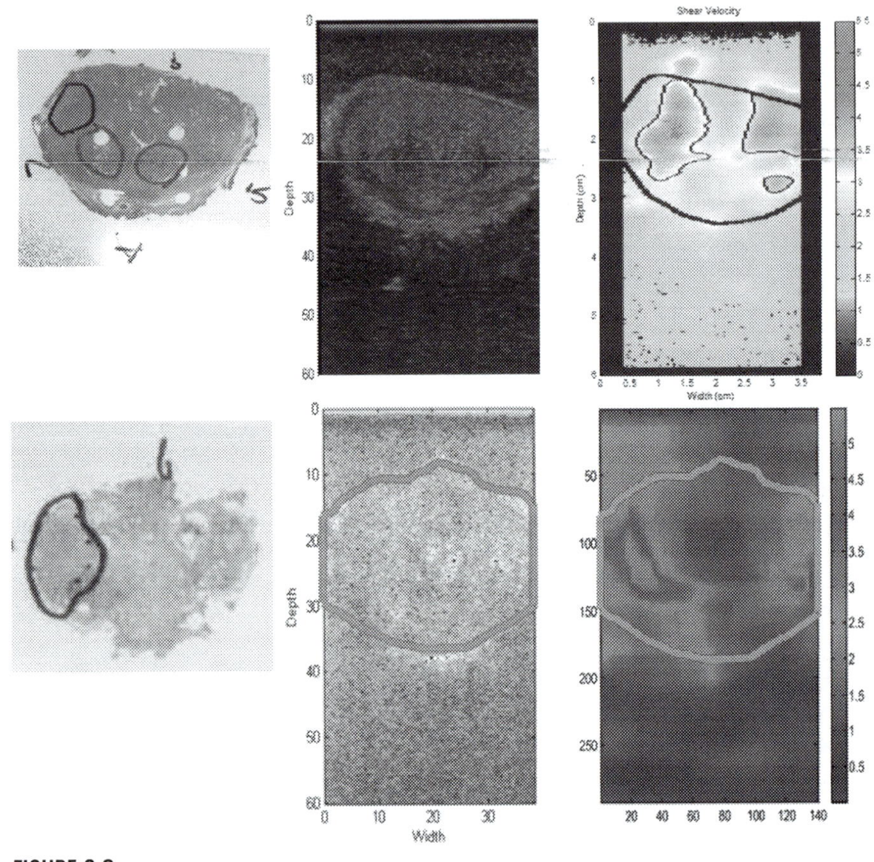


**FIGURE 3.2**

Two cases of crawling wave sonoelastography on ex vivo prostate glands, visualizing two cross sections close to the apex region. (A) Histology slice. (B) B-mode image. (C) Reconstructed shear wave speed image in meters per second.

*Modified from B. Castaneda et al., Prostate cancer detection using crawling wave sonoelastography, SPIE Med. Imaging 7265 (2009) 726513. 10: SPIE Medical Imaging.*

glands, in which cancerous tissue had higher SWS values than normal tissue as per histologic examination.

## 3.2 Vibroelastography

Proposed by Turgay et al. [52], vibroelastography (VE) is a dynamic elasticity method that combines multiple frequency components to excite the tissue simultaneously, measuring the displacement at multiple locations and time instants for a robust estimation of its biomechanical properties. Mahdavi et al. compared the performance of reconstructed volumes obtained from VE, MRI, and B-mode images of the prostate, reporting improved contrast-to-noise ratio as well as slightly lower error in volume reconstruction when compared with B-mode results. The total gland volume error was 8.8 ± 2.5% for VE versus MRI and 10.3 ± 4.6% for B-mode versus MRI [53]. Subsequently, they presented a fusion technique with B-mode and VE for automatic three-dimensional (3D) segmentation of 61 images of the prostate, reporting a reduced volume error of 10.2 ± 2.2% in comparison with manual segmentation (13.5% ± 4.1%) [54]. Lobo et al. [22] validated the technique with calibrated phantoms and performed experiments with prostatectomy samples, reporting a noticeable correlation of the VE images with histopathology slices, in which an area under receiver operating characteristic (ROC) curve of 0.82 ± 0.01 was achieved. Fig. 3.3 shows an example of transient transfer functions and absolute elasticity images of ex vivo prostate gland with PCa, both being featured modalities in VE. Transfer functions are qualitative maps that represent the tissue response at a certain frequency of excitation, where absolute elasticity images depict the shear modulus (in kilopascals) by the analysis of several transfer functions at different excitation frequencies.

## 3.3 Magnetic resonance elastography

This dynamic elasticity imaging technique uses mechanical waves to quantitatively measure the shear modulus of tissues [55]. The most popular clinical implementation is as an add-on to conventional MRI scanners. It applies shear waves with frequencies between 50 and 500 Hz through an external driver. The waves traveling

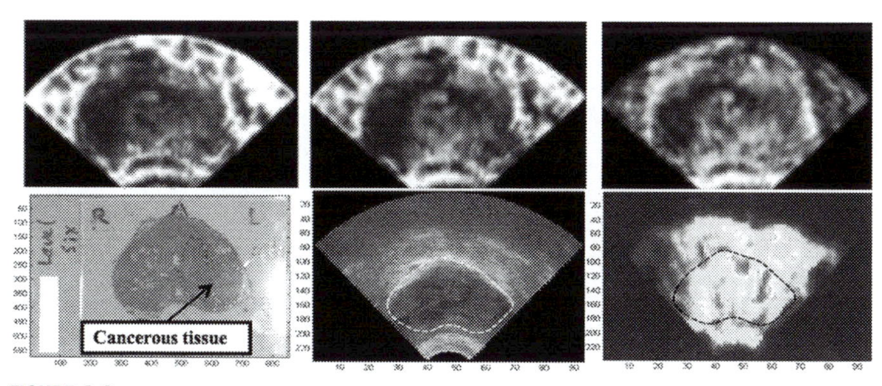


**FIGURE 3.3**

(A) Transfer function images at different excitation frequencies, resampled from axial transducer volumetric acquisition. Excitation amplitude, 0.5 mm; excitation frequency, 2–10 Hz; and radiofrequency data acquisition rate, 36 Hz. (B). Visualization of cancerous tissue in the prostate: (left) histopathology, (middle) B-mode, and (right) absolute elasticity image. Excitation frequency, 69–80 Hz.

*Modified from J. Lobo, A. Baghani, H. Eskandari, Prostate vibro-elastography: multi-frequency 1D over 3D steady-state shear wave imaging for quantitative elastic modulus measurement, Ultrasonics Symposium (IUS), 2015 IEEE International, 2015, pp. 1–4.*

inside the tissue are imaged using an MRI technique to generate quantitative data of tissue stiffness [56].

The first study assessing the technical feasibility of in vivo magnetic resonance elastography (MRE) of the prostate gland was performed using dynamic sinusoidal MRE in healthy volunteers [57]. In this study, Kemper and colleagues placed volunteers in the prone position and induced a mechanical wave via an external oscillator attached to the pubic bone. A 1.5-T magnetic resonance (MR) system was used to acquire data using a motion-sensitive spin echo MR sequence whose phase was locked to the mechanical oscillation. The images acquired were used to reconstruct the local distribution of elasticity inside the prostate gland, thus achieving enough penetration of the mechanical wave into the prostate gland in all volunteers. The reconstructed distribution of elasticity (shear modulus) inside the healthy prostate gland correlated with the zonal anatomy of the gland. The elasticity of the central portion ($2.2 \pm 0.3$ kPa) appeared to be lower than that of the peripheral prostatic portion ($3.3 \pm 0.5$ kPa).

More recently, a novel method for MRE was implemented at 3.0 T, in which a commercial endorectal coil was modified to dynamically generate mechanical stress in prostate phantoms with 6-mm lesions and in a porcine model. Tissue displacements were measured at actuation frequencies of 50–200 Hz to calculate maps of the shear modulus (G) from the measured phase-difference shear-wave patterns. Surprisingly, the authors found that the average G-values of simulated lesions were significantly higher ($8.2 \pm 1.9$ kPa) than those for background ($3.6 \pm 1.4$ kPa) but systematically lower than the values reported by the manufacturer (lesions: $13.0 \pm 1.0$ and background: $6.7 \pm 0.7$ kPa). In the porcine model, they reported shear moduli for muscle ($7.1 \pm 2.0$ kPa), prostate ($3.0 \pm 1.4$ kPa), and bulbourethral gland ($5.6 \pm 1.9$ kPa). These results encourage the technical feasibility of the modified technique [58]. In another study, transperineal prostate MRE was assessed for diagnostic power by correlation with histopathologic findings in cancer patients [59]. The 3D wave field of displacement was obtained using a fractionally encoded gradient echo sequence using a custom-made transducer. After quality assessment of the in vivo technique in healthy volunteers, patients were examined with MRE before RP. A windowed voxel-to-voxel technique compared the two-dimensional registered slides to the Gleason scores to calculate the areas under the ROC curves. The average elasticity of PCa patients was $8.2 \pm 1.7$ kPa in the prostate capsule, $7.5 \pm 1.9$ kPa in the peripheral zone, $9.7 \pm 3.0$ kPa in the central gland (CG), and $9.0 \pm 3.4$ kPa in the transition zone. Cancerous tissue with Gleason scores higher than $3 + 3$ was significantly different from healthy tissue in 10 out of 11 cases. However, cancerous tissue was not always stiffer than healthy tissue. Patient movement presented the biggest challenge for the measurements because it caused misalignment of the 3D wave field. Another strategy [20] was used to clinically assess whether fused 3D TRUS images and MR images of the prostate gland can provide high-quality information by combining statistical and biomechanical modeling. Data acquired from healthy men and 12 patients with suspected PCa show that this approach outperforms the model without patient-specific biomechanical parameters for higher than 15 kPa Young's modulus offsets. The study also shows that there is large variation in the average Young's modulus for healthy and PCa tissues and stresses the need to normalize data with biomechanical parameters on a patient basis.

Altogether, there is a lot of interest and potential to implement MRE techniques when MR equipment is available. However, the clinical protocols will need to deal with the noisy data provided by the mechanical waves to find prostate tumors.

## 4. Excitation by acoustic radiation force

### 4.1 Acoustic radiation force impulse imaging

The acoustic radiation force impulse (ARFI) method remotely palpates tissues through absorption of ultrasonic energy. ARFI imaging applies a focused US pulse (<1 ms) whose energy displaces the prostatic tissue to about 10 μm. The differential displacement with respect to the moment before applying the pulse is monitored using traditional B-mode US. Similar to SE images, the information provided is qualitative, where regions of decreased displacement suggest stiffer tissues. The clinical viability of this method was studied as an approach to detect local variations in the mechanical properties of soft tissues by applying radiation force to small volumes of tissue (2 $mm^3$) to exert displacements around 10 μm [21]. A decade later, the potential of this technique to guide prostate needle biopsy within specific structures of human prostates was shown in vivo [60]. This study used custom ARFI imaging sequences implemented with a scanner modified with a 3D wobbler, end-firing, transcavity transducer to image prostates of 19 patients before and after undergoing RP. PCa was present in most patients as bilaterally asymmetric stiff structures, while benign prostatic hyperplasia (BPH) appeared heterogeneous with a nodular texture. Unlike B-mode images, ARFI images were characterized by higher contrast of the prostate structures. However, the low resolution and limited depth penetration (22 mm) presented challenges when discriminating among PCa, BPH, and the discrete structures using the implemented qualitative ARFI. Recently, a comparison of prostatic volume estimation using B-mode/ARFI images and MR T2-weighted images, conducted by Palmeri et al.., reported that both US and ARFI volumes yielded a good correlation with MR results in the CG (R2 = 0.77 and 0.85, respectively). The CG volume differences were attributed to underestimation in apex-to-base regions (B-mode: $-10.8\% \pm 13.9\%$, ARFI: $-28.8\% \pm 9.4\%$) and overestimation of the lateral dimension (B-mode: $18.4\% \pm 13.9\%$, ARFI: $21.5\% \pm 14.3\%$) due to poor contrast caused by extraprostatic fat [61]. Fig. 3.4 shows histology slices from two patients with Gleason scores of 3 and 4, in which ARFI images feature better discrimination of the regions of suspicion than the B-mode images by depicting higher contrast due to decreased displacement.

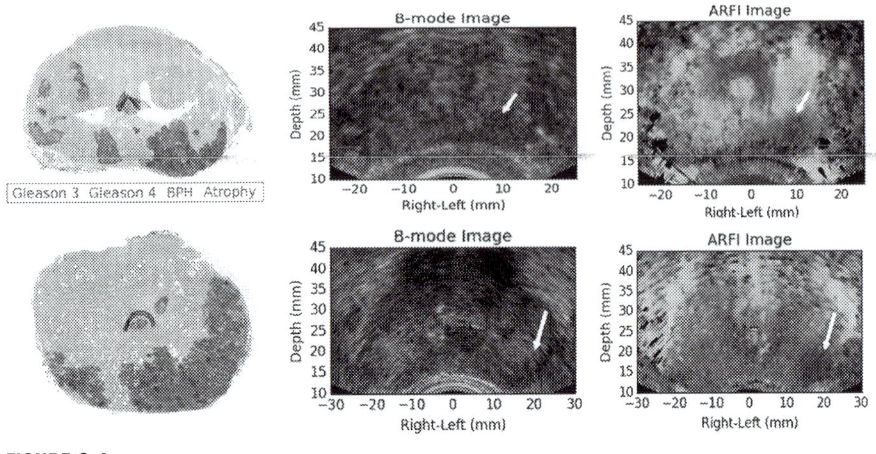


**FIGURE 3.4**

Examples of acoustic radiation force impulse (ARFI) imaging from regions of suspicion of two different study patients (A,B) that corresponded to large, posterior prostate cancer index lesions (*white arrows*). *BPH*, benign prostatic hyperplasia.

*Courtesy of M. Palmeri and Kathy Nightingale from the Duke University.*

## 4.2 Shear wave elastography

In quantitative ARFI or shear wave elasticity imaging (SWEI), first proposed by Sarvazyan et al. [62], acoustic focused beams are used to excite the examined tissue. Then shear waves are generated from the tissue displacement response and propagate away from the excitation region, where the behavior is tracked with US acquisitions in order to assess the shear modulus of the tissue. SWEI has been successfully implemented in several in vivo and ex vivo studies, such as in breast tissue and abdomen [63], as well as in quantifying the shear modulus of the liver [64]. Regarding prostate examination, Zhai et al. [19,65] conducted a study of 6 ex vivo prostate specimens that were imaged immediately after RP, reporting a shear modulus of 10.0 ± 1.0 kPa for PCa in contrast to the other healthy regions (peripheral zone: 4.1 ± 0.8 kPa, central zone: 9.9 ± 0.9 kPa, transition zone: 4.8 ± 0.6 kPa).

On the other hand, shear wave elastography (SWE) or supersonic shear imaging (SSI), proposed by Bercoff et al., is based on the utilization of ultrafast imaging to monitor the propagation of plane shear waves created by a series of focused US beams at different depths. The velocity of propagation of the tissue away from the exciting wave is monitored to calculate the stiffness of the soft tissue. For this purpose, SWE requires an ultrafast acquisition of approximately 5000 frames/s, using inversion algorithms to quantitatively map the shear elasticity of the tissue. In particular, SSI refers to the supersonic speed of the tissue excited by ultrasonic focused beams. Similar to the sonic boom originating from a supersonic aircraft, these shear waves interfere constructively along a Mach cone, creating two intense plane shear waves that propagate through the tissue and are progressively distorted by tissue's heterogeneities. The main assumption of the shear wave velocity estimation algorithm is that the medium is locally homogeneous. While the technique was first tested on healthy volunteers, the investigators proposed it initially for breast cancer detection [66].

SWE's quantitative assessment of prostatic tissues outperformed grayscale US imaging in a prospective study of 50 volunteers undergoing diagnostic biopsies because of suspected PCa. In this study the US image of the whole gland was arbitrarily divided into 12 zones analogously to the routine sampling for biopsies. Next, the regions were imaged by grayscale US and SWE, and quantitative data from SWE was compared to the histopathologic findings. Data analyzed per core for SWE was further grouped into patients with PSA <20 μg/L and >20 μg/L. The sensitivity and specificity for the low-PSA group were 0.9 and 0.88, respectively, whereas in the high-PSA group, they were 0.93 and 0.93, respectively. It was also observed that PCa had significantly higher stiffness values than those of benign tissues [67]. Fig. 3.5 depicts examples of stiffness differentiation in the prostate from healthy and PCa patients, while assessing the shear modulus in the region of examination.

Barr and colleagues compared the results of paired standard TRUS-guided biopsies and SWE techniques in 53 patients (318 sextants; mean age, 64.2 years) with elevated PSA levels (mean PSA = 5.05 ng/mL > 4.0 ng/mL) or abnormal DRE results. A second TRUS examination and sextant biopsy by a second physician blinded to SWE results was then performed. Using ROC curve analysis, a value of 37 kPa was determined as the cutoff between benign and malignant, with a sensitivity of 96.2% (25/26) and specificity of 96.2% (281/292). However, with a 40 kPa threshold value, all cases of PCa would have been detected. Of the 11 false-positive samples, 55% were secondary or benign calcifications. Altogether,

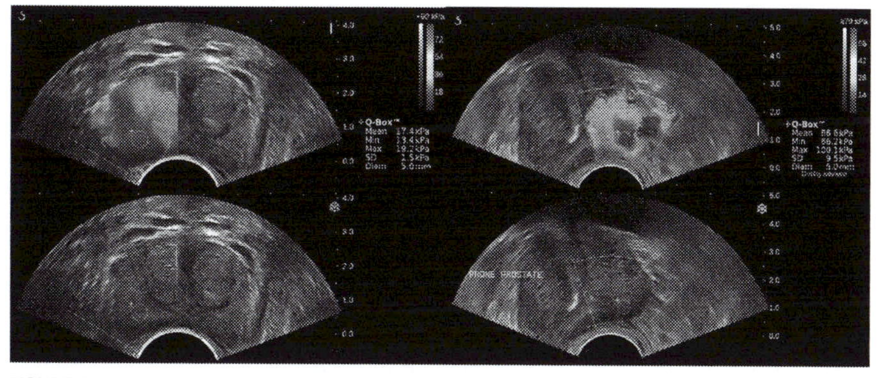


**FIGURE 3.5**

Examples of stiffness estimation using shear wave elastography (SWE). Left: SWE image from a 59-year-old patient with negative findings in biopsy and low stiffness value (17.4 ± 1.5 kPa). Right: SWE image from a 79-year-old patient with a G7 lesion at 1.5 cm depth (86.6 ± 9.5 kPa).

*Courtesy of Richard Barr from the Northeastern Ohio Medical University.*

these results suggest that SWE may help reduce the negative biopsy rate of PCa diagnosis [68]. A larger study including 184 men shows that a 35 kPa cutoff can help reduce the number of biopsies while ensuring detection of peripheral zone adenocarcinomas in patients with high and/or increasing PSA levels and/or abnormal DRE results [69]. In this study, TRUS SWE of the prostate was performed after a conventional TRUS examination followed by an US-guided 12-core sextant biopsy for pathologic evaluation. Elasticity values for 1040 peripheral zone sextants from the 184 men were matched to pathologic results at the patient and sextant levels. The reported sextant-level sensitivity and specificity were 96% and 85%, respectively, using the maximized SWE Youden index for the elasticity threshold. However, it is worth mentioning that the Youden index thresholding has proven not to be useful for making clinical decisions [70].

In summary, SWE imaging has recently been introduced commercially for prostate imaging [14] and the most encouraging results show that SWE may enable a substantial reduction in the number of biopsies while ensuring detection of peripheral zone adenocarcinomas [68].

## 5. Current status and future trends

SE feasibility has been widely validated in the literature for guidance of target biopsies. Currently, measurement errors due to the instability of the pressure applied by the user can be compensated with automatic balloon inflation techniques, which generate a uniform distribution and widen the effective area of examination. On the other hand, several studies have demonstrated a higher sensitivity for SE than B-mode TRUS images in the detection of PCa. Furthermore, a combination of SE with techniques, such as Color-Doppler Imaging (CDI) and PCA3 biomarker detection, has reported increased accuracy and better discrimination of prostatic regions. However, the quality of the images currently reached in SE is not sufficient for a definitive diagnosis. Hence, standard sextant or 10-core biopsies are still required in conjunction to SE.

Sonoelastography provides qualitative and quantitative information of the prostate using external vibrations of low frequency. Thus it features a wide penetration range while avoiding the risk of overheating the tissue unlike ARFI. In vivo and in vitro experiments have proven its capability to differentiate cancerous from benign tissue in the prostate. However, the implementation of quantitative sonoelastography presents several challenges due to the reflection of shear waves and the successful coupling of both actuators—opposing each other—into the examined tissue. Similarly, while VE has been implemented on real-time US systems focused on prostate visualization and segmentation, further studies are required for Young's modulus assessment of healthy and malignant tissue.

ARFI imaging is not yet commercially available for prostate imaging because it presents limitations such as low penetration range due to the attenuation of sound in the intervening tissue, apparent increase of the lesion with time after excitation due to shear wave reflections at the lesion boundaries, and the risk of tissue overheating. However, recent studies have demonstrated higher contrast of the prostate structures than conventional B-mode imaging, while showing promise for volume assessment of the gland when comparing with more sophisticated techniques such as T2-weighted MRI.

Due to the high estimation sensitivity and specificity for PCa, SWE recently became available on end-fire endocavity transducers, being already included in the World Federation for Ultrasound in Medicine and Biology guidelines of prostate examination with US. Despite the fact that technicians must be trained on the proper use of TRUS and US-guided biopsy procedures with SWE, the technique has a smaller learning curve than that for SE. Moreover, the addition of SWE to prostate care can improve staging over TRUS alone. However, there is still insufficient evidence to make a recommendation regarding SWE.

Individually, elastographic techniques have reported promising results on the detection accuracy and classification of PCa. However, the combination of shear-velocity-based approaches and quantitative strain elasticity methods applied on the same clinical case could potentially improve both the elasticity estimates and the cutoff values for detecting malignant lesions. Moreover, emerging shear-wave-based techniques such as reverberant elastography could benefit from the shear wave reflections at tissue boundaries, enhancing the signal-to-noise ratio and providing a more robust quantification of the tissue stiffness. Up till now, those reflections have created only problems for the estimation of SWS and tissue characterization. With the addition of other quantitative ultrasound (QUS) parameters that involve backscatter and attenuation coefficient estimation [71], probability maps of PCa findings could be achieved, which would benefit exploration time and diagnostic accuracy. Finally, enhancement of detection and classification of PCa could be achieved with the application of neural networks and deep learning algorithms [72], which are arguably the state of the art in image processing. For this, a well-structured data base involving subgroups of the aforementioned elastography and QUS techniques [71], prostate postoperative atlases [73], and patient's metadata [74] should be used as inputs to the training algorithm, while ground truth values (outputs) are obtained from prostatectomy and/or histopathologic findings.

## 6. Conclusions

In the past decade, several elastographic approaches have been developed for PCa diagnosis. These techniques bring new information regarding the tissue stiffness of the gland, which could aid technicians to locate suspicious regions for guiding biopsies. Compared to the conventional US, elastographic methods yield higher accuracy as well as high NPV in most of the prostate regions, suggesting that their performance can be further enhanced in combination with TRUS and/or MRI, as well as contrast approaches such as PCA33 imaging. Additionally, recent studies have demonstrated the possibility that elastographic methods are beneficial to estimate the volume of the prostate. However, there are still many challenges to overcome in the use of elastographic methods for PCa care, which is reflected in the broad range of reported accuracy values. Clinical applications of elastography are being only slowly incorporated into clinical practice mostly because of the mixed results regarding sensitivity and specificity for PCa diagnosis. Although this technique shows promise to help reduce the number of biopsies or to better guide biopsies, more work is required to change clinical practice and replace histopathologic analysis as the best way to diagnose PCa. A road map for clinical translation should include addressing user-dependence and proposing rigorous guidelines to properly assess elastograms of the prostate, with the final goal of improving the quality of life of patients.

## References

[1] J. Ferlay, et al., GLOBOCAN 2012 v1.0, Cancer Incidence and Mortality Worldwide: IARC CancerBase No. 11, June 29, 2013. Available: http://globocan.iarc.fr.

[2] J.M. Correas, A.M. Tissier, A. Khairoune, G. Khoury, D. Eiss, O. Hélénon, Ultrasound elastography of the prostate: state of the art (in Eng), Diagn. Interv. Imaging 94 (5) (2013) 551–560.

[3] F. Cornud, X. Belin, D. Piron, T. Flam, J. Casanova, O. Helenon, A. Mejean, N. Thiounn, J. Moreau, Color Doppler-guided prostate biopsies in 591 patients with an elevated serum PSA level: impact on Gleason score for nonpalpable lesions, Urology 49 (5) (1997) 709–715.

[4] H. Singh, et al., Predictors of prostate cancer after initial negative systematic 12 core biopsy (in Eng), J. Urol. 171 (5) (2004) 1850–1854.

[5] A.V. Taira, et al., Performance of transperineal template-guided mapping biopsy in detecting prostate cancer in the initial and repeat biopsy setting (in Eng), Prostate Cancer Prostatic Dis. 13 (1) (2010) 71–77.

[6] P. Pepe, F. Fraggetta, A. Galia, G. Grasso, F. Aragona, Prostate cancer detection by TURP after repeated negative saturation biopsy in patients with persistent suspicion of cancer: a case-control study on 75 consecutive patients (in Eng), Prostate Cancer Prostatic Dis. 13 (1) (2010) 83–86.

[7] H.F. Dvorak, Tumors: wounds that do not heal. Similarities between tumor stroma generation and wound healing (in Eng), N. Engl. J. Med. 315 (26) (1986) 1650–1659.

[8] J. Marx, How cells cycle toward cancer (in Eng), Science 263 (5145) (1994) 319–321.

[9] S. Gout, J. Huot, Role of cancer microenvironment in metastasis: focus on colon cancer (in Eng), Cancer Microenviron. 1 (1) (2008) 69–83.

[10] J.A. Tuxhorn, G.E. Ayala, M.J. Smith, V.C. Smith, T.D. Dang, D.R. Rowley, Reactive stroma in human prostate cancer: induction of myofibroblast phenotype and extracellular matrix remodeling (in Eng), Clin. Cancer Res. 8 (9) (2002) 2912–2923.

[11] M. Nilsson, H. Adamo, A. Bergh, S. Halin Bergström, Inhibition of lysyl oxidase and lysyl oxidase-like enzymes has tumour-promoting and tumour-suppressing roles in experimental prostate cancer (in Eng), Sci. Rep. 6 (2016) 19608.

[12] L.W. Chung, A. Baseman, V. Assikis, H.E. Zhau, Molecular insights into prostate cancer progression: the missing link of tumor microenvironment (in Eng), J. Urol. 173 (1) (2005) 10–20.

[13] T. Shiina, K. Nightingale, M. Palmeri, T. Hall, J. Bamber, R. Barr, L. Castera, B. Choi, Y. Chou, D. Cosgrove, C. Dietrich, H. Ding, D. Amy, A. Farrok, G. Ferraioli, C. Filice, M. Friedrich-Rust, K. Nakashima, M. Kudo, WFUMB guidelines and recommendations for clinical use fo ultrasound elastography: part 1: basic principles and terminology, Ultrasound Med. Biol. 41 (5) (2015) 1126–1147.

[14] R. Barr, D. Cosgrove, M. Brock, V. Cantisani, J. Correas, A. Postema, G. Salomon, M. Tsutsumi, H. Xu, C. Dietrich, WFUMB guidelines and recommendations on the clinical use of ultrasound elastography: part 5. Prostate, Ultrasound Med. Biol. 43 (1) (2017) 27–48.

[15] T.A. Krouskop, T.M. Wheeler, F. Kallel, B.S. Garra, T. Hall, Elastic moduli of breast and prostate tissues under compression (in Eng), Ultrason. Imaging 20 (4) (1998) 260–274.

[16] M. Zhang, et al., Quantitative characterization of viscoelastic properties of human prostate correlated with histology (in Eng), Ultrasound Med. Biol. 34 (7) (2008) 1033–1042.

[17] QIBA and QI/imaging biomarkers in the literature, J. Ultrasound Med. 33 (2014) S1–S124.

[18] B. Castaneda, et al., Prostate cancer detection using crawling wave sonoelastography, SPIE Med. Imaging 7265 (2009) 726513, 10: SPIE Medical Imaging.

[19] L. Zhai, et al., Characterizing stiffness of human prostates using acoustic radiation force (in Eng), Ultrason. Imaging 32 (4) (2010) 201–213.

[20] Y. Wang, D. Ni, J. Qin, M. Xu, X. Xie, P.A. Heng, Patient-specific deformation modelling via elastography: application to image-guided prostate interventions (in Eng), Sci. Rep. 6 (2016) 27386.

[21] K. Nightingale, M.S. Soo, R. Nightingale, G. Trahey, Acoustic radiation force impulse imaging: *in vivo* demonstration of clinical feasibility (in Eng), Ultrasound Med. Biol. 28 (2) (2002) 227–235.

[22] J. Lobo, A. Baghani, H. Eskandari, Prostate vibro-elastography: multi-frequency 1D over 3D steady-state shear wave imaging for quantitative elastic modulus measurement, in: Ultrasonics Symposium (IUS), 2015 IEEE International, 2015, pp. 1–4.

[23] E.D. Nelson, C.B. Slotoroff, L.G. Gomella, E.J. Halpern, Targeted biopsy of the prostate: the impact of color Doppler imaging and elastography on prostate cancer detection and Gleason score (in Eng), Urology 70 (6) (2007) 1136–1140.

[24] Japan Society of Ultrasonics in Medicine, Terminology and Diagnostic Criteria Committee, Clinical practice guidelines for ultrasound elastography: prostate (in Eng), J. Med. Ultrason. 43 (3) (2016) 449–455.

[25] D.L. Cochlin, R.H. Ganatra, D.F. Griffiths, Elastography in the detection of prostatic cancer (in Eng), Clin. Radiol. 57 (11) (2002) 1014–1020.

[26] K.J. Parker, M.M. Doyley, D.J. Rubens, Imaging the elastic properties of tissue: the 20 year perspective (in Eng), Phys. Med. Biol. 56 (1) (2011) R1–R29.

[27] M. Tsutsumi, T. Miyagawa, T. Matsumura, T. Endo, S. Kandori, T. Shimokama, S. Ishikawa, Real-time balloon inflation elastography for prostate cancer detection and initial evaluation of clinicopathologic analysis, Am. J. Roentgenol. 194 (6) (2010).

[28] M. Sumura, K. Shigeno, T. Hyuga, T. Yoneda, H. Shiina, M. Igawa, Initial evaluation of prostate cancer with real-time elastography based on step-section pathologic analysis after radical prostatectomy: a preliminary study (in Eng), Int. J. Urol. 14 (9) (2007) 811–816.

[29] M.M. Doyley, Model-based elastography: a survey of approaches to the inverse elasticity problem (in Eng), Phys. Med. Biol. 57 (3) (2012) R35–R73.

[30] C. Sumi, Spatially variant regularization for tissue strain measurement and shear modulus reconstruction (in Eng), J. Med. Ultrason. 34 (3) (2007) 125–131.
[31] J. Jiang, et al., Young's modulus reconstruction for radio-frequency ablation electrode-induced displacement fields: a feasibility study (in Eng), IEEE Trans. Med. Imaging 28 (8) (2009) 1325–1334.
[32] S. LeFloch, G. Cloutier, G. Finet, P. Tracqui, R. Pettigrew, J. Ohayon, Vascular imaging modulography: an experimental in vitro study, Comput. Methods Biomech. Biomed. Eng. 13 (2010) 89–90.
[33] A.A. Oberai, N.H. Gokhale, G.R. Feijoo, Solution of inverse problems in elasticity imaging using the adjoint method, Inverse Probl. 19 (2003) 297–313.
[34] R.A. Baldewsing, F. Mastik, J.A. Schaar, P.W. Serruys, A.F. van der Steen, Young's modulus reconstruction of vulnerable atherosclerotic plaque components using deformable curves (in Eng), Ultrasound Med. Biol. 32 (2) (2006) 201–210.
[35] M. Seitz, et al., Imaging procedures to diagnose prostate cancer (in Ger), Urologe A 46 (10) (2007) W1435–W1446. quiz W1447-8.
[36] K. König, U. Scheipers, A. Pesavento, A. Lorenz, H. Ermert, T. Senge, Initial experiences with real-time elastography guided biopsies of the prostate, J. Urol. 174 (1) (2005) 115–117.
[37] Y. Zhang, J. Tang, Y. Li, X. Fei, E. He, Y. Gao, Diagnostic value of strain index in the prostate peripherical zone lesions by real time tissue elastography, Zhongguo Yi Xue Ke Xue Yuan Xue Bao 32 (5) (2010) 549–552.
[38] L. Pallwein, et al., Sonoelastography of the prostate: comparison with systematic biopsy findings in 492 patients (in Eng), Eur. J. Radiol. 65 (2) (2008) 304–310.
[39] G. Salomon, et al., Evaluation of prostate cancer detection with ultrasound real-time elastography: a comparison with step section pathological analysis after radical prostatectomy (in Eng), Eur. Urol. 54 (6) (2008) 1354–1362.
[40] B. Zhang, et al., Real-time elastography in the diagnosis of patients suspected of having prostate cancer: a meta-analysis (in Eng), Ultrasound Med. Biol. 40 (7) (2014) 1400–1407.
[41] K. Kamoi, et al., The utility of transrectal real-time elastography in the diagnosis of prostate cancer (in Eng), Ultrasound Med. Biol. 34 (7) (2008) 1025–1032.
[42] S. Sarkar, S. Das, A review of imaging methods for prostate cancer detection (in Eng), Biomed. Eng. Comput. Biol. 7 (Suppl. 1) (2016) 1–15.
[43] D. Junker, et al., Real-time elastography of the prostate (in Eng), BioMed Res. Int. (2014) 180804.
[44] F. Aigner, et al., Comparison of real-time sonoelastography with T2-weighted endorectal magnetic resonance imaging for prostate cancer detection (in Eng), J. Ultrasound Med. 30 (5) (2011) 643–649.
[45] A. Pelzer, J. Heinzelbecker, C. Weiss, Real-time sonoelastography compared to magnetic resonance imaging using four different modalities at 3.0 T in the detection of prostate cancer: strength and weaknesses, Eur. J. Radiol. 82 (5) (2013) 814–821.
[46] D. Junker, et al., Comparison of real-time elastography and multiparametric MRI for prostate cancer detection: a whole-mount step-section analysis (in Eng), Am. J. Roentgenol. 202 (3) (2014) W263–W269.
[47] Y. Nygård, S. Haukaas, O. Halvorsen, K. Gravdal, J. Frugård, L. Akslen, C. Beisland, A positive Real-Time Elastography (RTE) combined with a Prostate Cancer Gene 3 (PCA3) score above 35 convey a high probability of intermediate- or high-risk prostate cancer in patient admitted for primary prostate biopsy, BMC Urol. 1 (39) (2016).

[48] K. Hoyt, B. Castaneda, M. Zhang, P. Nigwekar, P. di Sant'agnese, J. Joseph, J. Strang, D. Rubens, K. Parker, Tissue elasticity properties as biomarkers for prostate cancer (in Eng), Cancer Biomark. 4 (4–5) (2008) 213–225.

[49] B. Castaneda, J. Ormachea, P. Rodríguez, K.J. Parker, Application of numerical methods to elasticity imaging (in Eng), Mol. Cell. Biomech. 10 (1) (2013) 43–65.

[50] S. Taylor, D. Rubens, B. Porter, Z. Wu, R. Baggs, P. di Sant'Agnese, G. Nadasdy, D. Pasternack, E. Messing, P. Nigwekar, K. Parker, Prostate cancer: three-dimensional sonoelastography for in vitro detection, Genitourin. Imaging 237 (3) (2005).

[51] B. Castaneda, K. Hoyt, K. Westesson, L. An, J. Yao, L. Baxter, J. Joseph, J. Strang, D. Rubens, K. Parker, Performance of three-dimensional sonoelastography in prostate cancer detection: a comparison between *ex vivo* and *in vivo* experiments, Int. Ultrason. Symp. (2009) 519–522.

[52] E. Turgay, S. Salcudean, R. Rohling, Identifying the mechanical properties of tissue by ultrasound strain imaging (in Eng), Ultrasound Med. Biol. 32 (2) (2006) 221–235.

[53] S.S. Mahdavi, M. Moradi, X. Wen, W.J. Morris, S.E. Salcudean, Evaluation of visualization of the prostate gland in vibro-elastography images (in Eng), Med. Image Anal. 15 (4) (2011) 589–600.

[54] S.S. Mahdavi, M. Moradi, W.J. Morris, S.L. Goldenberg, S.E. Salcudean, Fusion of ultrasound B-mode and vibro-elastography images for automatic 3D segmentation of the prostate (in Eng), IEEE Trans. Med. Imaging 31 (11) (2012) 2073–2082.

[55] R. Muthupillai, D.J. Lomas, P.J. Rossman, J.F. Greenleaf, A. Manduca, R.L. Ehman, Magnetic resonance elastography by direct visualization of propagating acoustic strain waves (in Eng), Science 269 (5232) (1995) 1854–1857.

[56] Y.K. Mariappan, K.J. Glaser, R.L. Ehman, Magnetic resonance elastography: a review (in Eng), Clin. Anat. 23 (5) (2010) 497–511.

[57] J. Kemper, R. Sinkus, J. Lorenzen, C. Nolte-Ernsting, A. Stork, G. Adam, MR elastography of the prostate: initial *in vivo* application (in Eng), Röfo 176 (8) (2004) 1094–1099.

[58] G. Thörmer, M. Reiss-Zimmermann, J. Otto, K. Hoffmann, M. Moche, N. Garnov, T. Kahn, H. Busse, Novel technique for MR elastography of the prostate using a modified standard endorectal coil as actuator, J. Magn. Reson. Imaging 37 (6) (2013) 1480–1485.

[59] R.S. Sahebjavaher, et al., MR elastography of prostate cancer: quantitative comparison with histopathology and repeatability of methods (in Eng), NMR Biomed. 28 (1) (2015) 124–139.

[60] L. Zhai, et al., Acoustic radiation force impulse imaging of human prostates: initial *in vivo* demonstration (in Eng), Ultrasound Med. Biol. 38 (1) (2012) 50–61.

[61] M. Palmeri, Z. Miller, T. Glass, K. Garcia-Reyes, R. Gupta, S. Rosenzweig, C. Kauffman, T. Polascik, A. Buck, E. Kullbacki, J. Madden, S. Lipman, N. Rouze, K. Nightingale, B-mode and acoustic radiation force impulse (ARFI) imaging of prostate zonal anatomy: comparison with 3T T2-weighted MR imaging, Ultrasound Imaging 37 (1) (2015) 22–41.

[62] A. Sarvazyan, O. Rudenko, S. Swanson, J. Fowlkes, S. Emelianov, Shear wave elasticity imaging: a new ultrasonic technology for medical diagnostics, Ultrasound Med. Biol. 24 (9) (1998) 1419–1435.

[63] K. Nightingale, S. McAleavey, G. Trahey, Shear-wave generation using acoustic radiation force: *in vivo* and *ex vivo* results, Ultrasound Med. Biol. 29 (12) (2003) 1715–1723.

[64] M. Palmeri, M. Wang, J. Dahl, K. Frinkley, K. Nightingale, Quantifying hepatic shear modulus *In vivo* using acoustic radiation force, Ultrasound Med. Biol. 34 (4) (2008) 546–558.

[65] L. Zhai, J. Madden, V. Mouraviev, T. Polascik, K. Nightingale, Correlation between SWEI and ARFI image findings in *ex vivo* human prostates, Ultrason. symp. IEEE Int. (2009) 523–526.

[66] J. Bercoff, M. Tanter, M. Fink, Supersonic shear imaging: a new technique for soft tissue elasticity mapping (in Eng), IEEE Trans. Ultrason. Ferroelectr. Freq. Control 51 (4) (2004) 396–409.

[67] S. Ahmad, R. Cao, T. Varghese, L. Bidaut, G. Nabi, Transrectal quantitative shear wave elastography in the detection and characterisation of prostate cancer (in Eng), Surg. Endosc. 27 (9) (2013) 3280–3287.

[68] R.G. Barr, R. Memo, C.R. Schaub, Shear wave ultrasound elastography of the prostate: initial results (in Eng), Ultrasound Q. 28 (1) (2012) 13–20.

[69] J.M. Correas, et al., Prostate cancer: diagnostic performance of real-time shear-wave elastography (in Eng), Radiology 275 (1) (2015) 280–289.

[70] N. Perkins, E. Schisterman, The inconsistency of "optimal" cutpoints obtained using two criteria based on the receiver operating characteristic curve, Am. J. Epidemiol. 63 (7) (2006) 670–675.

[71] J. Mamou, M. Oelze, Quantitative Ultrasound in Soft Tissues, Springer, 2013.

[72] K.Diaz, B.Castaneda, Semi-automated segmentation of the prostate gland boundary in ultrasound images using a machine learning approach, Proc. SPIE 6914, Medical Imaging 2008: Image Processing.

[73] K. Diaz, B. Castaneda, M. Montero, J. Yao, J. Joseph, D. Rubens, K. Parker, Analysis of the spatial distribution of prostate cancer obtained from histopathological images, Proc. SPIE 8676, Medical Imaging 2013: Digital Pathology 86760V.

[74] X. Yang, B. Fei, 3D prostate segmentation of ultrasound images combining longitudinal image registration and machine learning, Proc. SPIE 8316, Medical Imaging 2012: Image-Guided Procedures, Robotic Interventions, and Modeling 83162O.

CHAPTER

# Cardiovascular elastography

**Elisa Konofagou**
*Columbia University, New York, NY, United States*

## 1. Cardiac imaging

### 1.1 Myocardial elastography

#### *1.1.1 Introduction*

Cardiovascular diseases remain America's primary killer by a large margin, claiming the lives of more Americans than the next two main causes of death combined (cancer and pulmonary complications). In particular, coronary artery disease (CAD) is by far the most lethal, causing 17% of all (cardiac-related or not) deaths every year. According to the latest report on Heart Disease and Stroke Statistics by the American Heart Association [1], in 2019, ≈720,000 Americans will have a new coronary event (defined as first hospitalized myocardial infarction (MI) or CAD death) and ≈335,000 will have a recurrent event [1]. Individuals self-reporting low income and low education have twice the incidence of CAD as those reporting high income and high education, thereby warranting a low-cost modality. However, there are currently no screening or early detection imaging techniques that can identify abnormalities prior to any symptoms or fatalities. Clinically available imaging techniques such as echocardiography or nuclear perfusion (radionuclide imaging) are typically used after a cardiac event has already occurred to determine the extent of damage. Despite the fact that over the past few years, therapeutic techniques, such as angioplasty, heart valve surgery and pacemakers, have experienced exponential growth in new procedures, the progress in the development of novel diagnostic techniques has stalled by comparison.

#### *1.1.2 Mechanical deformation of normal and ischemic or infarcted myocardium*

Detection of cardiac dysfunction through assessment of the mechanical properties of the heart, and more specifically, the left-ventricular muscle, has been a long-term goal in diagnostic cardiology. The increased stiffness could be due to myocardial remodeling, including elevated collagen and desmin expression as well as the titin isoform switch [1a]. Acute myocardial infarction caused by partial or total blockage of one or more coronary arteries can cause complex structural alterations of the left-ventricular muscle [1b]. These alterations may lead to collagen synthesis and scar formation, which can cause the myocardium to irreversibly change its mechanical properties.

Tissue Elasticity Imaging. https://doi.org/10.1016/B978-0-12-809662-8.00004-8

### *1.1.3 Imaging the deformation of the myocardium*

#### 1.1.3.1 Motion estimation

Echocardiography uses B-mode and/or Doppler to visually detect wall motion abnormalities that the physician subsequently rates on a scoresheet. Several methods have been developed to render these measurements quantitative and less subjective. Such motion estimation techniques include Doppler Myocardial Imaging (DMI) [2–4] and Strain Rate Imaging (SRI) [5–7] with several applications ranging from theoretical elasticity reconstruction models [8] and phantoms [9,10] to resynchronized heart monitoring [11] and stress echocardiography [12]. Despite their implementation on certain commercially available ultrasound systems (for Doppler, DMI and SRI), there have been several disadvantages reported that have stalled clinical acceptance including low axial resolution [13–16], aliasing (occurring at half the center frequency), increase in ambiguity with frequency [13], and attenuation [17]. "Speckle tracking" available on GE scanners employs such methods but the low frame rates used (~50 fps for 2D, ~10 fps for 3D) can only provide an approximate estimate of the total myocardial contraction. Other strain estimation techniques that have been developed for 3D ultrasound include elastic registration and model-based estimation [18], although these techniques involve intensive and lengthy computations and are largely confined to the research setting. Time-shift techniques, the time shift, or delay, between consecutively received radio frequency (RF) signals is directly equivalent to the amount of displacement that the tissue traveled between two consecutive acquisitions. In general, time-shift-based methods, such as cross-correlation and sum-of-absolute-difference (SAD) estimators, have been repeatedly shown to overcome the main limitations of low resolution and aliasing associated with the phase-shift methods while at the same time providing higher precision [13,13a][15,16]. Prior to our work, time-shift estimation techniques in cardiac applications in vivo had mainly been reported on M-mode data [19,20] or envelope-detected signals [21–23] due to the higher frame rate and previous unavailability of the RF signals, respectively.

#### 1.1.3.2 High frame rate ultrasound imaging

A fundamental tradeoff exists in ultrasound physics between depth, beam density, and frame rate. For cardiac applications in humans, the imaging depth is typically at least 12–14 cm in order to visualize the entire left ventricle. Increasing the frame rate in a conventional sequence comes, thus, at the cost of decreasing the number of lines for a given depth; i.e., increasing the temporal resolution decreases the spatial resolution. Indeed, early attempts to increase frame rate used smaller fields of view [6,24] or sparse sector scans [25]. Parallel beamforming using plane wave [26–28] or diverging beams [29,29a] has the advantage of reconstructing a large field of view at very high frame rates (up to 5500 frames/s for cardiac imaging).

### *1.1.4 Myocardial elastography*

Myocardial elastography has been under development for several years [24,30–36]; Konofagou et al., 2017; [37]. This technique encompasses imaging of mechanical (cumulative or systolic; Fig. 4.1) deformation to unveil the mechanical function of the myocardium. Myocardial elastography benefits from the development of techniques be used for high-precision 2D time-shift-based strain estimation [38] and high frame rates [24] to obtain a detailed map of the transmural strain in normal [36,39,40], ischemic [40], and infarcted [29] cases in vivo. These strain imaging

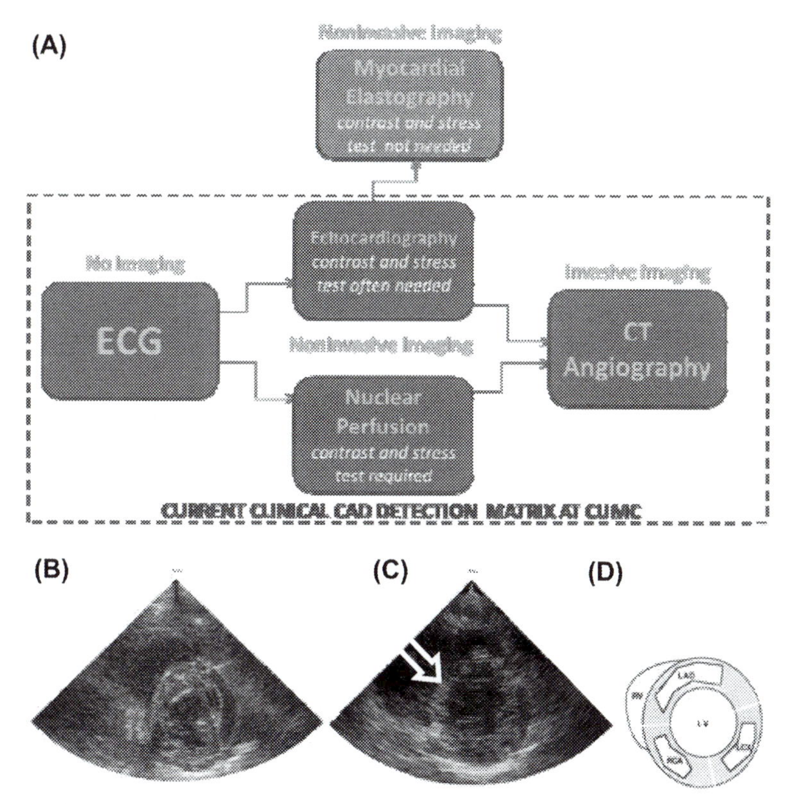


**FIGURE 4.1**

(A) Envisioned role and (B—C) initial findings of Myocardial Elastography in the current clinical routine to avoid false positives and/or thus unnecessary invasive procedures and false negatives by the currently used techniques; (B) Normal short-axis radial strain image (largely red (thickening) myocardium) in an echocardiography false positive case where CT angiography was administered only to confirm normal function; (C) Abnormal short-axis radial strain image containing an ischemic region (in blue or thinning; arrow) in a nuclear perfusion and subsequently CT angiography confirmed Myocardial Elastography findings regarding two occluded territories (64-year-old female, 40% LAD and 30% RCA occlusion); (D) Coronary territories given for Ref. [57]. CUMC: Columbia University Medical Center. LAD: Left-anterior descending artery. LCx: Left-circumflex artery. RCA: right coronary artery. Myocardial Elastography was capable of detecting normal function and identifying both compromised territories.

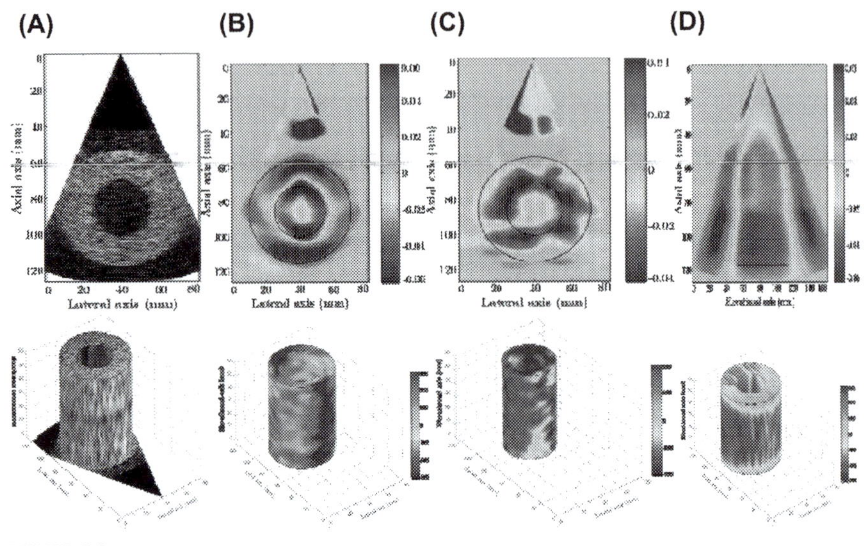


**FIGURE 4.2**

(A) B-mode, (B) radial, (C) circumferential and (C) longitudinal strain images of a cylindrical model undergoing radial deformation using PBME in phased array configuration in FIELD II for 2D (Top panel) and 3D (Bottom panel) Myocardial Elastography. In the proposed study, the canine LV geometry and electromechanical simulation model (Fig. 4.3) are used. The black boundaries in the top panel represent where the simulated myocardium. Strains outside the tissue are not taken into account.

techniques aim at achieving high precision estimates through correction techniques [38] and customized cross-correlation methods [41] and are thus successful in mapping the full 2D and 3D strain tensors [36,42,43]. A fast, normalized cross-correlation function is applied on channel data from consecutive frames to compute 2D motion and deformation [38,44]. The cross-correlation function uses a 1D kernel in a 2D (or, 3D) search to estimate both axial and lateral (and elevational) displacements (Eq. 1; Figs. 4.1 and 4.2).

### *1.1.5 Computational models*

A phased array simulation model for the quality assessment of parallel beamforming myocardial elastography (PBME) was used in order to establish performance. Field II, an established and publicly available ultrasound field simulation program [16,16a], is used to simulate the RF signals of the myocardium. The simulated mesh, including the myocardium, cavity and background are loaded into FIELD II [45,46]. Instead of a focused wave, a diverging wave sequence is employed for transmit by placing the focus 6.75 mm (half the size of the aperture) behind the array surface to achieve a 90 degrees angle insonification [47]. For 2D-PBME, the RF signals are obtained from 192 elements and 3.5-MHz center frequency with 60% bandwidth (at -6dB) phased-array similar to what are used in the canine study. Hanning apodization are

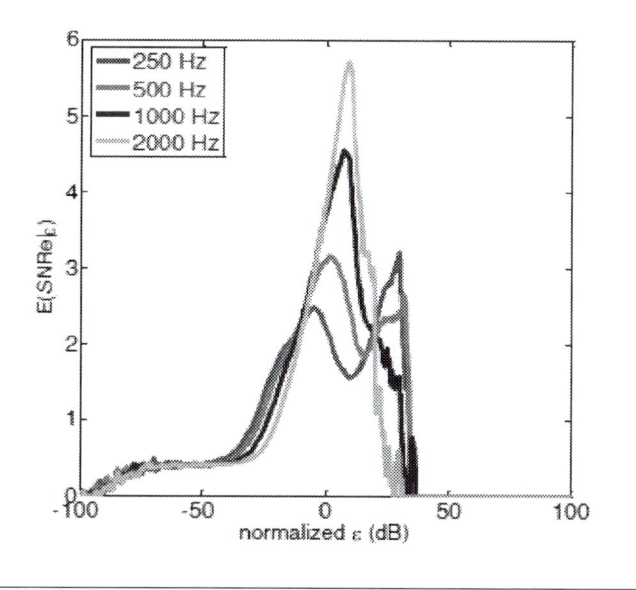


**FIGURE 4.3**

E(SNRe|ε) transverse strain curves increase with frame rate.

applied both during transmit and receive to reduce grating lobes. The width and height of the phased array are 43 and 7 mm, respectively. The RF signals are reconstructed by coherent summation of the signals received by all the elements using a delay-and-sum algorithm. For **3D-PBME**, RF signals are obtained from a 2-D array with 16x16, 32x32 and 64x64 elements and 3-MHz center frequency (Fig. 4.3).

### *1.1.6 Phantoms*

As in our previous studies [48], polyacrylamide (Acros Organics, Geel, Belgium) gel phantoms containing agar powder that produces full speckle are used to model the left ventricle of the human heart. The cylindrical phantom (Fig. 4.4) are molded as a thick-walled, hollow cylinder in order to mimic the geometry of the left ventricle in the short-axis view, and the center are left hollow to allow water to flow through the phantom and cause wall motion and deformation. Two phantoms are prepared for this study: one of uniform wall stiffness (non-inclusion; 25 kPa [48]; and one containing a stiff inclusion representing a region of reduced compliance (50 kPa). A combination of rotation and deformation are studied to better mimic the cardiac function using an existing setup. The pumping frequency are one pulse/sec, and the motor inducing rotation are synchronized for clockwise rotation to start with the pump stroke. One motor will rotate for 30 degrees clockwise at a speed of 112°/s (268 ms) while the other are fixed.

### *1.1.7 Myocardial ischemia and infarction detection in canines in vivo*

#### 1.1.7.1 Ischemic model

Myocardial ischemia was induced in 12 (n = 12) adult Mongrel male dogs (20–25 kg). All animals underwent a left thoracotomy, and the mid-proximal, left

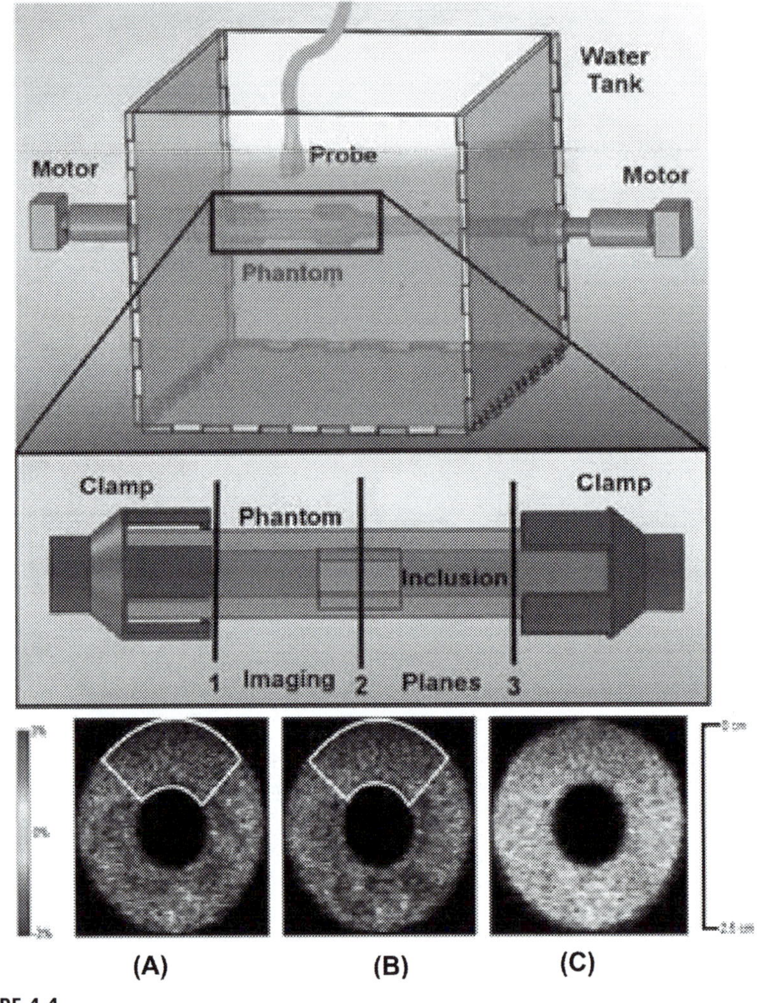

FIGURE 4.4

Top: Experimental setup used to impose motion schemes on a phantom. Bottom: peak cumulative (A) radial and (B) circumferential strain and (C) peak rotation during relaxation in an inclusion phantom. The inclusion is outlined in white. Note the regions of reduced strains within the region of the inclusion (ROI shown; [48]).

anterior descending coronary artery (LAD) was isolated and mounted with an Ameroid constrictor that was used to adjust the stenosis and thus the flow by 0%–100% of baseline, at intervals of 20% for 15 min each. Coronary flow was constantly monitored using a flow probe. Pressure-flow curves were also used to monitor the level of ischemia. Beamformed signals are continuously acquired to establish the strain baseline at normal coronary flow (Fig. 4.5).

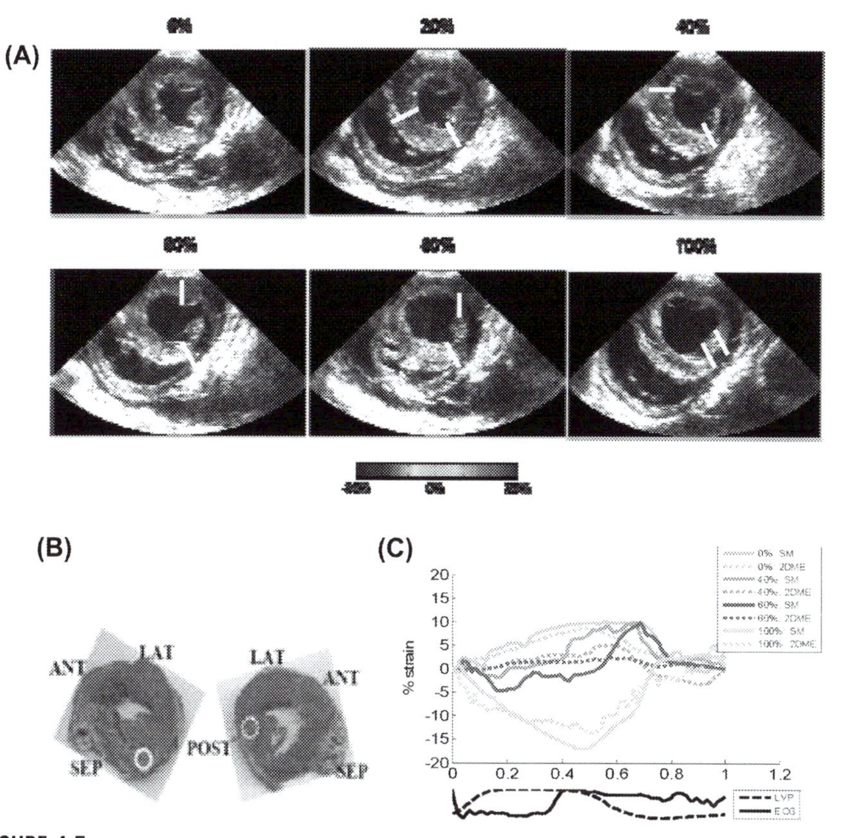


**FIGURE 4.5**

(A) Progression of ischemia (in 'blue' or thinning systolic strain as indicated compared to the surrounding 'red' or non-ischemic (thickening) strain) from 0% to 100% occlusion in the same dog using 2D-PBME and identifying the region and extent of ischemia in each case (shown with the ROI); (B) TTC pathology showing pale (unstained by TTC; ischemic) versus red (stained by TTC; normal) myocardium. (C) Temporal radial strain profiles in the anterior wall region of 3 × 3 mm$^2$ at 0%, 40%, 60%, and 100% occlusion levels with sonomicrometry (SM) and 2DME [40].

#### 1.1.7.2 Infarct model

Mongrel dogs underwent similar surgical procedure as in the ischemia model but survived for 4 days after complete ligation only in order to develop a fully-developed infarct. This protocol was used for our preliminary data shown in Fig. 4.6 that clearly depicts the infarcted region.

### *1.1.8 Validation of myocardial elastography against CT angiography (CTA)*

CT angiography is an established technique used for reliable detection of a coronary occlusion. We conducted a study where CTA was used to confirm the occlusion and

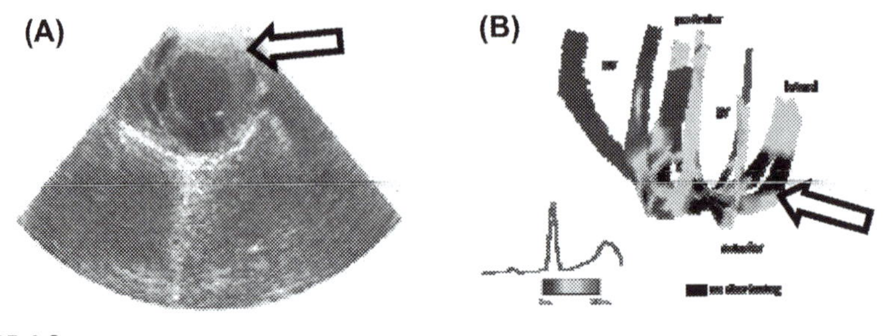

FIGURE 4.6

(A) Transthoracic systolic radial strain image of a canine left ventricle in vivo. Thinning (blue) strain region clearly indicates the extent of infarction. (B) electromechanical isochrones in 3D identifying spatial extent of MI. Arrow denotes MI region.

validate the location of the ischemia or infarct as detected by Myocardial Elastography. The study objectives were to show that 2-D myocardial strains can be imaged with diverging wave imaging and are different in average between healthy subjects and coronary artery disease (CAD) patients. In the study, 15 CAD patients and eight healthy subjects were imaged with ME. 50% occlusion or more was considered associated with obstructive CAD. Axial and lateral left-ventricular strains were imaged and translated to radial cumulative strains (Figs. 4.7 and 4.8). The end-systolic radial strain in normal left ventricles ($14.9 \pm 8.2\%$) was found to be significantly higher than in obstructive ($-0.9 \pm 7.4\%$, $P < .001$) and non-obstructive ($3.7 \pm 5.7\%$, $P < .05$) coronary disease. In the LAD territory, normal myocardium underwent an average strain of $16.9 \pm 12.9\%$, i.e., higher than in obstructed ($2.2 \pm 7.0\%$, $P < .05$) and non-obstructed LAD disease ($1.7 \pm 10.3\%$, $P < .05$) (Fig. 4.9).

### 1.1.9 Importance of myocardial elastography in the clinic

Consider a real-time, quantitative assessment of cardiac contractility and conductivity that can be wheeled into any noninvasive, interventional cardiology or emergency

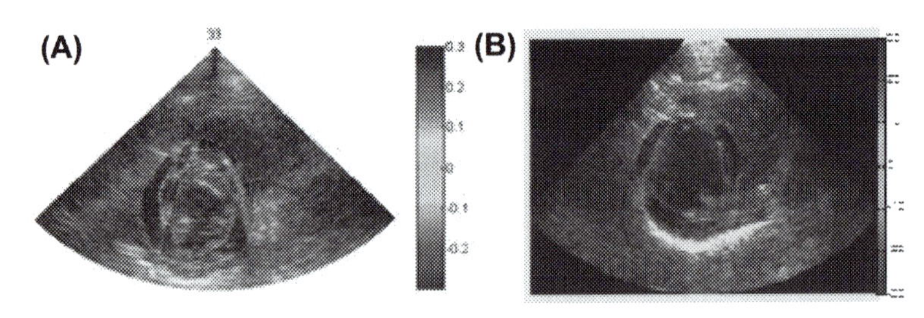

FIGURE 4.7

Comparison between (A) 2DME (conventional (focused) beamforming) using ECG gating and (B) 2D-PBME (parallel beamforming) systolic radial strain image in a normal patient. The quality is comparable without the arfifacts from sector matching and ECG gating in the former (A).

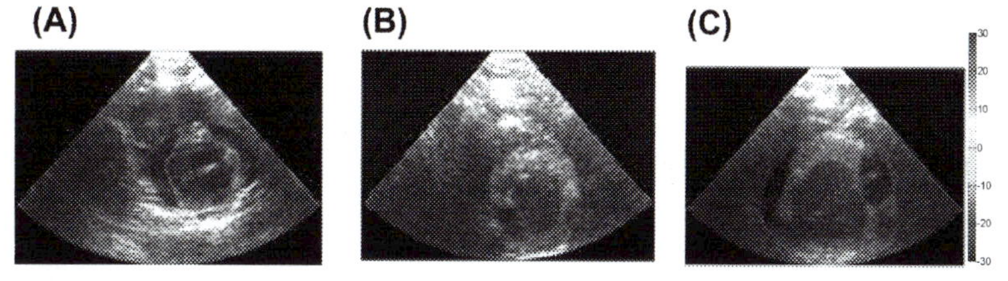


**FIGURE 4.8**

2D-PBME showing progression of coronary disease in three different patients: (A) Normal, (B) (RCA: 90%; LAD: 20%, LCX: 20%) occlusions and (C) (RCA: 99%; LAD: 100%, LCX: 100%) occlusions. Fig. 4.1D can be used for reference of coronaries.

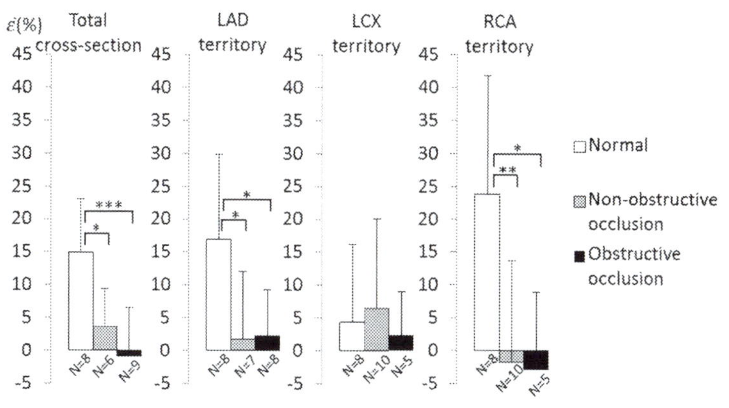


**FIGURE 4.9**

Preliminary findings in 25 human subjects (19 patients and six normals) using PBME. Non-severe CAD (stenosis<50%) and severe (>50% stenosis). LAD: Left-anterior descending artery. LCx: Left-circumflex artery. RCA: right coronary artery. Total: All territories. In all territories except LCx PBME was capable of differentiating normal from non-severe and from severe CAD.

room, and that can specifically inform a cardiologist or emergency room physician of the status of a patient's heart muscle in different regions of the heart well before the specialist can visually detect it and without requiring stress tests or contrast injection (Fig. 4.1A). The current clinical routine at the Columbia University Medical Center (CUMC) where 85% of patients undergo nuclear perfusion stress testing (vs. 15% for echocardiography) that entails radiation exposure, radionuclide (Tc99m) IV injection and stress induced by exercise (30%–40%) or drugs (60%–70%). Current limitations of echocardiography include lack of depiction of cardiac transient mechanical or electromechanical activation at the early onset of disease, and thus going beyond its existing potential. Potentially, this technology will not only diagnose ischemic regions of a patient's heart, but also inform caregivers of whether and how well their medical or interventional therapy or even heart transplant is working.

This new physiologic data could open up interactive therapy regimes that currently are not considered. The eventual goal of this technology is to become a specific method for estimating the position and severity of contraction defects in cardiac infarcts or angina, improving care and outcomes at little more cost or risk than that of a clinical ultrasound.

## 1.2 Electromechanical wave imaging

### 1.2.1 Cardiac arrhythmias

Cardiac arrhythmias can be separated into atrial (or, supraventricular) and ventricular. While ventricular arrhythmias, such as ventricular fibrillation (chaotic rhythm), ventricular tachycardia (rapid rhythm) and ventricular bradycardia (slow rhythm), incur the most episodes of sudden death, they are less common and easier to diagnose than atrial arrhythmias. Atrial arrhythmias include atrial fibrillation (A fib or AF) and atrial flutter (AFL), are the most common. Similar to the ventricular definition, atrial fibrillation denotes the chaotic rhythm while atrial flutter denotes the regular but abnormal (rapid or slow) rhythm. The number of individuals with atrial fibrillation in the United States is expected to reach 12 million by 2050 [48a] with the atrial flutters, often a result of treatment, also expected to rise as more of these treatments are administered. The study and methodology described herein have the dual purpose of both enhancing the diagnosis and treatment planning and guidance of arrhythmias that are the most common and most difficult to diagnose, i.e., the atrial arrhythmias. Below a short overview of the state-of-the-art diagnostic and treatment techniques is provided.

### 1.2.2 Clinical diagnosis of atrial arrhythmias

ECG recordings without imaging is typically used for the noninvasive identification of atrial arrhythmias. If ECG recordings indicate atrial flutter or fibrillation, RF ablation (see next section) is warranted that allows for catheterized cardiac mapping during the procedure. Cardiac mapping involves the insertion of a catheter containing a small number of electrodes in the heart chamber to contact the endocardium and measure times of activation. The procedure is usually minimally invasive and ionizing since it uses Computed Tomography (CT) guidance, but it can be a lengthy procedure requiring topical or general anesthesia and is therefore warranted only when the ECG recordings indicate that RF ablation is the appropriate course of treatment. Limitations include some inaccessible endocardial sites and the inability to map the mid-myocardium and epicardium. More importantly, potentials in only one or a few locations in the atrium are measured per heartbeat, it can be used only to study stable, repeatable arrhythmias.

### 1.2.3 Treatment of atrial arrhythmias

Radio-frequency (RF) ablation is a minimally invasive technique rapidly emerging as the most commonly used therapeutic modality for atrial flutter and atrial fibrillation. For this treatment, a catheter, whose tip carries an electrode, is inserted through the femoral vein and the tip of the electrode is positioned at the arrhythmic origin. Resistive heating of the tip is generated by intracellular ions moving in response to an alternating current. The electrode is connected to a function generator and the electrical current flows and raises the local temperature up to 95°C maintaining it for about 15 min, generating thermal lesions in the vicinity of the ablating catheter. The treatment consists thus in modifying or in blocking the circuits of electrical conduction in heart.

Current surgical procedures are invasive and have a moderate efficiency in the persistent forms of atrial fibrillation. AF in some patients may be due to focal activity originating in the pulmonary veins and 70% of these patients can be successfully treated by RF ablation of the focus inside the pulmonary veins [48b]. However, the treatment of AF using RF ablation can often lead to the development of other arrhythmias. For example, a sizable increase of atypical flutter is due to catheter ablation of atrial fibrillation.

### *1.2.4 Electromechanical wave imaging (EWI)*

The heart, being essentially an electrically driven mechanical pump, alters its mechanical and electrical properties to compensate for loss of normal mechanical and electrical function as a result of disease. During contraction, the electrical activation, or depolarization, wave propagates throughout all four chambers. During propagation, depolarization causes mechanical deformation yielding the electromechanical wave. The resulting regional muscle deformation is extremely rapid and completes within 15–20 ms following depolarization. Therefore, fast acquisition and precise estimation is extremely important in order to properly map and identify the transient and minute mechanical events that occur during depolarization. To this end, our group has pioneered Electromechanical Wave Imaging (EWI) that characterizes the electromechanical function throughout the entire myocardium. Our group was the first to demonstrate that EWI using parallel beamforming is feasible 1) at extremely high frame rates (up to 5500 fps) thus ensuring 2) high precision electromechanical activation maps that include transmural propagation, 3) imaging of transient cardiac events (electromechanical strains within ~0.2–1 ms) with 4) 3D capability, 5) real-time implementation and 6) without requiring ECG gating or breath holds, i.e., in a single heartbeat. . An important additional advantage of this methodology is that it can be routinely applied in the clinic through straightforward integration with any echocardiography system. It could thus be used both at the diagnostic and at the treatment guidance levels, either off- or on-line. Despite the fact that EWI can map both the atria and the ventricles, we chose to focus on the atrial arrhythmias in this study given that they are far more common, more salvageable and completely lacking of any noninvasive imaging modality that can map their conduction properties.

EWI can help guide treatment with ablation therapy: e.g., atrial fibrillation, atrial flutter, ventricular tachycardia, Wolff-Parkinson-White (WPW) syndrome, etc. WPW is a heart condition where an additional electrical pathway links the atria to the ventricles. It is the most common heart rate disorder in infants and children and is preferably treated using catheter ablation. EWI could be used in predicting the location of the accessory pathway, thus reducing the overall time of the ablation procedure. Since a sizable increase of atypical flutter is due to catheter ablation of atrial fibrillation, the prevalence of atrial flutter is also expected to rise. However, the relationship between AF and AFL is still not fully understood [48c]. Atrial flutter, a type of atrial tachycardia, has historically been defined exclusively from the ECG recordings. More specifically, differentiation between flutter and other tachycardias was based on the atrial rate and the presence or absence of isoelectric baselines between atrial deflections. Since then, electrophysiological studies and RF ablation brought new understanding into the atrial tachycardia mechanisms, which did not correlate well with these ECG-based definitions [48d]. In other words, ECG recordings offer only a limited value in the determination of specific atrial tachycardia mechanisms, and, ultimately, for the selection of the appropriate course of treatment. Cardiac mapping allowed different mechanisms of macro-reentrant atrial tachycardia to be identified, such as typical flutter, reverse typical flutter, left atrial flutter, etc., although a significant number of atypical flutters are still poorly understood [48d,48e]. While mapping the right atrium is routinely undertaken successfully using this procedure, mapping the left atrium is riskier since it requires a transseptal puncture. However, as it currently stands in the clinic, right-atrium arrhythmia cannot be distinguished from left-atrium one prior to treatment and only the latter warrants transseptal puncture, which is a riskier procedure. In some cases, the left atrium arrhythmia is not diagnosed until the entire right atrium is ablated (which may involve several unproductive hours of intervention), posing further risk to the patient. Consequently, left atrial activation sequences are not well characterized in many atrial tachycardia cases [48d]. Moreover, the surface ECG is insufficiently specific to distinguish left from right atrial flutters [48c,48e]. EWI could thus offer an important step for the localization of the right versus left atrial arrhythmia as well as localize the origin(s) at the treatment planning stage, i.e., prior to the catheterization of the patient rendering the treatment to be more efficient, much shorter in duration with unnecessary ablations in the wrong chamber or avoiding unnecessary transseptal punctures (Figs. 4.10 and 4.11).

#### 1.2.4.1 Treatment guidance capability of EWI

**Currently, there is simply no noninvasive electrical conduction mapping techniques of the heart that can be used in the clinic.** In addition, apart from being the currently available clinical electrical mapping methods are all catheter-based, and limited to mapping the endocardial or epicardial activation sequence; they are also time-consuming and costly. Even in a laboratory setting, mapping the 3D electrical activation sequence of the heart can be a daunting task [48f]. Studies of transmural electrical activation usually require usage of a large number of plunge electrodes to attain sufficient resolution [49,49a,49b], or are applied to small regions of interest in vivo [49c]. Non-contact methods to map the electrical activation sequence have also emerged but are not used in the clinic. Optical imaging techniques use voltage-sensitive dyes that bind to cardiac cell membranes and, following

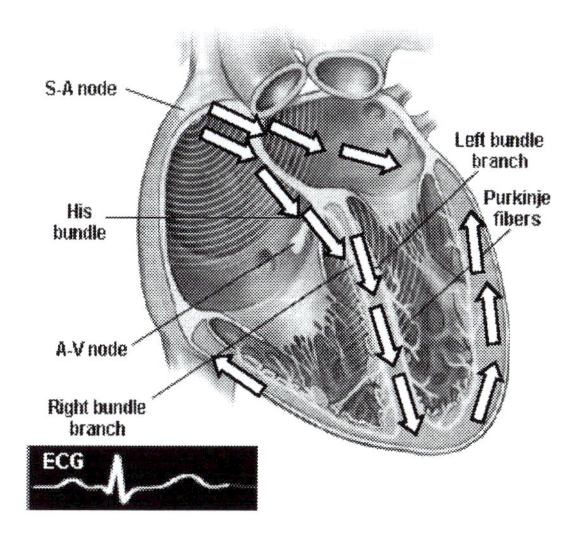


**FIGURE 4.10**

Illustration of the cardiac conduction system of the heart. The green arrows indicate the path of activation as the action potential propagates along the Purkinje fiber network (in yellow).

illumination, fluoresce if the cell undergoes electrical activation. Optical imaging methods can map the activation sequence of ex vivo tissue on the endo-an epicardial surfaces [49d,49e,49f] and transmurally [50,50a,50b] but cannot be applied in the clinic since they require the use of an electromechanical decoupler that inhibits cardiac contraction during imaging. Other newly developed methods to map the local electrical activity of the heart based on inverse problems are available: electrocardiographic imaging (ECGI) ([51,51a,51b]) and non-contact mapping [52,52a,52b]. The former is based on body surface potentials and CT or MRI scans and provides reconstructed epicardial action potentials, including the atria [52c]. The latter consists in reconstructing the transmural potentials from potentials measured in the heart chamber based on specific assumptions that may be susceptible to inverse problem errors [52d,52e,52f] (Fig. 4.12).

### *1.2.5 EWI sequences*

A block diagram of EWI is shown in Fig. 4.13. Two imaging sequences, the automated composite technique (ACT) [53] and the temporally-unequispaced acquisition sequences (TUAS) [26a] have been developed and implemented [53a,53b,53c,26a] as well as the single-heartbeat sequence ([47,54]).

#### 1.2.5.1 The ACT sequence

An automated composite technique (ACT) [53] was developed that solves the trade-off between frame rate and field of view by dividing the image into partially overlapping sectors from distinct cardiac cycles (Fig. 4.13A). A reference window at

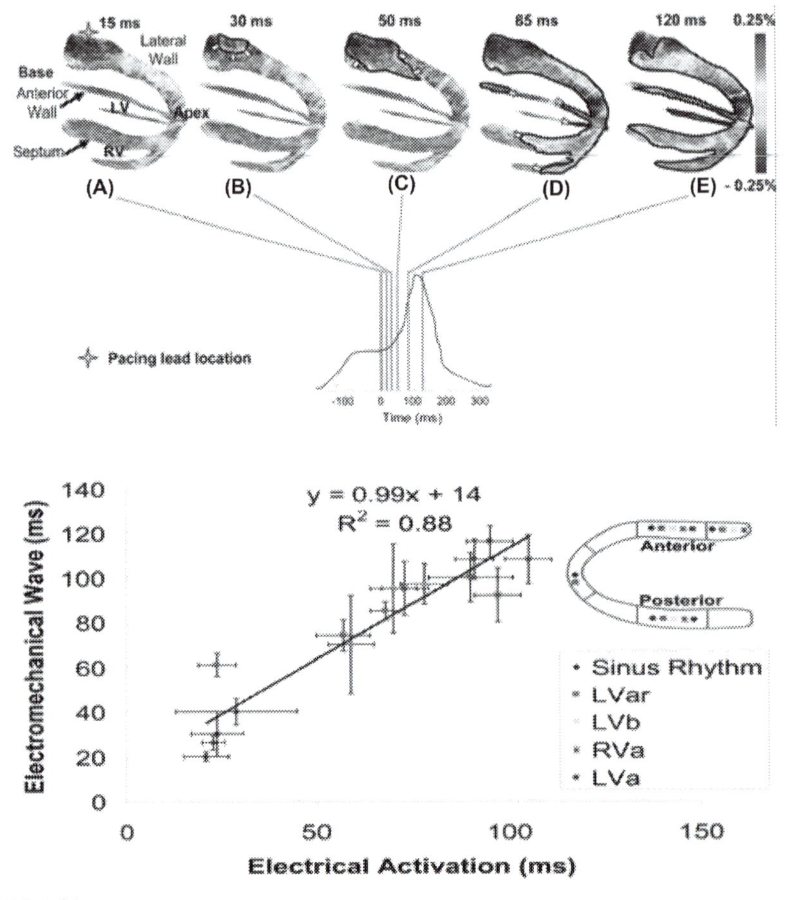


**FIGURE 4.11**

Top: Propagation of the electromechanical wave when paced from the lateral wall, near the base. Activation in this view corresponds to thickening of the tissue (red). Activated regions are traced at (A) 15 ms, (B) 30 ms, (C) 50 ms (D) 85 ms and (E) 120 ms and indicated on the (F) electrocardiogram. 0 ms corresponds to the pacing stimulus. (A–C) The EW propagates from the basal part of the lateral wall toward the apex. (D) Note that in the apical region, a transition from lengthening to shortening is observed rather than a transition from thinning to thickening. (D–E) In the anterior wall, the EW propagates from both the base and apex. The scale shows inter-frame strains. **Bottom**: Electrical and electromechanical activation times during the four pacing protocols and sinus rhythm in four different heart segments in the posterior and anterior walls.

time-point $t_1$ is correlated within the same beam at time-point $t_2$. The highest correlation identifies the axial displacement lapsing between the two time-points. The same process is repeated at multiple time-points and windows in order to obtain axial displacements along the ultrasound beam at each time-point. The full field-

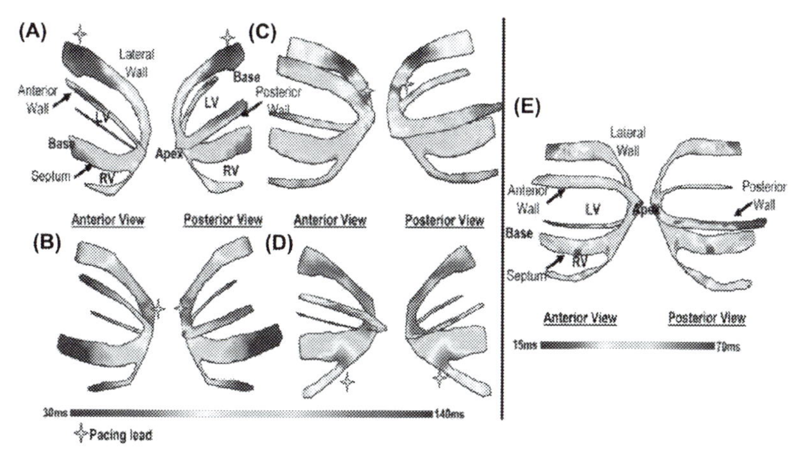


**FIGURE 4.12**

Isochrones showing the activation sequence under different pacing protocols as shown. Arrow indicates the pacing origin. (A) Pacing from the basal region of the lateral wall. (B) Pacing from the apex. (C) Pacing from the apical region of the lateral wall. (D) Pacing from the apical region of the right-ventricular wall. (E) Isochrones showing the EW activation sequence during sinus rhythm. The activation sequence exhibits early activation at the median level and late activation at the basal and apical levels. Activation of the right ventricular wall occurred after the activation of the septal and lateral walls. The blue cross indicates the pacing lead location.

of-view is then constructed using a motion-matching methodology (Fig. 4.13C) [53c] that compares all displacements estimated in the overlapping sectors at distinct cardiac cycles to match the sectors spatially and temporally. Each acquisition sequence includes at least one ultrasound beam which is contained in a neighboring sector. As a result, identical axial displacements are obtained in the corresponding sectors and cardiac cycles. The time delay corresponding to the maximum cross-correlation is thus obtained to synchronize neighboring sectors, which is repeated across all sectors eventually completing the reconstruction of the visualization of the entire chamber in question. It is important to note that this method is independent of the ECG, therefore not plagued by ECG gating limitations [53c]. Computation of the spatial derivative of the aforementioned displacement yields the strains in the axial direction using a least-squares methodology [55] (Fig. 4.13D). Segmentation of the myocardial wall is accomplished using an automated contour tracking technique [45] followed by strain estimates overlaid onto the B-mode images (Fig. 4.13E). Activation maps are computed by estimating the time-point at which the strain zero-crossing is incurred following the onset of the Q-wave. Spline and Delaunay interpolation increase the resolution of the corresponding activation maps. Two image planes imaged in the standard apical four- and two-chamber views are depicted across the long axis of the heart and temporally co-registered using

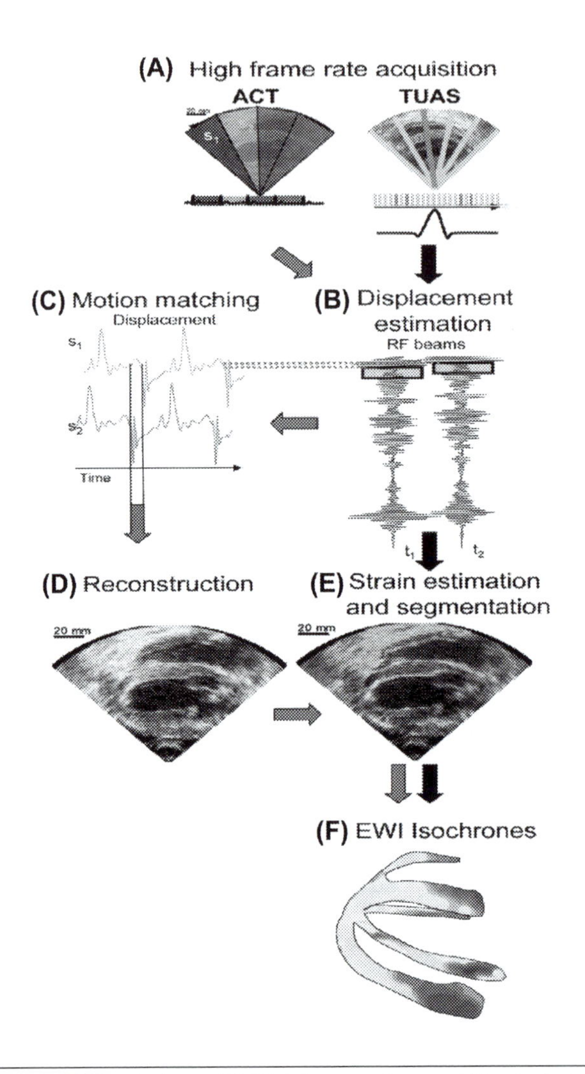


**FIGURE 4.13**

Block diagram of the EWI technique. (A) High frame-rate acquisition is first performed using either ACT (follow red arrows) or TUAS (black arrows). (B) High precision displacement estimation between two consecutively acquired RF beams (t1, t2) is then performed using very high frame rate RF speckle tracking. (C) In ACT only, a region of the heart muscle, common to two neighboring sectors, is then selected. By comparing the temporally varying displacements measured in neighboring sectors (s1, s2) via a cross-correlation technique, the delay between them is estimated. (D) In ACT only, a full-view cine-loop of the displacement overlaid onto the B-mode can then be reconstructed with all the sectors in the composite image synchronized. (E) In ACT and TUAS, the heart walls are then segmented, and incremental strains are computed to depict the EW. (F) By tracking the onset of the EW, isochrones of the sequence of activation are generated.

ECG, in a three-dimensional biplane view in Amira 4.1 (Visage Imaging) (Figs. 4.13F and 4.2).

### 1.2.5.2 The TUAS sequence

A temporally un-equispaced acquisition sequence (TUAS) (Fig. 4.14; Provost et al., 2011e), also a sector-based sequence [45a], was adapted to optimally estimate cardiac deformations where a wide range of frame rates are achieved for motion estimation, large beam density, and a large field of view in a single cardiac cycle, i.e., without the motion-matching and reconstruction steps in EWI (Fig. 4.10). As

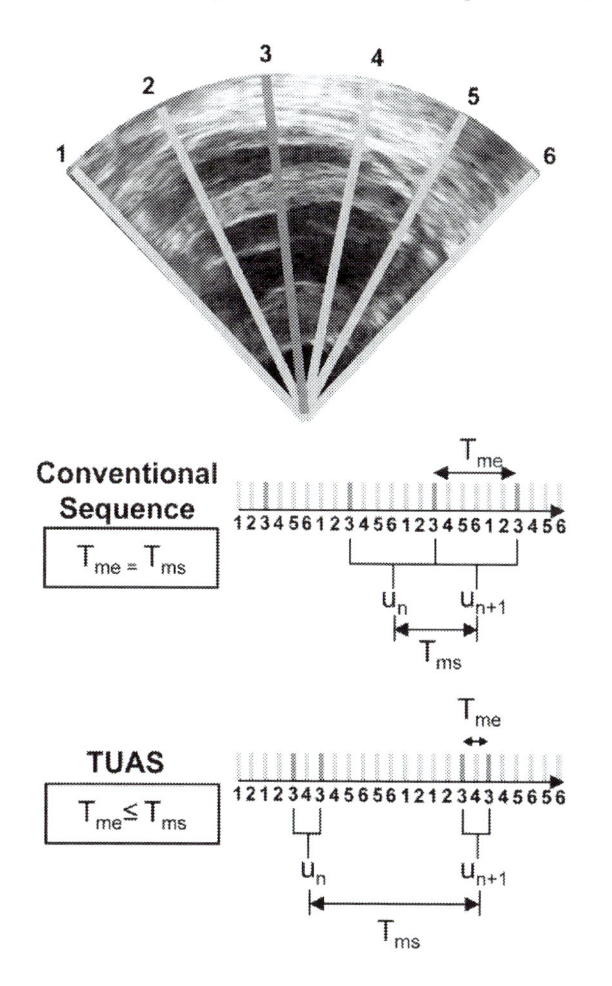


**FIGURE 4.14**

Illustration of an acquisition sequences in a simple case where only six lines form an image with each sector using two lines. In a conventional acquisition sequence, the time separating two acquisitions of the same line is the same. In TUAS, the time separating two acquisitions of the same line are modulated to optimize motion-estimation.

a result, for a given set of imaging parameters, motion can be estimated at frame rates varying from a few Hz to kHz at the minimum sampling rate, i.e., the Nyquist rate, of the motion over time at a given pixel. Typically, a finite number of beams, typically 64 or 128, over a 90 degrees angle are acquired while the process is repeated for each frame. A given beam, e.g., beam 3 (Fig. 4.14), is acquired at a fixed rate (Fig. 4.14, conventional sequence). However, In TUAS, the beam acquisition sequence is reordered to provide two distinct rates (Fig. 4.14): the motion-estimation rate (MER) and the motion-sampling rate (MSR). The MER is defined as the inverse of the time lapsing between the two RF frames used to estimate motion while the MSR denotes the inverse of the time lapsing between two consecutive displacement maps. In standard imaging sequences, MSR and MER are identical (Fig. 4.14) while TUAS can modify the MER. A TUAS frame is used only once for motion estimation, thus halving the MSR (Fig. 4.14). Thus, an acquisition performed at a 12-cm-depth with 64 beams with a standard sequence corresponds to a frame rate of 100 Hz, which can be insufficient for accurate motion tracking using RF cross-correlation. A higher frame rate of, e.g., 400 Hz typically used for EWI, is warranted which typically requires decreasing the number of beams by four, and thus reduction of beam density and/or field of view. On the other hand, TUAS provides an MSR of 50 Hz and a MER that can be varied, as follows: {6416, 3208, 1604, 802, 401, 201, 100} Hz with numerous advantages. Both the beam density and the field of view can be maintained while estimating the cardiac motion with an optimal frame rate, which could be, e.g., either 401 or 802 Hz, depending on the amplitude of the cardiac motion which results into halving of the MSR which has only little effect on the accuracy. An MSR above 120 Hz became negligible compared to the effect of the MER (Provost et al., 2011e).

### 1.2.5.3 Single-heartbeat EWI and optimal strain estimation

Previous Strain Filter reports [45b] indicate that the SNRe depends mostly on the value of the strain measured, when the imaging parameters are fixed. This framework defines upper limit on the $SNR_e$ as a function of the strain amplitude. The Ziv-Zakai Lower Bound (ZZLB) on the variance is a combination of the Cramér-Rao Lower Bound (CRLB) and the Barankin bound (BB) and transitions from the CRLB to the BB when decorrelation is very significant [45c]. In the correlation model reported [45d], this occurrence is noted only at very large strains. Finding an optimal strain value is thus equivalent to finding an optimal MER. The strain filter was previously developed for the analysis of strains occurring in static elastography, and is not adapted to the more complex motion of the heart. A new probabilistic framework based on experimental data has been introduced to not only establish an upper bound on the $SNR_e$, but to determine the probability of quality strains (Provost et al., 2011e). The optimal MER can thus be found by studying the link between the strains and the $SNR_e$, i.e., a conditional probability density function was constructed (Fig. 4.15B) and was found in agreement with the strain-filter theory. Fig. 4.15 shows that the conditional probability density function is contained within the CRLB up to approximately 4% strain before it transitions to BB, beyond which a

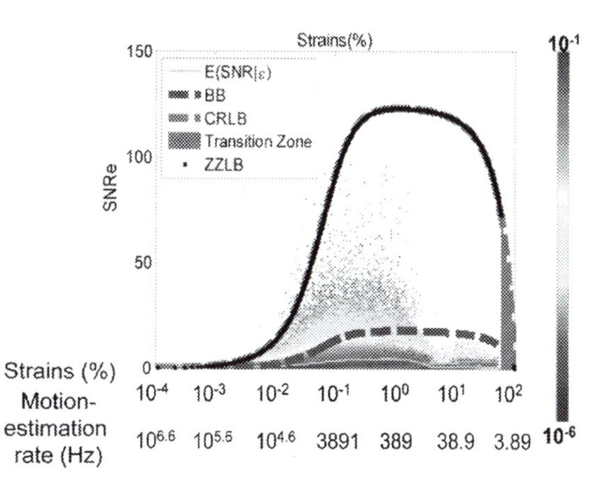


**FIGURE 4.15**

(A) Distribution of strains during the EW for different motion-estimation rates. (B) Conditional probabilty density function of the SNRe knowing the strain value (and the corresponding motion-estimation rate). The conditional expectation value of SNRe, the ZZLB, BB and CRLB are also displayed.

sharp decrease of the expected SNRe is noted. A high SNRe can thus be maintained but only above 350 Hz in this case (Fig. 4.15B). TUAS is also shown capable of accurately depicting non-periodic events at high temporal resolution at the optimal frame rate between 389 and 3891 Hz (Fig. 4.15).

### 1.2.6 Characterization of atrial arrhythmias in canines in vivo

Validation of EWI against electrical mapping using an electro anatomical mapping system in all four chambers of the heart is essential in order to determine the validity of the EWI maps. Electrical isochrones have been obtained in open-chest canine hearts in vivo in all four cardiac chambers at the endo- and epicardium using standard electroanatomic mapping (St. Jude Medical). EWI acquisition during normal sinus rhythm and pacing were acquired and electromechanical activation maps were computed. A linear regression between electrical and electromechanical maps indicated high correlation with slopes ranging from 0.77 to 1.83 and intercepts within 9 and 58 ms with $R^2$ within 0.71—0.92 (Fig. 4.16). This indicates that that the electromechanical activation is a reliable surrogate of the electrical activation and thus that EWI can characterize and localize arrhythmogenic sources.

### 1.2.7 EWI in normal human subjects and with arrhythmias

The objectives of the clinical studies [26—28,47]; Provost et al., 2015; [29]; were (1) to determine the potential for clinical role of EWI, by predicting activation patterns in normal subjects, (2) to determine the feasibility of EWI to identify the site of origin in subjects with tachyarrhythmia, and (3) to identify the myocardial activation sequence in patients undergoing CRT. In normal subjects (Figs. 4.17 and 4.18), the EW propagated, in both the atria and the ventricles, in accordance with the expected

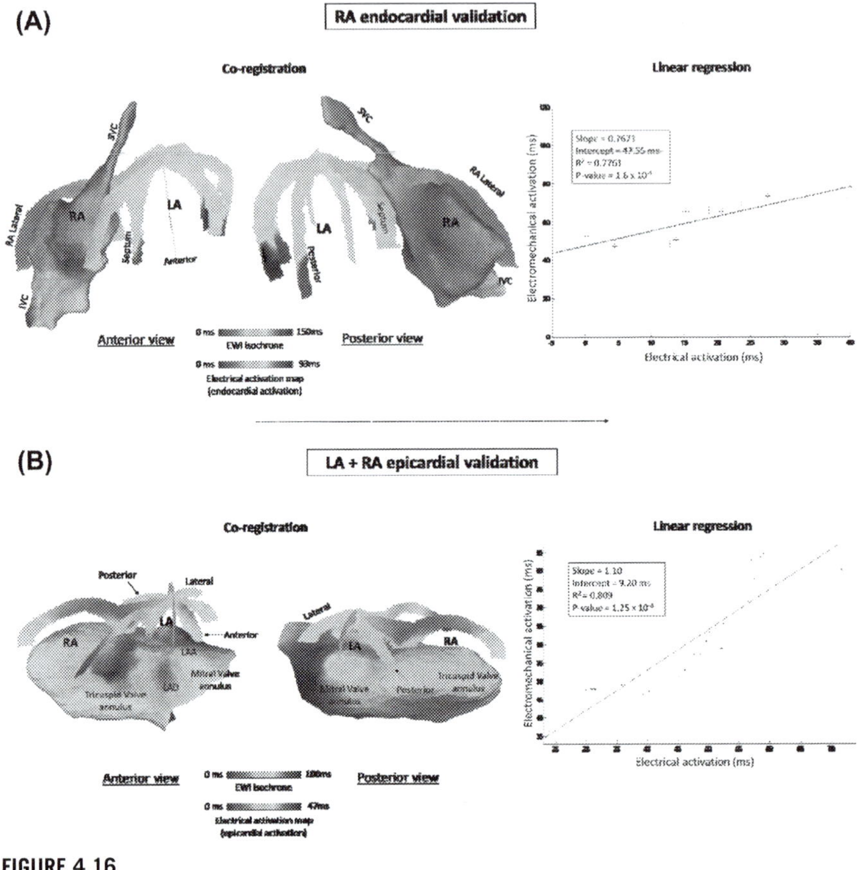


## FIGURE 4.16

Endocardial RA and epicardial RA + LA EWI validation in two canines during normal sinus rhythm. Doth endocardial and epicardial maps were acquired using a bipolar catheter. IVC = Inferior Véna Cava, LA = Left Atrium, LAD = Left Anterior Descending artery, LV = Left Ventricle, RA = Right Atrium, RV = Right Ventricle, SVC = Superior Vena Cava.

electrical activation sequences based on reports in the literature. In subjects with CRT, EWI successfully characterized two different pacing schemes, i.e., LV epicardial pacing and RV endocardial pacing versus sinus rhythm with conducted complexes. In two subjects with AFL (Figure 7-4), the propagation patterns obtained with EWI were in agreement with results obtained from invasive intracardiac mapping studies, indicating that EWI may be capable of distinguishing LA from RA flutters transthoracically. Finally, we have shown the feasibility of EWI to describe the activation sequence during a single heartbeat in a patient with AFL and RBBB ([47]; Fig. 4.19).The results presented demonstrate for the first time that mapping the transient strains occurring in response to the electrical activation, i.e., the

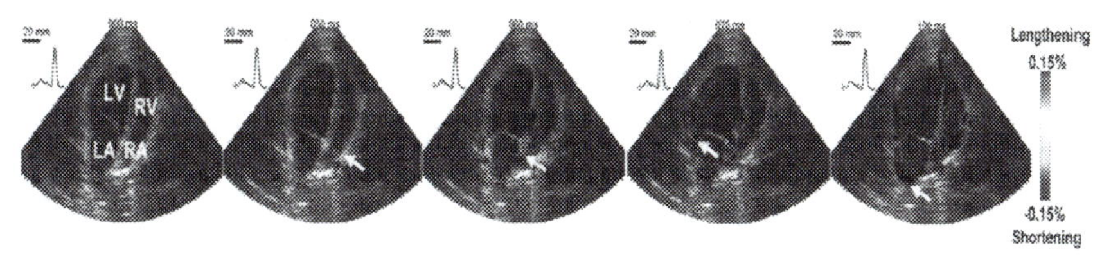


**FIGURE 4.17**

EWI using a flash sequence for motion estimation (motion sampling rate: 2000 Hz, motion- estimation rate: 500 Hz) overlaid on a standard 128-beams, 30-fps B-mode. All four chambers are mapped bu only atrial activation is shown here. Activation (shortening, in blue) was initiated in the right atrium (RA) (50 ms) and propagated toward the atrial septum (60 ms), the left atrium (LA) (70 ms) until complete activation of both atria (100 ms). RV: right ventricle, LV: left ventricle.

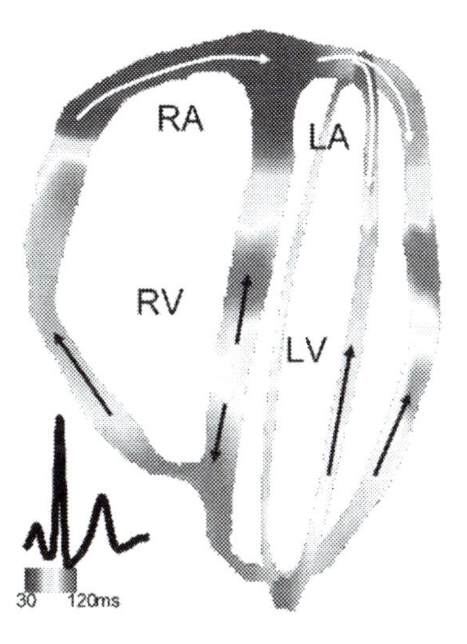


**FIGURE 4.18**

EWI isochrone in all four chambers in a healthy, 23-year-old male subject. Activation in this view corresponds to shortening of the tissue (blue). Activation is initiated in the right atrium and propagates in the left atrium. After the atrio-ventricular delay, activation is initiated in the ventricles from multiple origins, which are possibly correlated with the Purkinje terminals locations. Arrows (both white and black) indicate the direction of propagation. See Appendix 1 for cine-loop EWI.

electromechanical wave propagation, can be used to characterize both normal rhythm and arrhythmias in humans, in all four cardiac chambers transthoracically using multiple and single-heartbeat methodologies. EWI has the potential to noninvasively assist in clinical decision-making prior to invasive treatment, and to aid in optimizing and monitoring response to CRT.

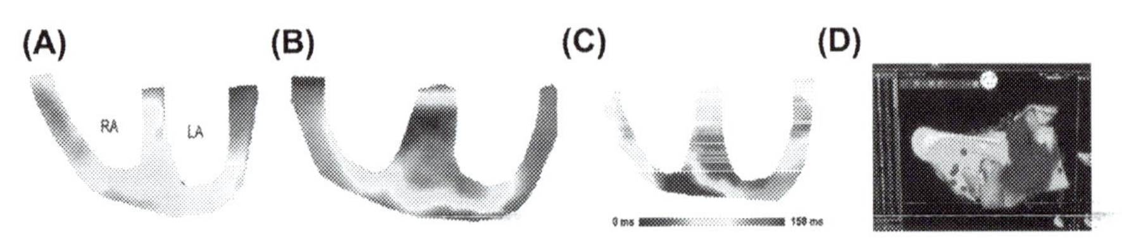


**FIGURE 4.19**

Noninvasively detecting and characterizing atrial flutter with EWI in three human subjects (RA: right atrium, LA: left atrium). (A) Normal subject (Activation is initiated in the right atrium and propagates toward the left atrium); (B) Unknown atrial flutter in a flutter patient with a prosthetic mitral valve, which was believed to cause the atrial flutter from the left side. Since the patient could not sustain the long procedure needed to construct a full activation map using cardiac mapping, ablation was attempted without complete information in two locations in the left atrium, which did not lead to cardioversion, i.e., did not return to sinus rhythm. The patient was sent home and a second ablation is scheduled in the near future. This case exemplifies the need for a faster mapping method, which could have led to the proper identification of the ablation site and potentially limit the number of ablation procedures needed to treat this patient. Indeed the EWI isochrones shown here display the initial activation in the left atrium, close to the septum; (C) Atypical left atrial flutter patient confirmed by cardiac mapping. EWI isochrones also show early activation is initiated in the left atrium. See Appendix 1 for the corresponding EWI cine-loops. (D) CARTO map of a left-sided flutter case.

## 2. Vascular imaging

### 2.1 Stroke

Stroke is the fifth leading cause of death and the primary cause of preventable disability in the U.S according to the most recent statistics by the American Heart Association [47a]. Each year, approximately 795,000 people continue to experience a new or recurrent stroke. On average, every 40 s, someone in the United States has a stroke, and someone dies of one approximately every 4 min. This high fatality rate is partly due to the fact that state-of-the-art methodologies such as CT, MRI and ultrasound are currently failing to inform timely intervention. Identification of patients with high-risk, asymptomatic carotid plaques remains thus an elusive but essential step in stroke prevention [47b]. In current clinical practice, the risk assessment of carotid plaque rupture is based on the degree of endoluminal stenosis [47c] despite it being shown not to necessarily correlate with vulnerable plaques [47d] or subsequent rupture [47b]. Current criteria thus relate to the vessel lumen properties but not directly to the carotid wall itself that bears the plaque, confirming the challenge at hand. A criterion that would reliably identify unstable plaques would ensure timely removal through carotid endarterectomy (CEA), especially in asymptomatic patients or those with low endoluminal stenosis whose stroke risk is currently poorly assessed.

## 2.2 Abdominal aortic aneurysm

Vascular stiffening happens naturally with aging. The elastin components of the wall are gradually reduced and the collagen that remains leads to overall hardening of the arteries. One of the results of vascular aging is also a higher probability of developing an abdominal aortic aneurysm (AAA), which is a focal, balloon-like dilation of the terminal aortic segment that occurs gradually over a span of years. This condition is growing in prevalence in the elderly population, with approximately 150,000 new cases being diagnosed every year[47e,47f] . An AAA may rupture if it is not treated, and this is ranked as the 13th most common cause of death in the US and the most common aneurysm type [47g]. Current AAA repair procedures are expensive and carry significant morbidity and mortality risks [47h,47i,47j,47k,47l,47m,47n]. The most standard diagnostic technique is abdominal ultrasound or CT imaging. In both cases, images of the abdominal aorta are obtained and the aneurysm is measured. If its transverse diameter is found to be higher than 5.5 cm, the chance of aneurysm rupture increases by 10%–20% within a year [47o] and the patient typically has to undergo surgery to remove the aneurysm. If left untreated, more than 30% of the aneurysms will rupture and the patient will die as a result.

## 2.3 Pulse-wave velocity (PWV)

Several techniques have been proposed for non-invasive measurement of the global PWV defined as the speed at which the arterial pulse propagates throughout the entire circulation tree which is directly linked to stiffness or Young's modulus through the Moens-Korteweg (MK) equation. In fact, the global PWV is widely considered as a surrogate to arterial stiffness and as a reliable indicator of cardiovascular mortality and morbidity in the case of hypertension [47p].

## 2.4 Pulse wave imaging

**Pulse Wave Imaging (PWI)** (Fig. 4.20) denotes an ultrasound-based method of imaging the ***pulse-wave induced displacement*** on the vascular wall during end-systole at very high framerates (>500 fps). As a result, unlike in alternative techniques, the pulse wave propagation (>1 m/s) can be sufficiently sampled and depicted as well as characterized. This entails the steps of: **1)** RF frame acquisition; **2)** motion estimation; **3)** segmentation and tracking; **4)** spatial-temporal mapping; **5)** PWV mapping and **6)** PWV and stiffness maps overlaid onto the B-mode images. The wall stiffness or, Young's modulus ($E$), is calculated in each vessel segment based on the regional PWV value.

PWI is thus a localized PWV and stiffness mapping modality that is thus highly suitable in characterizing focal disease such as plaques before the onset of pathology (normal, thin wall) to its later stages. The parameters *mapped* with PWI are the PWV (Fig. 4.20d), $r^2$ (coefficient of determination; Fig. 4.20c) and compliance (Fig. 4.20e), which indicate the magnitude of the v*elocity of the pulse wave*, its *uniformity of propagation* and the *underlying stiffness*, respectively, all of which can be mapped and used as biomarkers, highly complementary to the ultrasound and CT scan clinical standards.

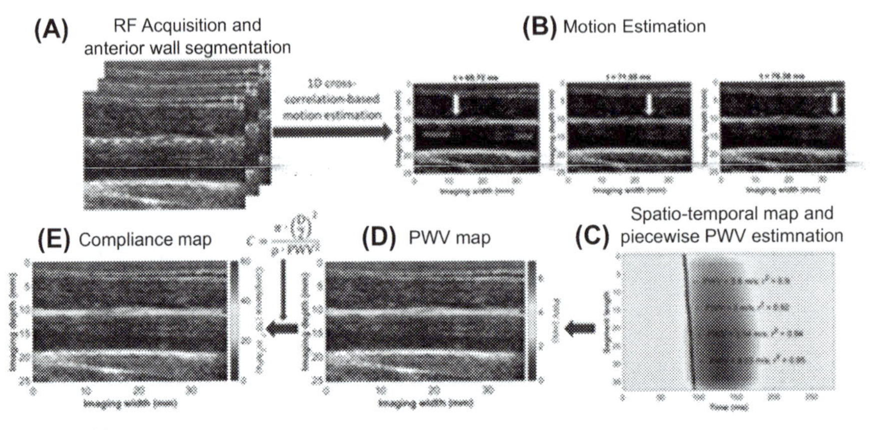


FIGURE 4.20

Block diagram of the Pulse Wave Imaging (PWI) technique on a normal carotid artery *in vivo*. (A) RF frame sequence and ciné-loop of RF frames. (B) PWI ciné-loop and frames (pulse wave induces upward (red) wall displacement along the anterior wall). The white arrows indicate the propagation of the wavefront along the anterior wall. (C) Spatio-temporal map of the pulse-wave-induced displacement along the vessel length and time with a single PWV value given in the entire segment imaged. (D) PWV and (E) stiffness maps overlayed on the anterior wall of the B-mode [47r].

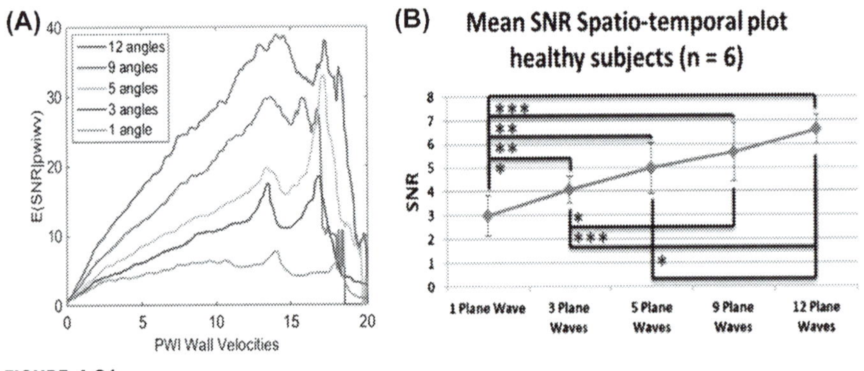


FIGURE 4.21

E(SNRdlδ) increase with number of angles in coherent compounding in (A) phantoms and (B) in vivo humans.

## 2.5 Methods

In order to accommodate both current clinical ultrasound practice but also ongoing methodologies developed by our group, a new PWI system are developed that will optimize parameters such as beam density and framerate. New PWI methodologies as those described below are applied in phantom studies. **The high framerates**

**facilitated by plane wave imaging will allow us to apply the following techniques that significantly enhance the quality and performance.**

### *2.5.1 PWI system using parallel beamforming*

An open-architecture system (Vantage, Verasonics, Redmond, WA) allows for software-based beamforming for simultaneous high framerate and high beam density. Similar to others in the field, we have found that the performance of speckle tracking is not compromised when such systems are used based on the advantages of differential, phase-based measurements (as opposed to amplitude-based approaches) [29,47]. The acoustic exposure is on the same level as that of a standard ultrasound scanner. We have implemented such a sequence [47] (Fig. 4.21B) that excites the transducer elements with specific delays allowing a large field of view and then uses the delay–and-sum method to reconstruct the RF signals from the channel data. Framerates are constrained by the image depth. In theory, the system can operate at up to 8333 fps. Parallel beamforming ensures that the highest beam density (128 beams) and framerate (up to 8333 fps) can simultaneously be achieved at a center frequency of 5.5 MHz or with a 128-element linear array at a center frequency of 17.5 MHz for higher spatial resolution, the latter allowing characterization of composition of plaques.

- ***Automated diameter and wall thickness*** methodologies allow for segmentation and contour tracking [47q].
- ***PWV mapping*** – Our fast 1-D normalized cross-correlation method are applied on the consecutively acquired RF signals to estimate the normalized wall displacement (Fig. 4.21) as in our previously published studies [44,56]. In order to obtain the regional PWV, all of the 50% upstroke points of the spatio-temporal plots are divided into overlapping segments. 50% upstroke was determined to yield highest SNR for PWV estimation in both human aortas and carotids [56a,56b]. Fixed regression windows have been used thus far to perform linear regression for PWV estimation, especially due to the low beam density afforded by conventional beamforming techniques. A piecewise, linear regression fit has thus been developed. Sub-regional PWVs are estimated within 2–4 mm segments along the length of the arterial wall and estimates the regional stiffness using the methodology described previously. Instead of single values for the entire vessel, this method provides maps of PWV along the entire wall imaged (Fig. 4.21) but also within the plaque itself).
- ***Automated PWV estimation*** – A dynamic programming algorithm are implemented to automatically select the size, location and number of windows according to a mean $r^2$ optimization routine with criteria based on the performance of the resulting linear regression. More specifically, the windows are appropriately selected to satisfy the maximization criterion: $\max_{NW,SWi,PWi} \frac{1}{NW} \cdot \left( \sum_{i=1}^{NW} r_i^2 \right) \max_{NW.SWi.PWi} \frac{1}{NW} \cdot \left( \sum_{i=1}^{NW} r_i^2 \right)$, where $NW$ is the number of windows used to model the stiffness of the imaged arterial wall, $SW_i$ is the

size of the ith window (namely the number of markers that it contains) and $PW_i$ is its position (namely the number of the starting marker).

- ***Stiffness/compliance mapping*** – The 1D system of the coupled governing pulse wave propagation are $\frac{dA}{dt} + \frac{d}{dx}(uA) = 0$ **(1a)**, and $\frac{du}{dt} + u\frac{du}{dx} + \frac{1}{\rho}\frac{dP}{dx} + K_R u = 0$ **(1b)**, where $u$ is the fluid velocity, $P$ is the pressure, $A$ is the cross-sectional area, $\rho$ the fluid density and $K_R$ is the resistance per unit length. The relationship between the PWV and vessel stiffness is derived by assuming a pressure-area relationship. Commonly, a thin-walled cylindrical elastic vessel is assumed, so that the wall displacement, $\Delta r = \frac{r^2(1-\nu^2)}{Eh}\Delta P$ **(2)** for a pressure increment $\Delta P$ in a vessel of Young's Modulus, $E$, Poisson's Ratio, $\nu$, radius $r$, and thickness, $h$, which gives the well-known Moens-Korteweg equation. In this study of carotid plaques, cylindrical geometry cannot be reasonably assumed, so we **use a generalized linear pressure-area relationship, which is valid for non-circular cross-sections**, $\Delta A = k_p \Delta P$ **(3).** The compliance, $k_p$, is related to the PWV through the Bramwell-Hill equation [56c], i.e., $k_p = \frac{\pi r^2}{4\rho PWV^2}.PWV = \sqrt{\frac{A}{\rho k_p}}$ **(3)** can be related to the effective Young's modulus as follows: $E_{eff} = \frac{2\pi r^3(1-\nu^2)}{k_p h}$. **(4)** that shows that compliance and stiffness are inversely proportional for a given geometry [56d]. As a result, as a first approximation, the longitudinal distribution of $k_p$ can be computed by regional PWV estimation map without homogeneity or symmetry assumptions (Fig. 4.21E).

### 2.5.2 3D PWI

A 2D matrix array (Sonic Concepts, Inc., Bothell, WA) was used in this study with a total of 1024 elements (32x32) and a 4.5 MHz center frequency and will be used with the Verasonics Vantage system for the highest framerate. A custom diverging beam sequence with a virtual focus placed behind the transducer face was implemented in order to interrogate the entire field of view in a single transmit sequence. Element data will be acquired, and reconstructed in a pixel-wise fashion. In sections of the common carotid prior to the bifurcation, inter-frame pulse-wave-induced axial displacements (not the full 3D vector) will be mapped in 3D given the lower SNR in lateral and elevational displacement estimation. Bicubic Hermite interpolation was implemented to render the vessel in 3D, which will provide better visualization of the complex anatomy and motion of non-axisymmetric plaques (Fig. 4.22). The framerate was similar to 2D, i.e., up to 5500 frames/s, and sufficient to visualize the pulse wave and estimate the PWI parameters. B- mode and PWV imaging is shown in Fig. 4.22.

## 2.6 PWI performance assessment in experimental phantoms

In order to ensure highest quality PWI in vivo, its performance is optimized in phantoms prior to application in vivo. Parallel beamforming provides images with reduced sonographic SNR; however, this issue has been addressed with coherent

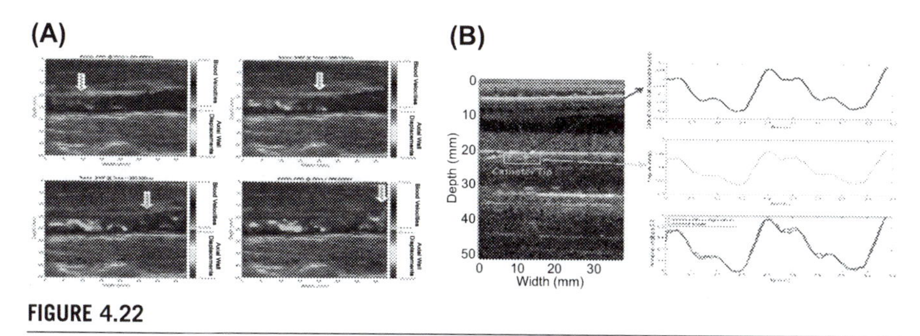


**FIGURE 4.22**

(A) PWI frames with Doppler flow; (B) Pressure/displacement curves revealing the phase lag amount and thus viscoelasticity of the vessel (phantom). The blue curve denotes the pressure, the green the cumulative displacement and the last curve the two overlaid.

compounding, at the cost, however, of framerate. Thus, by varying the number of angled plane wave acquisitions this tradeoff between framerate and image SNR can be fine-tuned and adjusted to the requirements of the application. This are verified in PVA cryogel [56a,56b] and silicone [56d] phantoms with preliminary results shown in Table 4.1. Carotid artery phantoms with a plaque mimicking inclusion are constructed from a mixture (w/w) of 87% deionized water, 10% polyvinyl alcohol (PVA) with a molecular weight of 56.140 g/mol (Sigma–Aldrich, St. Louis, MO, USA) and 3% graphite powder with particle size <50 μm (Merck KGaA) (Fig. 4.10(iii) [56e].

As in our reported studies, a peristaltic pump (Manostat Varistaltic, Barrington, IL) operating at 2 Hz are used. The parameters to be studied for optimizing the PWI performance are: 1) framerate, 2) number of transmitted plane waves (coherent compounding), 3) Motion-estimation rate (MER) and 4) Size of kernel in piecewise PWI (pPWI) (PWV is estimated within these kernels). A probabilistic framework has been developed by our group in order to compare the strain estimation quality between conventional and parallel beamforming [29,47]. The signal-to-noise-ratio of the displacement ($SNR_d$) are calculated for each sequence over the phase of systole. $SNR_e$ is computed for every point in an image using $SNR_d = \frac{\mu(\delta)}{\sigma(\delta)}$ **(3)** where $\mu(\delta)$ and $\sigma(\delta)$ refer to the mean and standard deviation of the displacement ($\delta$) magnitude

**Table 4.1** Preliminary validation of PWV values in a PVA and a silicone (with soft and stiff layer) phantoms. The static PWV is a mechanical testing method.

| Material | Static PWV estimates (m/s) | PWI derived PWV values (m/s) |
|---|---|---|
| P1 PVA | 1.57 | 1.53 |
| P2 soft | 2.31 | 1.86 |
| P2 stiff | 2.87 | 2.86 |

within a small 2D ROI (3.0 × 3.2 mm). Since both strain and $SNR_d$ are computed for each point in the vessel, a large number (>600,000) of displacement-$SNR_d$ pairs are generated for each sequence [29]. The conditional expected value of $SNR_d$ for each strain are calculated using $E(SNR_d|\delta) = \int_0^{+\infty} SNR_d \frac{f(SNR_d, \delta)}{f(\delta)} dSNR_d$ $E(SNR_e|\varepsilon) = \int_0^{+\infty} SNR_e \frac{f(SNR_e, \varepsilon)}{f(\varepsilon)} dSNR_e$ **(4)** [29,47]. $E(SNR_d|\delta)$ curves are generated for each sequence, which allows for a relatively easy comparison to be performed between different sequences for a wide range of strain values. Examples of $E(SNR_d|\delta)$ curves for displacement estimation are provided in Fig. 4.21A. The $SNR_d$ of both 2D PWI and 3D PWI are computed.

## 2.7 Mechanical testing

Three types of mechanical testing are performed to validate our in vivo PWI findings. First, in order to preserve the geometry of the sample and reproduce the in vivo pre-stretch, effect of surrounding tissue and loading conditions, the entire sample is kept intact and inflated at **static pressures** Fig. 4.21A. The compliance, $k_p = \frac{dA}{dP}$, can be measured experimentally by applying this static pressure and measuring the diameter change on the B-mode images. The PWV can then be estimated by $PWV = \sqrt{\frac{k_p r}{2\rho}}$. A phantom example showing good agreement between the PWV estimated through static testing and PWI is shown in Table 4.1.

## 2.8 PWI in aortic aneurysms and carotid plaques in human subjects in vivo

The clinical potential of PWI to provide complementary information on the mechanical properties of vessels in order to help stage disease and inform treatment are assessed. Therefore, the underlying hypothesis of this study is that imaging the regional pulse wave propagation at high temporal resolution is sufficient to determine regional elastic vessel wall properties and thus localize and characterize focal vascular disease at its very early stages, i.e., before changes in aortic diameter or other anatomical changes may occur or become detectable. It is also hypothesized that the regional stiffness changes will unveil the regions most susceptible to wall rupture.

### 2.8.1 Abdominal aortic aneurysms

The clinical feasibility of PWI was evaluated in normal and aneurysmal human aortas. PWI was performed in normal (N = 15, mean age 32.5 years ±10.2), and aneurysmal (N = 5, mean age 71.6 years ±11.8) human subjects [47r]. The PWV of the AAA aortas was significantly higher ($P < .001$) compared to that of the other two groups. The average $r^2$ in the AAA subjects was significantly lower ($P < .001$) than that in the normal subjects, very similar to what was shown in mice.

### 2.8.2 Carotid plaques

The feasibility of PWI in normal ([47q] and atherosclerotic (Fig. 4.23) common carotid arteries has been demonstrated in vivo. The pulse wave velocity (PWV) in eight (N = 8) normals varied between 4.0 and 5.2 m/s with high propagation uniformity ($r^2 > 0.90$) with excellent reproducibility (Figs. 4.23 and 4.24).

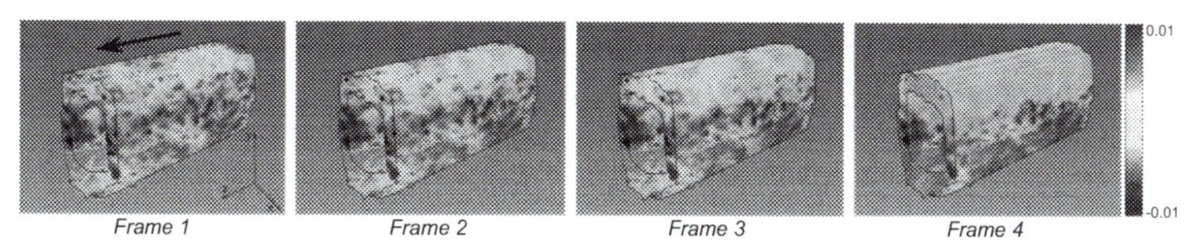


**FIGURE 4.23**

3D PWI frames with wave propagating from right to left in a human carotid in vivo on both the proximal (top) and distal (bottom) walls (Apostolakis et al. 2017).

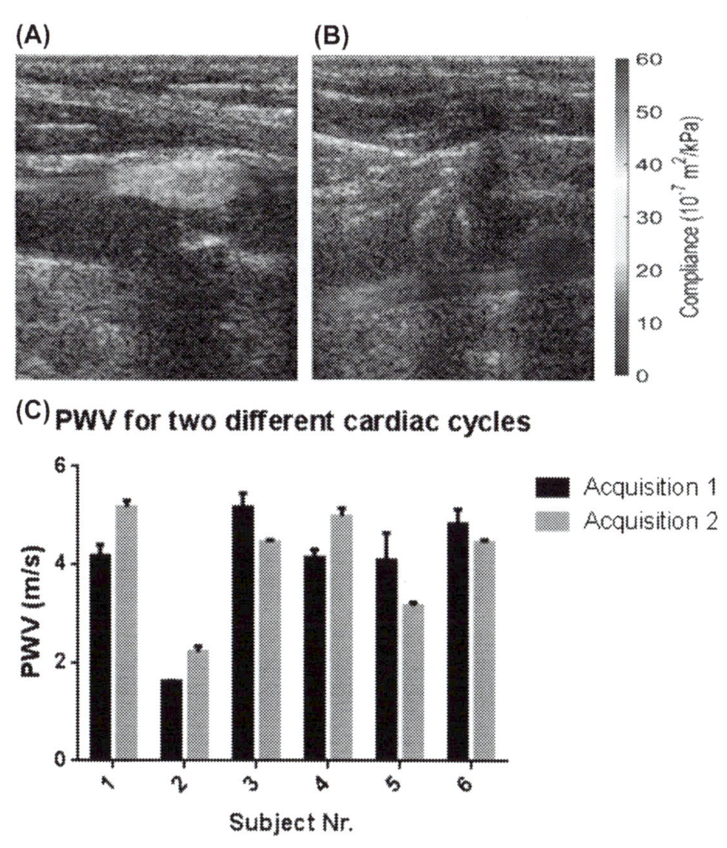


**FIGURE 4.24**

PWI compliance maps in human carotids in vivo: (A) lipid plaque (M, 75), (B) calcified plaque (F, 78); (C) PWI reproducibility in 6 normal subjects and different acquisitions.

## Acknowledgments

The results presented herein were produced by current and previous members of the Ultrasound and Elasticity Imaging Laboratory: Jean Provost, Ronny Li, Ethan Bunting, Alexandre Costet, Julien Grondin, Pierre Nauleau and Iason Apostolakis. The studies were in part supported by R01 EB006042, R01 HL098830 and R01 HL 114358.

## References

[1] E.J. Benjamin, et al., Heart disease and stroke statistics-2018 update: a report from the American heart association, Circulation 137 (12) (2018) e67–e492, https://doi.org/10.1161/CIR.0000000000000558. Epub 2018 Jan 31.

[1a] C. Neagoe, M. Kulke, F. del Monte, J.K. Gwathmey, P.P. de Tombe, R.J. Hajjar, W.A. Linke, Titin isoform switch in ischemic human heart disease, Circulation 106 (11) (2002) 1333–1341.

[1b] K.B. Gupta, M.B. Ratcliffe, M.A. Fallert, L.H. Edmunds, D.K. Bogen, Changes in passive mechanical stiffness of myocardial tissue with aneurysm formation, Circulation, vol, 89, no 5 (1994) 2315–2326.

[2] W.N. McDicken, G.R. Sutherland, C.M. Moran, L.N. Gordon, Colour Doppler velocity imaging of the myocardium, Ultrasound Med. Biol. 18 (1992) 651–654.

[3] G.R. Sutherland, M.J. Stewart, K.W. Groundstroem, C.M. Moran, A. Fleming, F.J. Guell-Peris, R.A. Riemersma, L.N. Fenn, K.A. Fox, W.N. McDicken, Color Doppler myocardial imaging: a new technique for the assessment of myocardial function, J. Am. Soc. Echocardiogr. 7 (1994) 441–458.

[4] J. Zamorano, D.R. Wallbridge, J. Ge, J. Drozd, J. Nesser, R. Erbel, Non-invasive assessment of cardiac physiology by tissue Doppler echocardiography – a comparison with invasive haemodynamics, Eur. Heart J. 18 (1997) 330–339.

[5] J. D'hooge, B. Bijnens, F. Jamal, C. Pislaru, S. Pislaru, J. Thoen, P. Suetens, F. Van de Werf, C. Angermann, F.E. Rademakers, M.C. Herregods, G.R. Sutherland, High frame rate myocardial integrated backscatter. Does this change our understanding of this acoustic parameter? Eur. J. Echocardiogr. 1 (2000) 32–41.

[6] J. D'hooge, E. Konofagou, F. Jamal, A. Heimdal, L. Barrios, B. Bijnens, J. Thoen, F. Van de Werf, G. Sutherland, P. Suetens, Two-dimensional ultrasonic strain rate measurement of the human heart in vivo, IEEE Trans. Ultrason. Ferroelectr. Freq. Control 49 (2002) 281–286.

[7] A. Heimdal, A. Stoylen, H. Torp, T. Skjaerpe, Real-time strain rate imaging of the left ventricle by ultrasound, J. Am. Soc. Echocardiogr. 11 (1998) 1013–1019.

[8] M. O'Donnell, A.R. Skovoroda, Prospects for elasticity reconstruction in the heart, IEEE Trans. Ultrason. Ferroelectr. Freq. Control 51 (2004) 322–328.

[9] K. Kaluzynski, X.C. Chen, S.Y. Emelianov, A.R. Skovoroda, M. O'Donnell, Strain rate imaging using two-dimensional speckle tracking, IEEE Trans. Ultrason. Ferroelectr. Freq. Control 48 (2001) 1111–1123.

[10] S. Langeland, J. D'hooge, T. Claessens, P. Claus, P. Verdonck, P. Suetens, G.R. Sutherland, B. Bijnens, RF-based two-dimensional cardiac strain estimation: a validation study in a tissue-mimicking phantom, IEEE Trans. Ultrason. Ferroelectr. Freq. Control 51 (2004) 1537–1546.

[11] O.A. Breithardt, C. Stellbrink, L. Herbots, P. Claus, A.M. Sinha, B. Bijnens, P. Hanrath, G.R. Sutherland, Cardiac resynchronization therapy can reverse abnormal myocardial strain distribution in patients with heart failure and left bundle branch block, J. Am. Coll. Cardiol. 42 (2003) 486–494.

[12] M. Kowalski, M.C. Herregods, L. Herbots, F. Weidemann, L. Simmons, J. Strotmann, C. Dommke, J. D'Hooge, P. Claus, B. Bijnens, L. Hatle, G.R. Sutherland, The feasibility of ultrasonic regional strain and strain rate imaging in quantifying dobutamine stress echocardiography, Eur. J. Echocardiogr. 4 (2003) 81–91.

[13] O. Bonnefous, P. Pesqué, Time domain formulation of pulse-Doppler ultrasound and blood velocity estimation by cross-correlation, Ultrason. Imaging 8 (1986) 73–85.

[13a] L.N. Bohs, G.E. Trahey, A Novel Method for Angle Independent Ultrasonic-Imaging of Blood-Flow and Tissue Motion, IEEE Trans. Biomed. Eng 38 (1991) 280–286.

[14] P.G.M. de Jong, T. Arts, A.P.G. Hoeks, R.S. Reneman, Determination of tissue motion velocity by correlation interpolation of pulsed ultrasonic echo signals, Ultrason. Imaging 12 (1990) 84–98.

[15] I.A. Hein, W.D. O'brien, Current time-domain methods for assessing tissue motion by analysis from reflected ultrasound echoes - a review, IEEE Trans. Ultrason. Ferroelectr. Freq. Control 40 (1993) 84–102.

[16] J.A. Jensen, Estimation of Blood Velocities Using Ultrasound, Cambridge University Press, New York, NY, 1996.

[16a] J.A. Jensen, N.B. Svendsen, Calculation of Pressure Fields from Arbitrarily Shaped, Apodized, and Excited Ultrasound Transducers. IEEE Trans. Ultrason. Ferroelectr. Freq, Control 39 (1992) 262–267.

[17] P.J. Brands, A.P.G. Hoeks, L.A.F. Ledoux, R.S. Reneman, A radio frequency domain complex cross-correlation model to estimate blood flow velocity and tissue motion by means of ultrasound, Ultrasound Med. Biol. 23 (1997) 911–920.

[18] X. Papademetris, A.J. Sinusas, D.P. Dione, R.T. Constable, J.S. Duncan, Estimation of 3-D left ventricular deformation from medical images using biomechanical models, IEEE Trans. Med. Imaging 21 (2002) 786–800.

[19] H. Kanai, H. Hasegawa, N. Chubachi, Y. Koiwa, M. Tanaka, Noninvasive evaluation of local myocardial thickening and its color coded imaging, IEEE Trans. Ultrason. Ferroelectr. Freq. Control 44 (1997) 752–768.

[20] H. Kanai, Y. Koiwa, J.P. Zhang, Real-time measurements of local myocardium motion and arterial wall thickening, IEEE Trans. Ultrason. Ferroelectr. Freq. Control 46 (1999) 1229–1241.

[21] J. Meunier, Analyse Dynamique des Textures D'Echographies Bidimensionelles Du Myocarde, Ph.D. thesis, Ecole Polytechnique Montreal, 1989.

[22] J. Meunier, M. Bertrand, G. Mailloux, A model for dynamic texture analysis in two-dimensional echocardiograms of the myocardium, Proc. SPIE 768 (1987) 193–200.

[23] J. Meunier, M. Bertrand, G. Mailloux, R. Peticlerc, A model for dynamic texture analysis in two-dimensional echocardiograms of the myocardium, Proc. SPIE 914 (1988) 20–29.

[24] E.E. Konofagou, J. D'Hooge, J. Ophir, Myocardial elastography – a feasibility study in vivo, Ultrasound Med. Biol. 28 (2002) 475–482.

[25] Kanai, Koiwa 2001.

[26] J. Provost, V. Gurev, N. Trayanova, E.E. Konofagou, Mapping of cardiac electrical activation with electromechanical wave imaging: an in silico-in vivo reciprocity study, Heart Rhythm 8 (5) (2011) 752–759. PMC3100212.

[26a] J. Provost, A. Gambhir, S. Thiébaut, V.T.-H. Nguyen, J. Vest, H. Garan, E. Konofagou, Non-invasive Electromechanical wave imaging of atrial, supraventricular and ventricular cardiac conduction disorders in canines and humans, In (Orlando, Florida), p. Accepted for presentation., 2011a.

[27] J. Provost, V.T. Nguyen, D. Legrand, S. Okrasinski, A. Costet, A. Gambhir, H. Garan, E.E. Konofagou, Electromechanical wave imaging for arrhythmias, Phys. Med. Biol. 56 (22) (2011) L1–L11. PMID: 22024555.

[28] J. Provost, W.-N. Lee, K. Fujikura, E.E. Konofagou, Imaging the electromechanical activity of the heart in vivo, Proc. Natl. Acad. Sci. 108 (21) (2011) 8565–8570. PMC3102378.

[29] E.A. Bunting, J. Provost, E.E. Konofagou, Stochastic precision analysis of 2D cardiac strain estimation in vivo, Phys. Med. Biol. 59 (22) (2014) 6841–6858. Epub 2014/10/22. NIHMS638316. PMCID: PMC Journal – In Process.

[29a] J. Grondin, A. Costet, E. Bunting, A. Gambhir, E.Y. Wan, E.E. Konofagou, Validation of Electro mechanical Wave Imaging in a canine model during pacing and sinus rhythm, Heart Rhythm 13 (11) (2016) 2221–2227.

[30] E. Konofagou, J. D'hooge, J. Ophir, Cardiac elastography – a feasibility study, Proc. IEEE Ultrason. Symp. (2000) 1273–1276.

[31] E.E. Konofagou, T. Harrigan, S. Solomon, Assessment of regional myocardial strain using cardiac elastography: distinguishing infarcted from non-infarcted myocardium, Proc. IEEE Ultrason. Symp. (2001) 1589–1592.

[32] E.E. Konofagou, W. Manning, K. Kissinger, S.D. Solomon, Myocardial elastography-Comparison to results using MR cardiac tagging, Proc. IEEE Ultrason. Symp. (2003) 130–133.

[33] E.E. Konofagou, S. Fung-Kee-Fung, J. Luo, M. Pernot, Imaging the mechanics and electromechanics of the heart, Proc. IEEE EMBS Conf. (2006) 6648–6651.

[34] E.E. Konofagou, J. Luo, K. Fujikura, D. Cervantes, J. Coromilas, Imaging the electromechanical wave activation of the left ventricle in vivo, Proc. IEEE Ultrason. Symp. (2006) 985–988.

[35] E.E. Konofagou, W.-N. Lee, J. Luo, Cardiovascular elasticity imaging, in: M. Fatemi, A. Al-Jumaily (Eds.), Biomedical Applications of Vibration and Acoustics in Imaging and Characterizations, ASME Press, New York, NY, 2008, pp. 93–117 (Chapter 6).

[36] E.E. Konofagou, J. Provost, Electromechanical wave imaging for noninvasive mapping of the 3D electrical activation sequence in canines and humans in vivo, J. Biomech. 45 (5) (2012) 856–864 [Invited] PMCID: PMC4005422.

[37] E.E. Konofagou, Intrinsic cardiovascular wave and strain imaging, in: M. Urban (Ed.), Ultrasound Elastography for Biomedical Applications and Medicine", Elsevier and Wiley Publishing, 2018.

[38] E. Konofagou, J. Ophir, A new elastographic method for estimation and imaging of lateral displacements, lateral strains, corrected axial strains and Poisson's ratios in tissues, Ultrasound Med. Biol. 24 (1998) 1183–1199.

[39] J. Provost, W.N. Lee, K. Fujikura, E.E. Konofagou, Electromechanical wave imaging of normal and ischemic hearts in vivo, IEEE Trans. Med. Imaging 29 (3) (2010a) 625–635. PMCID: PMC3093312.

[40] W.N. Lee, J. Provost, K. Fujikura, J. Wang, E.E. Konofagou, In vivo study of myocardial elastography under graded ischemia conditions, Phys. Med. Biol. 56 (4) (2011) 1155–1172. PMCID: PMC4005801.

[41] E.E. Konofagou, Estimation and Imaging of Three-Dimensional Motion and Poisson's Ratio in Elastography, Ph.D. dissertation, University of Houston, TX, 1999.

[42] I.K. Zervantonakis, S.D. Fung-Kee-Fung, W.N. Lee, E.E. Konofagou, A novel, view-independent method for strain mapping in myocardial elastography: eliminating angle and centroid dependence, Phys. Med. Biol. 52 (2007) 4063–4408.

[43] W.N. Lee, E.E. Konofagou, Angle-independent and multi-dimensional myocardial elastography - from theory to clinical validation, Ultrasonics 48 (2008) 563–567.

[44] J. Luo, E. Konofagou, A fast normalized cross-correlation calculation method for motion estimation, IEEE Trans. Ultrason. Ferroelectr. Freq. Control 57 (6) (2010) 1347–1357. PMCID: PMC4123965.

[45] J. Luo, E.E. Konofagou, High-frame rate, full-view myocardial elastography with automated contour tracking in murine left ventricles in vivo, IEEE Trans. Ultrason. Ferroelectr. Freq. Control 55 (2008) 240–248.

[45a] A. Baghani, A. Brant, S. Salcudean, R. Rohling, A high-frame-rate ultrasound system for the study of tissue motions, IEEE Trans Ultrason Ferroelectr Freq Control 57 (2010) 1535–1547.

[45b] T. Varghese, J. Ophir, A theoretical framework for performance characterization of elastography: the strain filter, IEEE Trans Ultrason Ferroelectr Freq Control 44 (1997) 164–172.

[45c] E. Weinstein, A. Weiss, Fundamental limitations in passive time-delay estimation—Part II: Wide-band systems. Acoustics, Speech and Signal Processing, IEEE, Transactions on 32 (1984) 1064–1078.

[45d] J. Meunier, M. Bertrand, Ultrasonic texture motion analysis: theory and simulation, IEEE Trans Med Imaging 14 (1995) 293–300.

[46] J. Luo, E.E. Konofagou, Effects of various parameters on lateral displacement estimation in ultrasound elastography, Ultrasound Med. Biol. 35 (8) (2009) 1352–1366.

[47] J. Provost, S. Thiebault, J. Luo, E.E. Konofagou, Single-heartbeat electromechanical wave imaging with optimal strain estimation using temporally unequispaced acquisition sequences, Phys. Med. Biol. 57 (4) (2012) 1095–1112. NIHMS375306 [PMCID in Process].

[47a] M.D. Mozaffarian, et al., Heart Disease and Stroke Statistics—2016 Update, A Report From the American Heart Association, Circulation, Circulation 133 (4) (Jan 26 2016) e38–60.

[47b] G.C. Makris, A.N. Nicolaides, X.Y. Xu, G. Geroulakos, Introduction to the biomechanics of carotid plaque pathogenesis and rupture: review of the clinical evidence, The British journal of radiology 83 (993) (2010) 729–735, https://doi.org/10.1259/bjr/49957752. PubMed PMID: 20647514; PubMed Central PMCID: PMCPMC3473420.

[47c] T. Saam, T.S. Hatsukami, N. Takaya, B. Chu, H. Underhill, W.S. Kerwin, J. Cai, M.S. Ferguson, C. Yuan, The vulnerable, or high-risk, atherosclerotic plaque: noninvasive MR imaging for characterization and assessment, Radiology 244 (1) (2007) 64–77.

[47d] H.H. Hansen, C.L. De Korte, Nonivasive carotid elastography, in: A.N. Nicolaides, K.W. Beach, E. Kyriacou, C.S. Pattichis (Eds.), Ultrasound and Carotid Bifurcation Atherosclerosis, Springer, London, 2012, pp. 341–353.

[47e] K. Ouriel, R.M. Green, C. Donayre, C.K. Shortell, J. Elliott, J.A. DeWeese, An evaluation of new methods of expressing aortic aneurysm size: relationship to rupture, Journal of vascular surgery 15 (1) (1992) 12–18.

[47f] H. Bengtsson, B. Sonesson, D. Bergqvist, Incidence and prevalence of abdominal aortic aneurysms, estimated by necropsy studies and population screening by ultrasound, Annals of the New York Academy of Sciences 800 (1996) 1–24.
[47g] M.I. Patel, D.T. Hardman, C.M. Fisher, M. Appleberg, Current views on the pathogenesis of abdominal aortic aneurysms, Journal of the American College of Surgeons 181 (4) (1995) 371–382.
[47h] R.C. Darling, C.R. Messina, D.C. Brewster, L.W. Ottinger, Autopsy study of unoperated abdominal aortic aneurysms. The case for early resection, Circulation 56 (3 Suppl) (1977) II161–I164.
[47i] R.A. Wain, M.L. Marin, T. Ohki, L.A. Sanchez, R.T. Lyon, A. Rozenblit, W.D. Suggs, J.G. Yuan, F.J. Veith, Endoleaks after endovascular graft treatment of aortic aneurysms: classification, risk factors, and outcome, Journal of vascular surgery 27 (1) (1998) 69–78.
[47j] W.D. Turnipseed, S.C. Carr, G. Tefera, C.W. Acher, J.R. Hoch, Minimal incision aortic surgery, Journal of vascular surgery 34 (1) (2001) 47–53.
[47k] O.C. Velazquez, R.A. Larson, R.A. Baum, J.P. Carpenter, M.A. Golden, M.E. Mitchell, A. Pyeron, C.F. Barker, R.M. Fairman, Gender-related differences in infrarenal aortic aneurysm morphologic features: issues relevant to Ancure and Talent endografts, Journal of vascular surgery 33 (2 Suppl) (2001) S77–84.
[47l] L. Gabrielli, A. Baudo, A. Molinari, M. Domanin, Early complications in endovascular treatment of abdominal aortic aneurysm, Acta chirurgica Belgica 104 (5) (2004) 519–526.
[47m] J.N. Ghansah, J.T. Murphy, Complications of major aortic and lower extremity vascular surgery, Seminars in cardiothoracic and vascular anesthesia 8 (4) (2004) 335–361.
[47n] E.D. Dillavou, S.C. Muluk, M.S. Makaroun, A decade of change in abdominal aortic aneurysm repair in the United States: Have we improved outcomes equally between men and women? Journal of vascular surgery 43 (2) (2006) 230–238.
[47o] D.C. Brewster, J.L. Cronenwett, J.W. Hallett Jr., K.W. Johnston, W.C. Krupski, J.S. Matsumura, Joint Council of the American Association for Vascular S, Society for Vascular S. Guidelines for the treatment of abdominal aortic aneurysms. Report of a subcommittee of the Joint Council of the American Association for Vascular Surgery and Society for Vascular Surgery, Journal of vascular surgery 37 (5) (2003) 1106–1117.
[47p] G. Mancia, G. De Backer, A. Dominiczak, R. Cifkova, R. Fagard, G. Germano, G. Grassi, A.M. Heagerty, S.E. Kjeldsen, S. Laurent, K. Narkiewicz, L. Ruilope, A. Rynkiewicz, R.E. Schmieder, H.A. Boudier, A. Zanchetti, A. Vahanian, J. Camm, R. De Caterina, V. Dean, K. Dickstein, G. Filippatos, C. Funck-Brentano, I. Hellemans, S.D. Kristensen, K. McGregor, U. Sechtem, S. Silber, M. Tendera, P. Widimsky, J.L. Zamorano, S. Erdine, W. Kiowski, E. Agabiti-Rosei, E. Ambrosioni, L.H. Lindholm, M. Viigimaa, S. Adamopoulos, E. Agabiti-Rosei, E. Ambrosioni, V. Bertomeu, D. Clement, S. Erdine, C. Farsang, D. Gaita, G. Lip, J.M. Mallion, A.J. Manolis, P.M. Nilsson, E. O'Brien, P. Ponikowski, J. Redon, F. Ruschitzka, J. Tamargo, P. van Zwieten, B. Waeber, B. Williams, Management of Arterial Hypertension of the European Society of H, European Society of C. 2007 Guidelines for the Management of Arterial Hypertension: The Task Force for the Management of Arterial Hypertension of the European Society of Hypertension (ESH) and of the European Society of Cardiology (ESC), Journal of hypertension 25 (6) (2007) 1105–1187.

[47q] J. Luo, R. Li, E.E. Konofagou, Pulse Wave Imaging (PWI) of the Human Carotid Artery: An In Vivo Feasibility Study, IEEE Trans. Ultras. Ferroel. Freq Control 59 (1) (2012) 174–181.

[47r] R. Li, J. Luo, S. Balaram, F. Chaudhry, J. Lantis, D. Shahmirzadi, E.E. Konofagou, Pulse wave imaging in normal, hypertensive and aneurysmal human aortas in vivo: a feasibility study, Phys. Med. Biol. 58 (2013) 4549–4562.

[48] S.J. Okrasinski, B. Ramachandran, E.E. Konofagou, Assessment of myocardial elastography performance in phantoms under combined physiologic motion configurations with preliminary in vivo feasibility, Phys. Med. Biol. 57 (17) (2012) 5633–5650. PMCID: PMC3704133.

[48a] D. Lloyd-Jones, M. Carnethon, G. De Simone, T.B. Ferguson, K. Flegal, E. Ford, K. Furie, A. Go, K. Greenlund, N. Haase, et al., Heart Disease and Stroke Statistics—2009 Update: A Report From the American Heart Association Statistics Committee and Stroke Statistics Subcommittee, Circulation 119 (2009) e21–181.

[48b] M. Haïssaguerre, P. Jaïs, D.C. Shah, A. Takahashi, M. Hocini, G. Quiniou, S. Garrigue, A. Le Mouroux, P. Le Métayer, J. Clémenty, Spontaneous initiation of atrial fibrillation by ectopic beats originating in the pulmonary veins. N. Engl, J. Med 339 (1998) 659–666.

[48c] H. Garan, Atypical atrial flutter, Heart Rhythm 5 (2008) 618–621.

[48d] N. Saoudi, F. Cosío, A. Waldo, S.A. Chen, Y. Iesaka, M. Lesh, S. Saksena, J. Salerno, W. Schoels, A classification of atrial flutter and regular atrial tachycardia according to electrophysiological mechanisms and anatomical bases; a Statement from a Joint Expert Group from The Working Group of Arrhythmias of the European Society of Cardiology and the North American Society of Pacing and Electrophysiology, Eur. Heart J 22 (2001) 1162–1182.

[48e] A. Bochoeyer, Y. Yang, J. Cheng, R.J. Lee, E.C. Keung, N.F. Marrouche, A. Natale, M.M. Scheinman, Surface Electrocardiographic Characteristics of Right and Left Atrial Flutter, Circulation 108 (2003) 60–66.

[48f] M.P. Nash, A.J. Pullan, Challenges Facing Validation of Noninvasive Electrical Imaging of the Heart, The Annals of Noninvasive Electrocardiology 10 (2005) 73–82.

[49] N. Chattipakorn, R.E. Ideker, The vortex at the apex of the left ventricle: a new twist to the story of the electrical induction of rotors? J. Cardiovasc. Electrophysiol. 14 (2003) 303–305.

[49a] D. Durrer, R.T. Van Dam, G.E. Freud, M.J. Janse, F.L. Meijler, R.C. Arzbaecher, Total Excitation of the Isolated Human Heart, Circulation 41 (1970) 899–912.

[49b] D.R. Sutherland, Q. Ni, R.S. MacLeod, R.L. Lux, B.B. Punske, Experimental measures of ventricular activation and synchrony, Pacing Clin Electrophysiol 31 (2008) 1560–1570.

[49c] H. Ashikaga, B.A. Coppola, B. Hopenfeld, E.S. Leifer, E.R. McVeigh, J.H. Omens, Transmural Dispersion of Myofiber Mechanics: Implications for Electrical Heterogeneity In Vivo, Journal of the American College of Cardiology 49 (2007) 909–916.

[49d] M.W. Kay, P.M. Amison, J.M. Rogers, Three-dimensional surface reconstruction and panoramic optical mapping of large hearts, IEEE Trans Biomed Eng 51 (2004) 1219–1229.

[49e] F. Qu, C.M. Ripplinger, V.P. Nikolski, C. Grimm, I.R. Efimov, Three-dimensional panoramic imaging of cardiac arrhythmias in rabbit heart, J Biomed Opt 12 (2007) 044019.

[49f] C.M. Ripplinger, Q. Lou, W. Li, J. Hadley, I.R. Efimov, Panoramic imaging reveals basic mechanisms of induction and termination of ventricular tachycardia in rabbit heart with chronic infarction: implications for low-voltage cardioversion, Heart Rhythm 6 (2009) 87–97.

[50] E.M.C. Hillman, O. Bernus, E. Pease, M.B. Bouchard, A. Pertsov, Depth-resolved optical imaging of transmural electrical propagation in perfused heart, Opt. Express 15 (2007) 17827–17841.

[50a] D.A. Hooks, I.J. LeGrice, J.D. Harvey, B.H. Smaill, Intramural multisite recording of transmembrane potential in the heart, Biophys. J 81 (2001) 2671–2680.

[50b] W. Kong, R.E. Ideker, V.G. Fast, Transmural optical measurements of Vm dynamics during long-duration ventricular fibrillation in canine hearts, Heart Rhythm 6 (2009) 796–802.

[51] C. Ramanathan, R.N. Ghanem, P. Jia, K. Ryu, Y. Rudy, Noninvasive electrocardiographic imaging for cardiac electrophysiology and arrhythmia, Nat. Med. 10 (2004) 422–428.

[51a] X. Zhang, I. Ramachandra, Z. Liu, B. Muneer, S.M. Pogwizd, B. He, Noninvasive three-dimensional electrocardiographic imaging of ventricular activation sequence, Am J Physiol Heart Circ Physiol 289 (2005) H2724–H2732.

[51b] C. Ramanathan, P. Jia, R. Ghanem, K. Ryu, Y. Rudy, Activation and repolarization of the normal human heart under complete physiological conditions. Proceedings of the National Academy of Sciences 103 (2006) 6309–6314.

[52] R.J. Schilling, N.S. Peters, W. Davies, Simultaneous endocardial mapping in the human left ventricle using a noncontact catheter – comparison of contact and reconstructed electrograms during sinus rhythm, Circulation 98 (1998) 887–898.

[52a] C.-T. Tai, T.-Y. Liu, P.-C. Lee, Y.-J. Lin, M.-S. Chang, S.-A. Chen, Non-contact mapping to guide radiofrequency ablation of atypical right atrial flutter, J. Am. Coll. Cardiol 44 (2004) 1080–1086.

[52b] B. Taccardi, G. Arisi, E. Macchi, S. Baruffi, S. Spaggiari, A new intracavitary probe for detecting the site of origin of ectopic ventricular beats during one cardiac cycle, Circulation 75 (1987) 272–281.

[52c] Y. Wang, R.B. Schuessler, R.J. Damiano, P.K. Woodard, Y. Rudy, Noninvasive electrocardiographic imaging (ECGI) of scar-related atypical atrial flutter, Heart Rhythm 4 (2007) 1565–1567.

[52d] F.B. Belgacem, Why is the Cauchy problem severely ill-posed? Inverse Problems 23 (2007) 823–836.

[52e] R.L. Lux, Noninvasive Assessment of Cardiac Electrophysiology for Predicting Arrhythmogenic Risk: Are We Getting Closer? Circulation 118 (2008) 899–900.

[52f] B. Olshansky, Electrocardiographic imaging: Back to the drawing board? Heart Rhythm 8 (2011) 700–701.

[53] S.G. Wang, W.N. Lee, J. Provost, J. Luo, E.E. Konofagou, A composite high-frame-rate system for clinical cardiovascular imaging, IEEE Trans. Ultrason. Ferroelectr. Freq. Control 55 (2008) 2221–2233.

[53a] J. Provost, W.-N. Lee, K. Fujikura, E.E. Konofagou, Electromechanical Wave Propagation in the Normal and Ischemic Canine Myocardium in Vivo (2008). In (Lake Travis, Austin, Texas, USA.).

[53b] J. Provost, W.-N. Lee, K. Fujikura, E.E. Konofagou, Electromechanical Wave Imaging: Non-invasive Localization and Quantification of Partial Ischemic Regions In Vivo (2009). In (Roma, Italy).

[53c] J. Provost, S. Okrasinski, W.-N. Lee, E.E. Konofagou, Abstract AB15-6 A Noninvasive, Direct Method for Cardiac Mapping: Electromechanical Wave Imaging, Heart Rhythm 7 (2010b) S32.

[54] A. Costet, J. Provost, A. Gambhir, Y. Bobkov, P. Danilo Jr., G.J. Boink, M.R. Rosen, E.E. Konofagou, Electromechanical wave imaging of biologically and electrically paced canine hearts in vivo, Ultrasound Med. Biol. 40 (1) (2014) 177–187. PMCID: PMC3897195.

[55] F. Kallel, J. Ophir, A least-squares strain estimator for elastography, Ultrason. Imaging 19 (1997) 195–208.

[56] J. Luo, W.-N. Lee, E.E. Konofagou, Fundamental performance assessment of 2-D myocardial elastography in a phased array configuration, IEEE Trans. Ultrason. Ferroelectr. Freq. Control 56 (10) (2009) 2620–2627.

[56a] R.X. Li, W.W. Qaqish, E.E. Konofagou, "Performance assessment of Pulse Wave Imaging using conventional ultrasound in canine aortas ex vivo and normal human arteries in vivo.", Artery Research Volume 11 (2015) 19–28.

[56b] R.X. Li, I.Z. Apostolakis, P. Kemper, M.D.J. McGarry, A. Ip, E.S. Connolly, J.F. McKinsey, E.E. Konofagou, Pulse Wave Imaging in Carotid Artery Stenosis Human Patients in Vivo, Phys Med Biol 64 (2) (2019 Jan 10) 025013.

[56c] M. McGarry, R. Li, I. Apostolakis, P. Nauleau, E.E. Konofagou, An inverse approach to determining spatially varying arterial compliance using ultrasound imaging, Physics in Medicine and Biology 61 (15) (2016) 5486–5507.

[56d] J.J. Westenberg, et al., Bramwell-Hill modeling for local aortic pulse wave velocity estimation: a validation study with velocity-encoded cardiovascular magnetic resonance and invasive pressure assessment, J Cardiovasc Magn Reson (2012).

[56e] E. Widman, E. Maksuti, D. Larsson, M.W. Urban, A. Bjallmark, M. Larsson, Shear wave elastography plaque characterization with mechanical testing validation: a phantom study, Phys Med Biol. 60 (8) (2015) 3151–3174.

[57] K. Lauerma, et al., Multislice MRI in assessment of myocardial perfusion in patients with single-vessel proximal left anterior descending coronary artery disease before and after revascularization, Circulation 96 (1997) 2859–2867.

[58] J. Provost, W.-N. Lee, K. Fujikura, E.E. Konofagou, Abstract 21142: Electromechanical wave imaging for noninvasive mapping of the 3D electrical activation sequence in vivo. Circulation 122 (2011c) A21142.

## Further reading

[1] J. Amano, J.X. Thomas, M. Lavallee, I. Mirsky, D. Glover, W.T. Manders, W.C. Randall, S.F. Vatner, Effects of myocardial-ischemia on regional function and stiffness in conscious dogs, Am. J. Physiol. 252 (1) (1987) H110–H117.

[2] M.E. Bertrand, M.F. Rousseau, J.M. Lefebvre, J.M. Lablanche, P.H. Asseman, A.G. Carre, J.P. Lekieffre, Left-ventricular compliance in acute transmural myocardial-infarction in man, Eur. J. Cardiol. 7 (1978) 179–193.

[3] B. Byram, G. Holley, D. Giannantonio, G. Trahey, 3-D phantom and in vivo cardiac speckle tracking using a matrix array and raw echo data, IEEE Trans. Ultrason. Ferroelectr. Freq. Control (2010) 839–854.

[4] E. Carmeliet, Cardiac ionic currents and acute ischemia: from channels to arrhythmias, Physiol. Rev. 79 (1999) 917–1017.
[5] R. Coronel, Heterogeneity in extracellular potassium concentration during early myocardial ischemia and reperfusion: implications for arrhythmogenesis, Cardiovasc. Res. 28 (1994) 770–777.
[6] N. Chattipakorn, B.H. KenKnight, J.M. Rogers, R.G. Walker, G.P. Walcott, D.L. Rollins, W.M. Smith, R.E. Ideker, Locally propagated activation immediately after internal defibrillation, Circulation 97 (1998) 1401–1410.
[7] N.A. Trayanova, J. Constantino, V. Gurev, Electromechanical models of the ventricles, Am J Physiol Heart Circ Physiol 301 (2) (2011 Aug) H279–H286.
[8] B. Denarie, T. Bjastad, H. Torp, Multi-line transmission in 3D with reduced crosstalk artifacts: a proof of concept study, IEEE Trans. Ultrason. Ferroelectr. Freq. Control 60 (2013) 1708–1718.
[9] C.H. Edwards, J.S. Rankin, P.A. Mchale, D. Ling, R.W. Anderson, Effects of ischemia on left-ventricular regional function in the conscious dog, Am. J. Physiol. 240 (3) (1981) H413–H420.
[10] J.M. Ferrero, J. Saiz, J.M. Ferrero, et al., Simulation of action potentials from metabolically impaired cardiac myocytes. Role of ATP-sensitive K$+$ current, Circ. Res. 79 (1996) 208–221.
[11] J.M. Ferrero, B. Trenor, B. Rodriguez, et al., Electrical activity ad reentry during acute regional myocardial ischemia: insights from simulation, Int. J. Bifurcation Chaos 13 (2003) 1–13.
[12] R.J. Gibbons, G.J. Balady, J.T. Bricker, B.R. Chaitman, G.F. Fletcher, V.F. Froelicher, et al., Acc/aha 2002 guideline update for exercise testing: summary article: a report of the american college of cardiology/american heart association task force on practice guidelines (committee to update the 1997 exercise testing guidelines), Circulation 106 (2002) 1883–1892.
[13] A.S. Go, et al., Heart disease and stroke statistics—2014 update: a report from the American heart association, Circulation 129 (2014) e28–e292.
[14] J. Grondin, E.E. Konofagou, Intracardiac myocardial elastography in vivo, IEEE Trans. Ultrason. Ferroelectr. Freq. Control 62 (2) (2015) 337–349. PMCID: PMC Journal – In Process.
[15] J. Guccione, K. Costa, A. McCulloch, Finite element stress analysis of left ventricular mechanics in the beating dog heart 28 (1995) 1167–1177.
[16] V. Gurev, J. Constantino, J.J. Rice, N.A. Trayanova, Three-dimensional Activation Sequence Determines Transmural Changes in Electromechanical Delay, 2009.
[17] J.H. Haga, A.J. Beaudoin, J.G. White, J. Strony, Quantification of the passive mechanical properties of the resting platelet, Ann. Biomed. Eng. 26 (1998) 268–277.
[18] J.W. Holmes, T.K. Borg, J.W. Covell, Structure and mechanics of healing myocardial infarcts, Annu. Rev. Biomed. Eng. 7 (2005) 223–253.
[19] G. Hou, J. Provost, J. Grondin, S. Wang, F. Marquet, E. Bunting, E. Konofagou, Sparse matrix beamforming and image reconstruction for real-time 2D HIFU monitoring using Harmonic Motion Imaging for Focused Ultrasound (HMIFU) with in vitro validation, IEEE Trans. Med. Imaging 33 (11) (2014) 2107–2117.
[20] M.J. Janse, J. Cinca, Morena He, et al., The "Border zone" in myocardial ischemia: an electrophysiological, metabolic, and histochemical correlation in the pig heart, Circ. Res. 44 (1979) 576–588.

[21] J.A. Jensen, in: Field: a program for simulating ultrasound systems. Paper presented at the 10th Nordic-Baltic Conference on Biomedical imaging vol. 34, Medical & Biological Engineering & Computing, 1996, pp. 1351–1353.

[22] X. Jie, V. Gurev, N. Trayanova, Mechanically-induced Spontaneous Arrhythmias in Acute Regional Ischemia, 2009 submitted.

[23] Y. Kagiyama, J.L. Hill, L.S. Gettes, Interaction of acidosis and increased extracellular potassium on action potential characteristics and conduction in Guinea pig ventricular muscle, Circ. Res. 51 (1982) 614–623.

[24] A.G. Kléber, M.J. Janse, F.J.L. van Capelle, et al., Mechanism and time course of s-t and t-q segment changes during acute regional myocardial ischemia in the pig heart determined by extracellular and intracellular recordings, Circ. Res. 42 (1978) 603–613.

[25] E.E. Konofagou, W.-N. Lee, C.M. Ingrassia, A theoretical model for myocardial elastography and its in vivo validation, Proc. IEEE EMBS Conf. (2005) 985–988.

[26] E.E. Konofagou, W.-N. Lee, S.D. Fung-Kee-Fung, Angle-independent myocardial elastography: theoretical analysis and clinical validation, Proc. SPIE 6513 (2007), 65130G-1-65130G-10.

[27] E.E. Konofagou, J. Luo, D. Saluja, K. Fujikura, D. Cervantes, J. Coromilas, Noninvasive electromechanical wave imaging and conduction velocity estimation in vivo, Proc. IEEE Ultrason. Symp. (2007) 969–972.

[28] E.E. Konofagou, J. Luo, D. Saluja, D. Cervantes, J. Coromilas, K. Fujikura, Noninvasive Electromechanical Wave Imaging and Conduction-Relevant Velocity Estimation in Vivo, Ultrasonics, 2009.

[29] E.E. Konofagou, J. Luo, D. Saluja, D.O. Cervantes, J. Coromilas, K. Fujikura, Noninvasive electromechanical wave imaging and conduction-relevant velocity estimation in vivo, Ultrasonics 50 (2) (2010) 208–215 [Invited] PMCID: PMC4005418.

[30] E.E. Konofagou, W.-N. Lee, J. Luo, J. Provost, J. Vappou, Physiologic cardiovascular strain and intrinsic wave imaging, Annu. Rev. Biomed. Eng. 13 (2011) 477–505 [Invited] PMID:21756144. PMC Exempt – invited review.

[31] E.E. Konofagou, J. Provost, Evolving concepts in measuring ventricular strain in the human heart: non-invasive imaging, in: P. Kohl (Ed.), Cardiac Mechano-Electric Coupling and Arrhythmias, second ed., 2014.

[32] W.N. Lee, C.M. Ingrassia, S.D. Fung-Kee-Fung, K.D. Costa, J.W. Holmes, E.E. Konofagou, Theoretical quality assessment of myocardial elastography with in vivo validation, IEEE Trans. Ultrason. Ferroelectr. Freq. Control 54 (2007) 2233–2245.

[33] W.N. Lee, Z. Qian, C.L. Tosti, T.R. Brown, D.N. Metaxas, E.E. Konofagou, Preliminary validation of angle-independent myocardial elastography using mr tagging in a clinical setting, Ultrasound Med. Biol. 34 (2008) 1980–1997.

[34] Lee W-N, Parker K., Luo J, Holmes J.W., Konofagou E.E. Frame rate dependence of myocardial elastography estimates using a physiologic 3D biventricular finite-element model of the heart with preliminary in vivo validation, 2009 IEEE International Ultrasonics Symposium, Rome, Italy, Sept. 18–21, 2009.

[35] W.N. Lee, Myocardial Elastography: A Strain Imaging Technique for the Reliable Detection and Localization of Myocardial Ischemia in Vivo, Ph.D. dissertation, Columbia University, 2010.

[36] X. Liu, K.Z. Abd-Elmoniem, M. Stone, E.Z. Murano, J. Zhuo, R.P. Gullapalli, J.L. Prince, Incompressible deformation estimation algorithm (IDEA) from tagged

MR images, IEEE Trans. Med. Imaging 31 (2) (2012) 326–340, https://doi.org/10.1109/TMI.2011.2168825. Epub 2011 Sep 19.

[37] J. Luo, K. Fujikura, S. Homma, E.E. Konofagou, Myocardial elastography at both high temporal and spatial resolution for the detection of infarcts, Ultrasound Med. Biol. 33 (2007) 1206–1223.

[38] J. Luo, E.E. Konofagou, Automated contour tracking for high frame-rate, full-view myocardial elastography in vivo, IEEE Ultrason. Symp. Proc. (2007) 1929–1932.

[39] J. Luo, E.E. Konofagou, Imaging of wall motion coupled with blood flow velocity in the heart and vessels in vivo: a feasibility study, Ultrasound Med. Biol. 37 (6) (2011) 980–995. PMCID: PMC4009734.

[40] W.G. O'Dell, C.C. Moore, W.C. Hunter, E.A. Zerhouni, E.R. McVeigh, 3-Dimensional myocardial deformations - calculation with displacement field fitting to tagged mr-images, Radiology 195 (1995) 829–835.

[41] N.F. Osman, J.L. Prince, Visualizing myocardial function using HARP MRI, Phys. Med. Biol. 45 (2000) 1665–1682.

[42] M. Pernot, E.E. Konofagou, Electromechanical imaging of the myocardium at normal and pathological states, Proc. IEEE Ultrason. Symp. (2005) 1091–1094.

[43] D.P. Perrin, N.V. Vasilyev, G.R. Marx, P.J. del Nido, Temporal enhancement of 3D echocardiography by frame reordering, JACC Cardiovasc. Imaging 5 (2012) 300–304.

[44] J. Provost, A. Gambhir, J. Vest, H. Garan, E.E. Konofagou, A clinical feasibility study of atrial and ventricular electromechanical wave imaging, Heart Rhythm 10 (6) (2013) 856–862. PMCID: PMC4005774.

[45] J. Provost, C. Papadacci, J. Arango, M. Imbault, M. Fink, et al., 3D ultrafast ultrasound imaging, Phys. Med. Biol. 59 (2014).

[46] N. Reichek, MRI myocardial tagging, J. Magn. Reson. Imaging 10 (1999) 609–616.

[47] J.J. Rice, F. Wang, D.M. Bers, P.P. de Tombe, Approximate model of cooperative activation and crossbridge cycling in cardiac muscle using ordinary differential equations, Biophys. J. 95 (2008) 2368–2390.

[48] R. Righetti, J. Ophir, P. Ktonas, Axial resolution in elastography, Ultrasound Med. Biol. 28 (1) (2002) 101–113.

[49] R.M. Shaw, Y. Rudy, Electrophysiologic effects of acute myocardial ischemia. A mechanistic investigation of action potential conduction and conduction failure, Circ. Res. 80 (1997) 124–138.

[50] N. Varma, F.R. Eberli, C.S. Apstein, Increased diastolic chamber stiffness during demand ischemia - response to quick length change differentiates rigor-activared from calcium-activated tension, Circulation 101 (18) (2000) 2185–2192.

[51] N. Varma, F. R. i Eberl, C.S. Apstein, Left ventricular diastolic dysfunction during demand ischemia: rigor underlies increased stiffness without calcium-mediated tension. Amelioration by glycolytic substrate, J. Am. Coll. Cardiol. 37 (8) (2001) 2144–2153.

[52] J.N. Weiss, K.L. Shine, Accumulation and electrophysiological alterations during early myocardial ischemia, Am. J. Physiol. 243 (1982) H318–H327.

[53] J.N. Weiss, N. Venkatesh, S.T. Lamp, ATP-sensitive K+ channels and cellular K+ loss in hypoxic and ischemic mammalian ventricle, J. Physiol. 447 (1992) 649–673.

[54] A.L. Wit, M.J. Janse, Experimental models of ventricular tachycardia and fibrillation caused by ischemia and infarction, Circulation 85 (Suppl. I) (1992) I32–I42.

[55] G.X. Yan, Kl'eber AeG, Changes in extracellular and intracellular pH in ischemic rabbit papillary muscle, Circ. Res. 71 (1992) 460–470.

[56] A.M. Yue, J.R. Paisey, S. Robinson, T.R. Betts, P.R. Roberts, J.M. Morgan, Determination of human ventricular repolarization by noncontact mapping - validation with monophasic action potential recordings, Circulation 110 (2004) 1343–1350.

[57] A.V. Zaitsev, P.K. Guha, F. Sarmast, et al., Wavebreak formation during ventricular fibirllation in the isolated, regionally ischemic pig heart, Circ. Res. 92 (2003) 546–553.

CHAPTER

# Ultrasound-based liver elastography

**Ioan Sporea, Roxana Şirli**

*Department of Gastroenterology and Hepatology, Victor Babeş University of Medicine and Pharmacy, Timişoara, Romania*

## 1. Introduction to chronic liver disease: etiology, screening, and diagnosis

There are many causes of chronic liver disease and, all told, chronic liver disease is a significant burden on our modern medical system. Common causes of chronic liver disease include viruses, alcohol consumption, and diet. Regardless of cause, they all lead to the common pathologic changes of fibrosis and eventually cirrhosis.

Among them, chronic viral hepatitis is still frequent in clinical practice, especially in some areas and in certain populations, such as intravenous drug users. Despite the fact that interferon-free treatment regimens for hepatitis C virus (HCV) chronic infection can cure the vast majority of cases, new cases are being continuously discovered. Hepatitis B virus (HBV) chronic infection is rarely cured with currently available treatment, and hence, patients should be followed up throughout their lives.

Nonalcoholic fatty liver disease (NAFLD) is a form of chronic liver disease that is the result of metabolic derangements secondary to the modern western lifestyle. Contributing conditions include obesity, type 2 diabetes mellitus, hypertriglyceridemia, and lack of physical exercise, which are more and more common in developed countries, hence the increase in NAFLD prevalence [1]. In patients with nonalcoholic steatohepatitis, fibrosis is present and can evolve toward advanced fibrosis and cirrhosis. In these cases, continuous monitoring of fibrosis is mandatory for prognostic assessment and for tailoring treatment.

Alcoholic liver disease is also a real problem of the modern world in many societies. A report from World Health Organization concluded that 3.3 million deaths (6% of all global deaths) are attributable to alcohol use and that alcohol abuse is a risk factor in about 50% of cases of cirrhosis [2]. The disease must be continuously monitored for progression, especially in those who are still drinking.

Other chronic causes of liver diseases, such as cholestatic liver disease (primary biliary cholangitis and primary sclerosing cholangitis), hemochromatosis, and autoimmune hepatitis, are less common than the previously mentioned diseases but are still a significant concern. Thus, in totality, the liver disease burden is sufficiently high and the need for monitoring is sufficiently critical to warrant investigations into inexpensive, effective, and noninvasive monitoring techniques.

**Tissue Elasticity Imaging. https://doi.org/10.1016/B978-0-12-809662-8.00005-X**

Classically, liver biopsy (LB) has been referred to as the "gold standard" for liver fibrosis assessment [3]. Published studies demonstrated significant limitations in fibrosis assessment by LB. These limitations include insufficient specimen size [4,5], intra- and interobserver variabilities [6,7] complications following the procedure [8,9], and patients' reluctance to undergo LB [10]. Histopathologic examination of liver tissue provides qualitative information regarding the severity of necroinflammation and fibrosis, as well as regarding the cause of chronic liver disease [11]. Commonly used pathologic scoring systems include the Knodell score/histology activity index [12], the METAVIR score [13] and the Ishak score (modified Knodell score) [14].

There are several reasons why a reference gold standard, LB, should be replaced by newer methods. First, some chronic liver diseases such as NAFLD are very frequent in some areas and performing LBs for the assessment of all these patients is unrealistic. The estimated prevalence of NAFLD ranges from 6.3% to 33% in the general population [15], higher rates being reported in obese patients, diabetic patients, and patients with dyslipidemia. According to the National Health and Nutrition Examination Survey, 33.8% of adult Americans are obese [16]. NAFLD is a lifelong disease and the surveillance of fibrosis progression is mandatory. Second, there is the reluctance of some patients to undergo an invasive procedure such as LB, especially those with alcoholic steatohepatitis (ASH). Third, nowadays, with highly effective curative treatments available, it is not so important to have a LB for chronic hepatitis C, when our fight is to cure the infection, not to find fibrosis severity, which is used in some areas only for treatment prioritization. Finally, in HBV chronic infection, in which the currently available treatment only suppresses viral replication (without virus disappearance) and the treatment is lifelong, fibrosis progression should be followed up and repetitive biopsies are also unrealistic.

Thus noninvasive evaluation of liver fibrosis can be an answer. Noninvasive methods are well accepted by patients, have no complications, are inexpensive, and available in many centers. Also, only a short training of the personnel is needed. Over the past decades, noninvasive tests for liver fibrosis assessment were developed, including biological tests as well as ultrasound-based elastographic techniques. Biological tests incorporate serum-based laboratory measurement into various mathematical formulas (some of them proprietary), resulting in an estimate of disease severity.

*Elastographic methods* can be divided into *ultrasound-based elastographic methods* and *magnetic resonance elastography* (MRE). Both these modalities can be used to measure liver stiffness (LS), which is considered to be an indicator of liver fibrosis severity. Liver disease progression leads to changes in tissue stiffness measurable by elastography (as fibrosis progresses, liver tissue becomes less elastic or stiffer). An external stimulus deforms the liver, generating shear waves that propagate through the liver. The speed of the shear waves is then measured in meters per second. The shear wave speed can be converted to a tissue stiffness estimate, Young's modulus, given in kilopascals (kPa) [17].

MRE evaluates LS by assessing the propagation of shear waves generated by an external vibration device, using a special magnetic resonance imaging technique. Images depicting the propagation of the induced shear waves are captured and processed, so that quantitative maps of LS are generated. It is an expensive yet precise method to assess liver fibrosis severity [18].

*Ultrasound-based elastographic methods* began to be used more than 10 years ago with strain elastography, mainly for breast lesion assessment [19], and with transient elastography (TE) dedicated to liver evaluation [20]. During this period, other ultrasound-based elastographic techniques were developed, all using the acoustic radiation force imaging (ARFI) technology: point shear wave elastography (pSWE) and two-dimensional shear wave elastography (2D-SWE).

*Ultrasound-based liver elastographic techniques can be divided* into the following:

**A.** *Shear wave elastography (SWE)* methods:
   **a)** TE (FibroScan from Echosens).
   **b)** pSWE (including Virtual Touch Quantification [VTQ] from Siemens and ElastPQ from Philips, more recently becoming available in Esaote and Hitachi systems).
   **c)** 2D-SWE (currently available on the Aixplorer system from Supersonic Shear Wave Imaging, on the LOGIQ systems from General Electric, and more recently on the Toshiba, Philips, and Mindray systems).

**B.** *Strain elastography*: First implemented as Real-time Tissue Elastography (RTE) in Hitachi systems and is now available in many other ultrasound machines, even if used for liver fibrosis assessment only by Hitachi systems.

Before discussing in detail the different types of elastography, we will present some general information on shear wave elastography that is important for clinical practice. TE values are expressed in kilopascals, whereas in systems using pSWE and 2D-SWE the results of LS measurements (LSMs) can be displayed in both meters per second and kilopascals. The choice of whether to display speed or modulus (kPa) may be under user control, or such a choice may be unavailable (sometimes determined by regulatory authorities for the region of the world in which the scanner or measurement system is manufactured or sold) [21]. Operators performing SWE must have some experience in these procedures. For TE, 50 or 100 evaluations are enough for training [22–24]. For pSWE and 2D-SWE (Supersonic Shear Wave Imaging), ultrasound experience is especially needed [21,25]. Another important fact is that intra- and interobserver reproducibilities of SWE methods are very good [26].

Regarding the *examination technique*, liver elastography should be performed in fasting conditions (at least 3–4 h after a meal) because meal increases the splanchnic flow and leads to an increase in LS [21–24,27]. For all the elastographic techniques, measurements are performed with the patient in supine position, with the right hand in maximal abduction [21]. The probe is placed in an intercostal space (V-VIII) and a visual inspection of the liver (for ARFI techniques) is performed to choose the measurement place. For TE an ultrasound image of the liver is not provided, only an A-mode image is used to avoid large vessels. The patient is asked to breath normally and then to stop breathing for a few seconds in a neutral position (not in deep inspiration, which increases the LS), during which the measurements are performed. According to the technique, a certain number of valid measurements should be obtained and their median (M) value is calculated. About 10 valid measurements should be obtained for TE, VTQ, and 2D-SWE.GE and 5 for the system by Supersonic Shear Wave Imaging [21]. Most consensus groups recommend basing measurement quality on the M of 10 measurements and an interquartile range (IQR)/M <30% [21,28].

An important fact to know is that some conditions such as obstructive jaundice [29], right-sided cardiac failure [30], nonfasting condition [27], and flares in chronic hepatitis B [31] are confounding factors for liver elastography, as they increase LS values irrespective of fibrosis severity, leading to an overestimation of the fibrosis stage. High levels of aminotransferases are indicators of inflammation, and in some conditions, such as acute alcoholic hepatitis, confident values of LS by TE to predict fibrosis severity are obtained only when aminotransferase levels are lower than 100 IU/mL (no more than twice the upper limit of normal) [32]. In chronic HBV infection, LS cutoff values by VTQ were calculated in connection with aminotransferase levels and it was demonstrated that they increase with the aminotransferase levels [33].

As a general recommendation, the results obtained by liver elastography should be carefully interpreted by physicians, knowing the clinical data of the patient (history, physical examination, laboratory results) [22,34].

## 2. Transient elastography

TE is the oldest ultrasound-based elastographic method used for LS evaluation. Because many of the theoretic and physical aspects of TE were covered in the general chapters, we shall discuss only the clinical aspects in this chapter.

This elastographic method was studied for more than 10 years in clinical practice and more than 1000 papers were published (1269 titles concerning TE were found on PubMed when we wrote this chapter). Thus the body of evidence regarding this method is quite large. At least three large meta-analyses evaluated TE, all of them showing that TE accuracy increases with an increase in the fibrosis stage (from 0.80 in moderate fibrosis to 0.95 in cirrhosis) [35–37]. TE is not able to discriminate between narrow stages of fibrosis, but it is quite good for discriminating between the absence and presence of significant fibrosis (F2 according to the Metavir score).

Regarding TE values *in healthy subjects*, in a study published on 152 subjects (144 with valid measurements), the mean LS value was 4.8 ± 1.3 kPa [38]. No significant difference was observed between the age groups. LS value in women was significantly lower than that in men (4.6 ± 1.2 kPa vs. 5.1 ± 1.2 kPa, $P = .0082$). In a study performed by Roulot [39] on 429 consecutive apparently healthy subjects, the mean LS value was 5.49 ± 1.59 kPa, with higher values in men versus women. Similar results were obtained by Corpechot et al. [40] and Kim et al. [41]. Overall, the upper limit of normal is accepted as 5.3 kPa [39,41].

TE was extensively studied in patients with HCV chronic infection, where it was compared with LB. In these comparative studies, TE showed good accuracy, with LS values increasing with fibrosis. The good results of several prospective studies [20,42–45] compelled the European Association for the Study of the Liver (EASL) to recommend TE for the evaluation of patients with chronic hepatitis C [46]. TE can be combined with biological tests to increase confidence, and when the results are concordant, LB can be avoided. However, when the results are discordant, LB is recommended [42]. By applying this strategy, more than 70% of the LBs can be avoided.

The most used cutoff values are those proposed by Tsochatzis et al. [37] in their meta-analysis. Due to overlap of cutoff values between different fibrosis stages in various studies, a meta-analysis was only possible for cutoff values ≥7, ≥9.5, and ≥12 kPa in stages 2, 3, and 4, respectively [37].

For patients with chronic hepatitis B, in the meta-analysis by Tsochatzis et al. [37], the pooled cutoff for F ≥ 2 (Metavir score) was 7 kPa, with 0.84 pooled sensitivity (Se) and 0.78 pooled specificity (Sp). In another meta-analysis performed by Chon et al. [47], which included 2772 patients, the mean area under the receiver operating characteristic (AUROC) curve values for the diagnosis of significant fibrosis (F2), severe fibrosis (F3), and cirrhosis (F4) were 0.859, 0.887, and 0.929, respectively. The proposed cutoff for F2 was 7.9 kPa (Se 74.3% and Sp 78.3%) and for F3 it was 8.8 kPa (Se 74.0% and Sp 63.8%), whereas for F4 it was 11.7 kPa (Se 84.6% and Sp 81.5%).

A study by our group showed that there were no significant differences among the mean LS values for different stages of fibrosis in patients with chronic hepatitis B and chronic hepatitis C: F1, 6.5 ± 1.9 versus 5.8 ± 2.1 ($P = .0889$); F2, 7.1 ± 2 versus 6.9 ± 2.5 ($P = .3369$); F3, 9.1 ± 3.6 versus 9.9 ± 5 ($P = .7038$); and F4, 19.8 ± 8.6 versus 17.3 ± 6.1 ($P = .6574$) [48].

Given the successes of TE in the setting of staging liver fibrosis in viral hepatitis, TE was evaluated as a predictor of fibrosis severity in other chronic hepatopathies, such as NAFLD [49–51], cholestatic hepatopathies [52–54], ASH [55,56], HCV-HIV coinfection [57,58], and after transplant [59–61]. In all these diseases, TE showed similar predictive value for fibrosis severity (expressed as AUROC) as in chronic hepatitis C.

Initially, TE was performed only with a standard M probe (3.5 MHz), with a feasibility of 90%–95%, with reliable measurements being obtained only in approximately 80% of cases [62] or less in some populations [63]. In both the mentioned studies, reliable measurements were considered, those which fulfilled the recommended technical qualitative criteria: success rate >60% and interquartile range (IQR) <30% of the M value [64]. Both feasibility and the rate of reliable measurements decreased with the increase in body mass index (BMI). Other factors that diminish the reliability of, and limit the applicability of, TE are narrow intercostal spaces and large skin/liver distance. These factors were partially overpassed by the introduction of the XL probe, whose lower frequency transducer (2.5 MHz) allows the examination of deeper regions [65]. In a large real-life cohort of more than 3200 successive subjects, using both probes (M and XL), reliable measurements were obtained in 92% of cases [66]. But even using the XL probe, the rate of reliable measurements decreases with the increase in BMI, so that in severely obese patients (BMI >40 kg/m$^2$), reliable measurements were obtained only in 53.2% of the cases [66].

Despite the fact that M and XL probes are used on the same machine, LS values differ slightly according to the probe used, being lower by approximately 1 kPa when the XL probe is used [65].

Promising results on TE were obtained in pediatric population, first using the M probe and then using the S probe, with higher frequency (5 MHz) [67,68].

In the past years, the FibroScan device has been improved with the inclusion of a module that allows the quantitative evaluation of liver steatosis: controlled attenuation parameter (CAP). CAP measurements showed a good correlation with steatosis severity on LB and is currently available on both M and XL probes [69–71]. A meta-analysis [72] that included 11 studies calculated the following cutoff values of CAP for predicting the presence of different grades of steatosis: 232.5 dB/m for $S \geq 1$, 255 dB/m for $S \geq 2$, and 290 dB/m for S3. The hierarchical summary receiver operating characteristic curve values were 0.85, 0.88, and 0.87, respectively.

Promising results have been obtained for predicting not only the complications of liver cirrhosis by TE, especially for portal hypertension [73–76], but also the occurrence of hepatocellular carcinoma [77–79].

There are a number of weaknesses to the TE method that limit its applicability. These weaknesses include the following: it cannot be used in the presence of ascites, it is unable to visualize the liver and thus one cannot pick a place for LSM, it is used only for LS evaluation, it gives unreliable results regarding fibrosis severity in cases with obstructive jaundice or cardiac failure (also true in other types of elastography), and sometimes repetitive examinations can obtain different results [80]. Another weak point is the fact that following successful treatment in chronic viral hepatitis, the LS values decrease owing to the disappearance of inflammatory lesions in the liver, so that noncirrhotic values are obtained in subjects with cirrhosis [81,82]. The same observation has been made in other types of elastography. Yet another drawback is that the FibroScan probes must be calibrated twice a year (once a year in newer models), with supplementary costs.

## 3. Point shear wave elastography

Systems that use pSWE and 2D-SWE are newer on the market. All these systems use ARFI technology for elastographic measurements and are integrated into an ultrasound machine (usually high end). These complex ultrasound machines can be used not only for LS evaluation but also for other purposes such as stiffness assessment by SWE and strain elastography of other organs (breast, thyroid, prostate), real-time ultrasound examination in B-mode, Doppler mode examination, and contrast-enhanced ultrasonography. Multiple probes (convex or linear) are available and no calibration of the systems is needed for elastography. pSWE modules are now available on multiple systems.

During pSWE measurements, real-time ultrasound is used to choose the area in which LS will be assessed, by placing a measurement box in the desired place, more than 1 cm below the liver capsule, thus avoiding large vessels [21,83]. The patient is in supine position, with the right arm in maximal abduction. Subsequently the patient is asked to stop breathing in an intermediate position and the acquisition button is pressed by the operator. The results of LSMs can be displayed in both meters per second and kilopascals. The choice of whether to display speed or modulus (kPa) may be under user control or determined by regulatory authorities for the region of the world in which the scanner is manufactured or sold [21]. Up to 10 measurements are performed, and the LS value is given by their M value. pSWE is quite simple to perform, but a good acoustic window of the liver as well as patient cooperation are needed. The feasibility of pSWE is quite high, being usable in more than 95% of patients [83,84].

## 3.1 Virtual touch quantification

The first group of published papers regarding pSWE evaluated VTQ (referred to as ARFI in the earliest publications). In studies that evaluated healthy volunteers, the mean VTQ values ranged from 1.05 to 1.19 m/s [85–88]. Early results revealed a good correlation between VTQ measurements and fibrosis severity on LB. In a study performed by Friedrich-Rust [88], the AUROC curve value for significant fibrosis ($F \geq 2$) was 0.82 and for the prediction of cirrhosis ($F = 4$), it was 0.91. In another early study [83], the cutoff values for different stages of fibrosis were as follows: for $F \geq 2$, 1.37 m/s (Fig. 5.1) and for $F = 4$, 1.75 m/s (Fig. 5.2).

In a large international multicenter study, comprising 1095 patients (181 with chronic hepatitis B and 914 with chronic hepatitis C) from 10 centers in 5 countries, in each patient, LB and VTQ measurements were performed [89]. The correlation of VTQ measurements and histologic fibrosis was significantly better in patients with chronic hepatitis C than that in those with chronic hepatitis B: $r = 0.653$, $P < .0001$ versus $r = 0.511$, $P < .0001$ ($P = .007$). The mean LS values depending on the stage of fibrosis were similar in patients with chronic hepatitis B and C.

The performance of VTQ was confirmed by meta-analyses. The first one, based on 13 studies and 1163 patients, compared the performance of VTQ and TE to diagnose significant fibrosis ($F \geq 2$) and cirrhosis (F4). The diagnostic odds ratios of VTQ and TE for $F \geq 2$ and F4 showed no significant differences: mean difference in relative diagnostic odds ratio was 0.27 for $F \geq 2$ and 0.12 for F4 [90].

In the largest published meta-analysis, including 36 studies with 3951 patients, the mean diagnostic performance of VTQ expressed as AUROC curve value was 0.84 for the diagnosis of significant fibrosis ($F \geq 2$), 0.89 for the diagnosis of severe

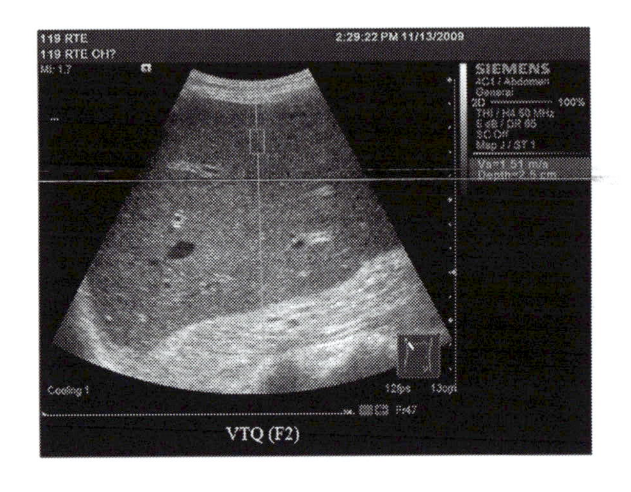


**FIGURE 5.1**

VTQ measurement in a patient with significant fibrosis.

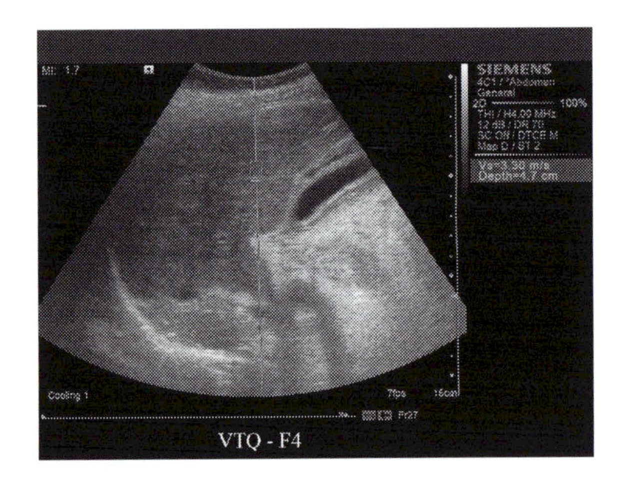

**FIGURE 5.2**

VTQ measurement in a patient with cirrhosis.

fibrosis ($F \geq 3$), and 0.91 for the diagnosis of cirrhosis ($F = 4$) [91]. The cutoff values suggested by this meta-analysis are 1.35 m/s for $F \geq 2$, 1.61 m/s for $F \geq 3$, and 1.87 m/s for F4 [91]. In another paper that compared the diagnostic value of VTQ and TE in a cohort of 241 patients with chronic hepatitis C, in which LB was considered the "gold standard," similar performance of VTQ and TE was found as measured by AUROC curve: 0.81 versus 0.85 for $F \geq 2$ ($P = .15$), 0.88 versus 0.92 for $F \geq 3$ ($P = .11$), and 0.89 versus 0.94 for $F = 4$ ($P = .19$) [92].

## 3.2 ElastPQ technique

For another pSWE technique, ElastPQ, there are fewer published papers, as it was more recently launched. In a pilot study on a mixed etiology cohort, the M LS values assessed by ElastPQ for different stages of liver fibrosis were 4.6 kPa for F0–F1, 5.9 kPa for F2, 7 kPa for F3, and 12 kPa for F4, with AUROC curve values for predicting significant fibrosis ($F \geq 2$), severe fibrosis ($F \geq 3$), and cirrhosis ($F = 4$) similar to those of TE [93].

In a paper published by Ferraioli [84], TE, ElastPQ, and LB were performed in the same session in a cohort of 134 patients with chronic hepatitis C. Similar AUROC curve values were obtained for ElastPQ and TE: 0.80 versus 0.82 ($P = .42$) for significant fibrosis ($F \geq 2$), 0.88 versus 0.95 for severe fibrosis ($F \geq 3$), and 0.95 versus 0.92 ($P = .30$) for cirrhosis ($F = 4$). In 116 subjects the intraobserver agreement for ElastPQ ranged from 0.83 to 0.96 and the interobserver agreement ranged from 0.83 to 0.93 [94].

In another study that included 228 consecutive subjects with chronic hepatopathies (26% HBV, 74% HCV), LS was evaluated in the same session by TE and ElastPQ, with TE being considered as the reference method [95]. Reliable LSMs were obtained in 90.7% subjects by TE and in 98.7% by ElastPQ, with a very strong correlation between LSMs by TE and ElastPQ ($r = 0.85$, $P < .001$). Based on the TE cutoff values published by Tsochatzis et al. [37], the AUROC curve values of ElastPQ were 0.90 for patients with mild fibrosis (F1), 0.95 for those with moderate fibrosis (F2), 0.96 for those with severe fibrosis (F3), and 0.94 for those with cirrhosis (F4). The best cutoff values for discriminating F1, F2, F3, and F4 were 6.4, 7.2, 8.5, and 9.9 kPa, respectively [95] (Figs. 5.3 and 5.4).

In a study that evaluated only patients with hepatitis C, the optimal cutoff values of ElastPQ measurements to predict significant fibrosis, advanced fibrosis, and

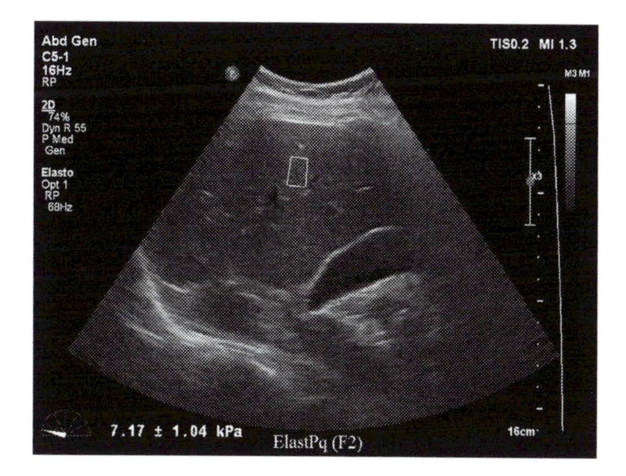

**FIGURE 5.3**

ElastPQ measurement in a patient with significant fibrosis.

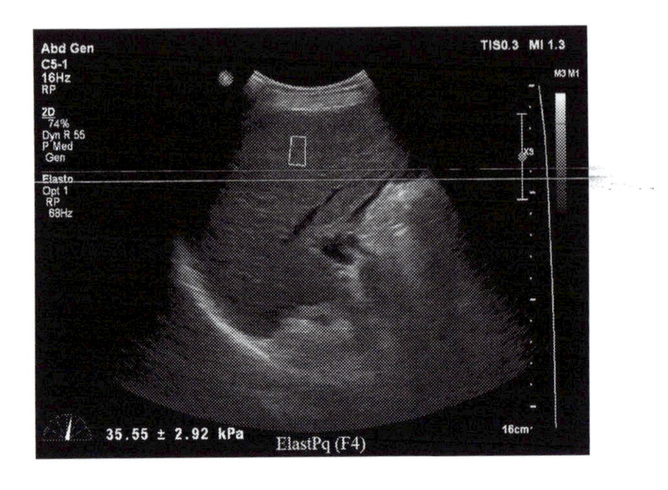


**FIGURE 5.4**

ElastPQ measurement in a patient with cirrhosis.

cirrhosis were 6.43, 9.54, and 11.34 kPa, respectively [94]. In patients with HBV chronic infection, the best LS cutoff values for predicting the presence of significant fibrosis ($F \geq 2$) and cirrhosis ($F = 4$) were 6.99 kPa (AUROC curve value = 0.94) and 9 kPa (AUROC curve value = 0.89), respectively [96]. Liver fibrosis and necroinflammatory activity were significantly correlated with ElastPQ measurements [96].

In a study comparing ElastPQ to VTQ in a cohort of 176 subjects [97], valid measurements were obtained in all cases by VTQ and in 97.7% by ElastPQ ($P = .12$). The mean LS values obtained by VTQ were significantly higher than those obtained with ElastPQ: 1.46 versus 1.32 ($P = .0004$).

## 3.3 Point shear wave elastography from Hitachi

A pSWE technique is now available from Hitachi. The technique showed good repeatability in quantitative measurements in a liver fibrosis phantom and excellent inter- and intraclass correlations [98]. In a cohort of 445 patients with HCV this method had very good results for diagnosing $F \geq 2$ (AUROC curve value, 0.92) and cirrhosis (AUROC curve value, 0.94). The cutoff values to rule in and rule out significant fibrosis were 6.78 kPa (76.9% Se and 90.3% Sp) and 5.55 kPa (90.6% Se, 72.2% Sp), respectively, whereas to rule in and rule out cirrhosis, they were 9.15 kPa (83.3% Se, 90.1% Sp) and 8.41 kPa (90.6% Se, 82.2% Sp), respectively [99].

There are many advantages in using pSWE. The main advantage of pSWE techniques is the fact that the region of interest in which measurements are made can be chosen by the operator, while the liver is visualized in real time. Other advantages are the systems are easy to manipulate by the operator, 10 measurements are acquired rapidly, only short training is needed for the operator, and good inter- and intraoperator reproducibilities [100–106]. Also, pSWE techniques can be used in patients with ascites and the feasibility is high. Comparative studies with TE showed similar results, but better feasibility.

## 4. Two-dimensional shear wave elastography

2D-SWE also uses the ARFI technology, but has the advantage of providing both a color-coded image and a numeric quantification of tissue elasticity. 2D-SWE technique was firstly implemented by Supersonic Shear Wave Imaging on the Aixplorer system, followed by General Electric on the LOGIQ systems (2D-SWE.GE) and by Toshiba on the Aplio systems.

### 4.1 Supersonic shear wave imaging

Early results regarding supersonic shear wave imaging (SSI) showed good correlation with histology or TE [107–109].

The largest cohort of patients was evaluated by 2D-SWE (supersonic shear wave imaging) during a multicenter international study that included retrospective data from 1340 patients (chronic hepatitis C, $n = 470$; chronic hepatitis B, $n = 420$; NAFLD, $n = 172$; others, $n = 278$) evaluated in the same session by SSI and LB, with data available on TE in a subsample of 972 patients [109]. The fibrosis severity on LB was found to be as follows: 40.8% of the patients had F0–F1, 19.3% had F2, 14.0% had F3, and 26.0% had F4. The overall performance of SSI was good to excellent in patients with chronic hepatitis C, chronic hepatitis B, and NAFLD, with AUROC curve values of 86.3%, 91.6%, and 85.9%, respectively, for diagnosing significant fibrosis ($F \geq 2$) and 96.1%, 97.1%, and 95.5% for diagnosing cirrhosis (F4), respectively. The optimal cutoff values were 7.1 kPa for diagnosing $F \geq 2$ in all patients, 13.5 kPa for diagnosing cirrhosis in patients with chronic hepatitis C and NAFLD, and 11.5 kPa for diagnosing cirrhosis in patients with chronic hepatitis B (Figs. 5.5 and 5.6). Based on the AUROC curve value comparison, 2D-SWE was equivalent to or superior than TE.

Other studies that compared SSI with TE showed at least the noninferiority of SSI and its good reproducibility [110–112].

### 4.2 2D-SWE.GE

Regarding the other 2D-SWE technique (2D-SWE.GE), in extenso publications are still relatively few, though they are being incorporated into clinical practice. The reproducibility of 2D-SWE.GE was very good, both inter- and intraobserver reproducibilities [113]. In healthy volunteers, the mean LSM assessed by 2D-SWE.GE was $5.1 \pm 1.3$ kPa, which is significantly higher than that by TE ($4.3 \pm 0.9$ kPa, $P < .0001$). Also, the mean LSMs were significantly higher for men than for women [114]. In a study including 331 subjects, a strong correlation was found between the LS values obtained by 2D-SWE.GE and TE: $r = 0.83$, $P < .0001$. The mean LS

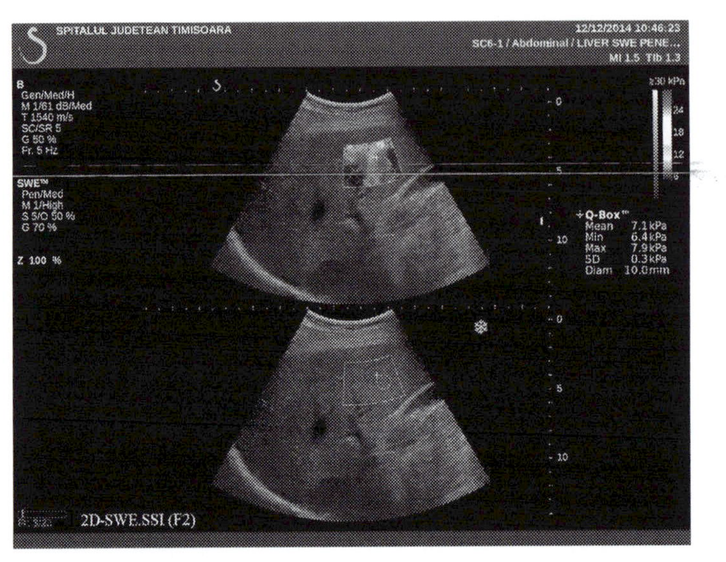


**FIGURE 5.5**

SSI measurement in a patient with significant fibrosis.

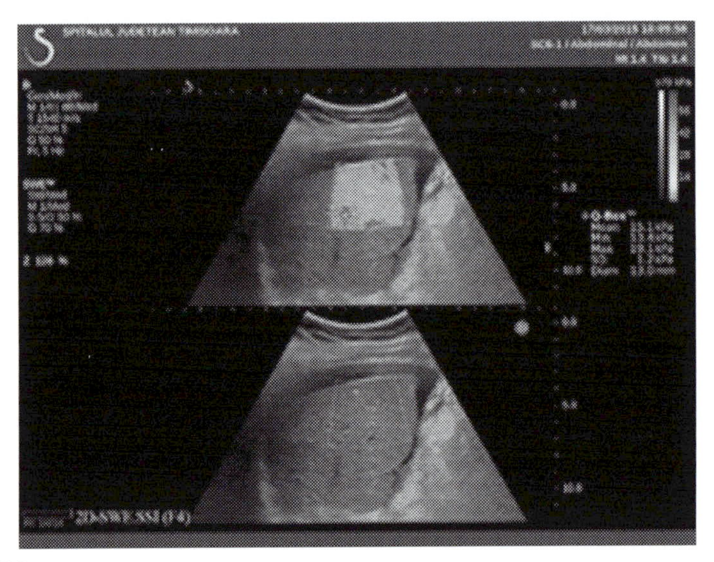

**FIGURE 5.6**

SSI measurement in a patient with cirrhosis and ascites.

values obtained by 2D-SWE.GE were significantly lower than those obtained by TE: $10.14 \pm 4.24$ kPa versus $16.72 \pm 13.4$ kPa ($P < .0001$). The best 2D-SWE.GE cut-off values for $F \geq 2$, $F \geq 3$, and for $F = 4$ were 6.7, 8.2, and 9.3 kPa, respectively [115] (Figs. 5.7 and 5.8).

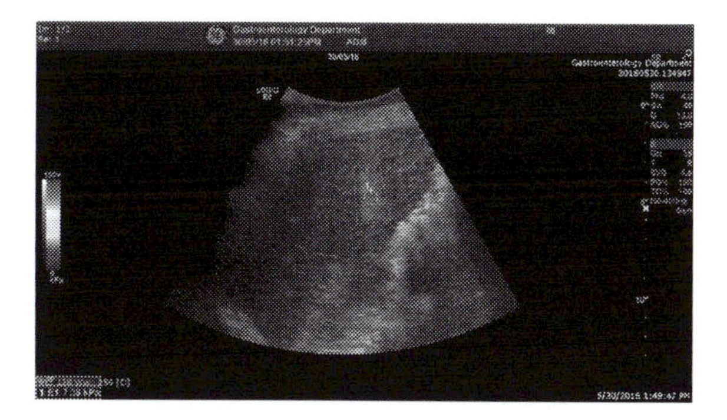

**FIGURE 5.7**

2D-SWE.GE measurement in a patient with significant fibrosis.

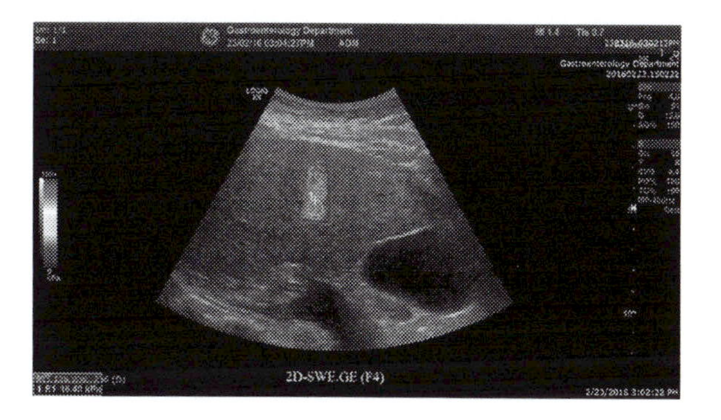


**FIGURE 5.8**

2D-SWE.GE measurement in a patient with cirrhosis and ascites.

## 4.3 Two-dimensional shear wave elastography from Toshiba

This system provides a display of shear waves traveling within the measuring box, allowing selection of areas not affected by artifacts for analysis (Fig. 5.9) [21]. Only a few studies have been published, showing promising results [116–118]. In a study on 115 patients, 2D-SWE Toshiba was compared to TE [116]. 2D-SWE proved to be highly reproducible, with the intraclass correlation test between an experienced radiologist and a third-year radiology resident being 0.878. The best cutoff values predicting significant hepatic fibrosis and liver cirrhosis by 2D-SWE were >1.78 m/s (AUROC curve value = 0.777) and >2.24 m/s (AUROC curve value = 0.935), respectively [116].

The weak points of 2D-SWE methods seem to be that some ultrasound experience is required in order to obtain reliable elastograms, good cooperation of the patient is needed for the same purpose, a good acoustic window is mandatory, and only a few papers have been published. The need of qualitative technical parameters is still under evaluation.

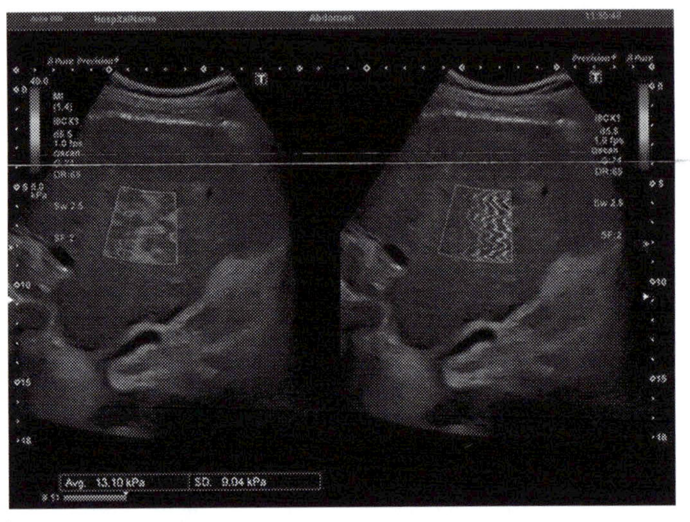

FIGURE 5.9

2D-SWE.Toshiba measurement in a patient with significant fibrosis.

## 5. Comparative studies

When newer elastographic techniques were developed (pSWE and 2D-SWE), many comparative studies were proposed, trying to see if there was a difference between them. Initially, the new techniques were compared with LB as a reference method, but later studies considered TE as a reference standard. In the meta-analysis performed by Bota et al. [90], no significant difference was found between VTQ and TE for predicting significant fibrosis and cirrhosis. In two studies performed by Ferraioli et al. [84,93], the AUROC curve values for predicting significant fibrosis ($F \geq 2$), severe fibrosis ($F \geq 3$), and cirrhosis ($F = 4$) were similar for TE and ElastPQ. Also, in the study performed by Hermann et al. [109], no significant differences were found between the TE and SSI results.

Cassinotto et al. [108] evaluated a cohort of 349 patients with chronic hepatitis who underwent LB and LS assessment by SSI, TE (M probe when BMI $<30$ kg/m$^2$ and XL probe when BMI $\geq 30$ kg/m$^2$), and VTQ. AUROC curve values were calculated and compared for each stage of fibrosis. SSI, TE, and VTQ correlated significantly with histologic fibrosis score ($r = 0.79$, $P < .00001$; $r = 0.70$, $P < .00001$; and $r = 0.64$, $P < .00001$, respectively). SSI had a higher accuracy than TE for the diagnosis of severe fibrosis (F3) ($P = .0016$) and a higher accuracy than VTQ for the diagnosis of significant fibrosis (F2) ($P = .0003$). But in the end, no significant difference between the methods was observed for the diagnosis of mild fibrosis and cirrhosis. The same group performed a similar comparison in a cohort of 291 patients with NAFLD [107]. In this study, the AUROCs for SSI, TE, and VTQ were 0.86, 0.82, and 0.77 for diagnosing $F \geq 2$; 0.89, 0.86, and 0.84 for $F \geq 3$; and 0.88, 0.87, and 0.84 for cirrhosis, respectively.

In a study that included 151 subjects with or without chronic hepatopathies evaluated in the same session by TE, VTQ, ElastPQ, and SSI, the feasibility and the performance of these methods were evaluated [119]. In this study, TE was considered as the reference method. Reliable LSMs were obtained in a significantly higher proportion of patients by means of ElastPQ as compared with TE, SSI, and VTQ: 99.3% versus 86.7% ($P < .0001$), 99.3% versus 86% ($P < .0001$), and 99.3% versus 92.7% ($P = .009$), respectively. VTQ, ElastPQ, and SSI had similar accuracies for diagnosing at least significant fibrosis (F2/F3), 83% (VTQ) versus 79% (ElastPQ) versus 78% (SSI), and also for diagnosing liver cirrhosis, 95% (VTQ) versus 94% (ElastPQ) versus 94% (SSI) [119].

Despite the fact that not so many comparative studies have been published, the new ultrasound-based elastographic methods have offered promising results [120]. Further prospective multicenter studies are required to confirm these preliminary studies.

## 6. Strain elastography

Strain elastography (RTE) has been used in clinical practice for more than 10 years (especially for breast or thyroid nodules) [17,121], but the results obtained in clinical research regarding LS assessment are contradictory. Strain elastography was implemented for the first time in Hitachi machines, and initially, manual compression with the probe was necessary for LSMs. In the latest models, heart beats are used for tissue compression to obtain an elastogram, which is superposed on a normal ultrasound image, so that the region of interest can be chosen in real time (Fig. 5.10).

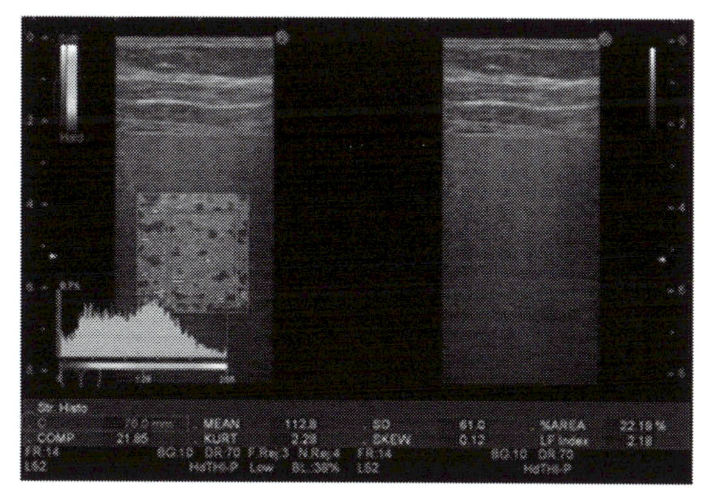

**FIGURE 5.10**

Strain elastography in a patient with mild fibrosis.

The results of strain elastography are expressed as ratios or indexes (such as the liver fibrosis index [LFI]).

The first published paper regarding strain elastography in the liver in the European population was performed by Friedrich-Rust et al. in a cohort of 79 patients with chronic viral hepatitis [122]. The accuracy of strain elastography was 0.75 for significant fibrosis ($F \geq 2$), 0.73 for severe fibrosis ($F \geq 3$), and 0.69 for cirrhosis ($F = 4$). The same author later evaluated a cohort of 134 patients with histology-proven chronic hepatitis ($n = 112$) or liver cirrhosis ($n = 20$) and concluded that strain elastography cannot replace TE for noninvasive assessment of liver fibrosis [123].

Better results were obtained in Japanese studies than in European ones. In a cohort of 310 patients with hepatitis C, in which the LFI was calculated based on strain elastography, LFI had a 78.4% accuracy to discriminate between F0-2/F3-4 and an 80.3% accuracy to discriminate between F0-3/F4 [124]. Another Japanese study using histograms from which 11 parameters were extracted obtained even better results [125].

A meta-analysis published in 2015 included 15 studies that had evaluated strain elastography for staging liver fibrosis [126]. This meta-analysis demonstrated that strain elastography is not highly accurate for diagnosing any stage of fibrosis (summary Se and Sp are roughly 0.80). When this meta-analysis compared strain elastography with TE and VTQ, the overall accuracy of strain elastography seems to be nearly identical for the diagnosis of significant fibrosis ($F \geq 2$), but less accurate for the diagnosis of cirrhosis ($F \geq 4$) [126].

The weak points of strain elastography are nonvalidation of the method's performance in a non-Japanese population and accuracy that seems to be inferior to other elastographic techniques.

After more than 15 years of clinical use, ultrasound-based liver elastography has accumulated a lot of information and hundreds of scientific papers. This large body of evidence is now included in several guidelines on liver elastography: national guidelines—Japanese [127] and Romanian [24], the European Federation of Societies for Ultrasound in Medicine and Biology guidelines [21,128], the World Federation for Ultrasound in Medicine and Biology guidelines [23], and, more recently, the consensus statement of the American Society of Radiologists in Ultrasound [28] and the guidelines of the EASL in cooperation with the Latin American Association for the Study of the Liver [19]. These guidelines can assist practitioners performing liver elastography to choose the best approach for daily practice, knowing the value and the limits of this method.

## References

[1] Z.M. Younossi, A.B. Koenig, D. Abdelatif, Y. Fazel, L. Henry, M. Wymer, Global epidemiology of nonalcoholic fatty liver disease-Meta-analytic assessment of prevalence, incidence, and outcomes, Hepatology 64 (2016) 73–84.

[2] Y.H. Yoon, C.M. Chen, Surveillance Report #105. Liver Cirrhosis Mortality in the United States: National, State, and Regional Trends, 2000–2013, 2016 (cited 31 May 2018). Available at:https://pubs.niaaa.nih.gov/publications/surveillance105/Cirr13.htm.

[3] S. Sherlock, Aspiration liver biopsy, technique and diagnostic application, Lancet (1945) 397.

[4] P. Bedossa, D. Dargere, V. Paradis, Sampling variability of liver fibrosis in chronic hepatitis, Hepatology 38 (2003) 1449–1457.

[5] G. Colloredo, M. Guido, A. Sonzogni, G. Leandro, Impact of liver biopsy size on histological evaluation of chronic viral hepatitis: the smaller the sample, the milder the disease, J. Hepatol. 39 (2003) 239–244.

[6] A. Regev, M. Berho, L.J. Jeffers, C. Milikowski, E.G. Molina, N.T. Pyrsopoulos, et al., Sampling error and intraobserver variation in liver biopsy in patients with chronic HCV infection, Am. J. Gastroenterol. 97 (2002) 2614–2618.

[7] M. Persico, B. Palmentieri, R. Vecchione, R. Torella, I. de Sio, Diagnosis of chronic liver disease: reproducibility and validation of liver biopsy, Am. J. Gastroenterol. 97 (2002) 491–492.

[8] J. West, T.R. Card, Reduced mortality rates following elective percutaneous liver biopsies, Gastroenterology 139 (2010) 1230–1237.

[9] L.B. Seeff, G.T. Everson, T.R. Morgan, T.M. Curto, W.M. Lee, M.G. Ghany, , et al.HALT–C Trial Group, Complication rate of percutaneous liver biopsies among persons with advanced chronic liver disease in the HALT-C trial, Clin. Gastroenterol. Hepatol. 8 (2010) 877–883.

[10] E.B. Tapper, A.S. Lok, Use of liver imaging and biopsy in clinical practice, N. Engl. J. Med. 377 (2017) 756–768.

[11] D.C. Rockey, S.H. Caldwell, Z.D. Goodman, R.C. Nelson, A.D. Smith, American association for the study of liver diseases. Liver biopsy, Hepatology 49 (2009) 1017–1044.

[12] R.G. Knodell, K.G. Ishak, W.C. Black, T.S. Chen, R. Craig, N. Kaplowitz, et al., Formulation and application of a numerical scoring system for assessing histological activity in asymptomatic chronic active hepatitis, Hepatology 1 (1981) 431–435.

[13] P. Bedossa, T. Poynard, An algorithm for the grading of activity in chronic hepatitis C. The METAVIR cooperative study group, Hepatology 24 (1996) 289–293.

[14] K. Ishak, A. Baptista, L. Bianchi, F. Callea, J. De Groote, F. Gudat, et al., Histological grading and staging of chronic hepatitis, J. Hepatol. 22 (1995) 696–699.

[15] G. Vernon, A. Baranova, Z.M. Younossi, Systematic review: the epidemiology and natural history of non-alcoholic fatty liver disease and nonalcoholic steatohepatitis in adults, Aliment. Pharmacol. Ther. 34 (2011) 274–285.

[16] K.M. Flegal, M.D. Carroll, C.J. Ogden, L.R. Curtin, Prevalence and trends in obesity among US adults, 1999-2008, J. Am. Med. Assoc. 303 (2010) 235–241.

[17] J. Bamber, D. Cosgrove, C.F. Dietrich, J. Fromageau, J. Bojunga, F. Calliada, V. Cantisani, et al., EFSUMB guidelines and recommendations on the clinical use of ultrasound elastography. Part 1: basic principles and technology, Ultraschall der Med. 34 (2013) 169–184.

[18] Y.K. Mariappan, K.J. Glaser, R.L. Ehman, Magnetic resonance elastography: a review, Clin. Anat. 23 (2010) 497–511.

[19] A. Itoh, E. Ueno, E. Tohno, H. Kamma, H. Takahashi, T. Shiina, et al., Breast disease: clinical application of US elastography for diagnosis, Radiology 239 (2006) 341–350.

[20] L. Castéra, J. Vergniol, J. Foucher, B. Le Bail, E. Chanteloup, M. Haaser, et al., Prospective comparison of transient elastography, Fibrotest, APRI, and liver biopsy for the assessment of fibrosis in chronic hepatitis C, Gastroenterology 128 (2005) 343–350.
[21] C.F. Dietrich, J. Bamber, A. Berzigotti, S. Bota, V. Cantisani, L. Castera, et al., EFSUMB Guidelines and Recommendations on the Clinical Use of Liver Ultrasound Elastography, Update 2017 (Long Version), Ultraschall Med 38 (2017) e16–e47.
[22] L. Castéra, H.L. Chan, M. Arrese, N. Afdhal, P. Bedossa, M. Friedrich-Rust, et al., EASL-ALEH clinical practice guidelines: non-invasive tests for evaluation of liver disease severity and prognosis, J. Hepatol. 63 (2015) 237–264.
[23] G. Ferraioli, C. Filice, L. Castéra, B.I. Choi, I. Sporea, S.R. Wilson, et al., WFUMB guideline and recommendations for clinical use of ultrasound elastography: part 3: liver, Ultrasound Med. Biol. 41 (2015) 1161–1179.
[24] I. Sporea, S. Bota, A. Săftoiu, R. Şirli, O. Grădinaru-Taşcău, A. Popescu, et al., Romanian national guidelines and practical recommendations on liver elastography, Med. Ultrason. 16 (2014) 123–138.
[25] O. Grădinaru-Taşcău, I. Sporea, S. Bota, A. Jurchiş, A. Popescu, M. Popescu, et al., Does experience play a role in the ability to perform liver stiffness measurements by means of supersonic shear imaging (SSI)? Med. Ultrason. 15 (2013) 180–183.
[26] J.H. Yoon, J.M. Lee, J.K. Han, B.I. Choi, Shear Waves elastography for liver stiffness measurement in clinical sonographic examinations: evaluation of intraobserver reproducibility, technical failure, and unreliable stiffness measurements, J. Ultrasound Med. 33 (2014) 437–447.
[27] I. Mederacke, K. Wursthorn, J. Kirschner, K. Rifai, M.P. Manns, H. Wedemeyer, et al., Food intake increases liver stiffness in patients with chronic or resolved hepatitis C virus infection, Liver Int. 29 (2009) 1500–1506.
[28] R.G. Barr, G. Ferraioli, M.L. Palmeri, Z.D. Goodman, G. Garcia-Tsao, J. Rubin, et al., Elastography assessment of liver fibrosis: society of radiologists in ultrasound consensus conference statement, Radiology 276 (2015) 845–861.
[29] G. Millonig, F.M. Reimann, S. Friedrich, H. Fonouni, A. Mehrabi, M.W. Büchler, et al., Extrahepatic cholestasis increases liver stiffness (FibroScan) irrespective of fibrosis, Hepatology 48 (2008) 1718–1723.
[30] G. Millonig, S. Friedrich, S. Adolf, H. Fonouni, M. Golriz, A. Mehrabi, et al., Liver stiffness is directly influenced by central venous pressure, J. Hepatol. 52 (2010) 206–210.
[31] B. Coco, F. Oliveri, A.M. Maina, P. Ciccorossi, R. Sacco, P. Colombatto, et al., Transient elastography; a new surrogate marker of liver fibrosis influenced by major changes of transaminases, J. Viral Hepat. 14 (2007) 360–369.
[32] S. Mueller, G. Millonig, L. Sarovska, S. Friedrich, F.M. Reimann, M. Pritsch, et al., Increased liver stiffness in alcoholic liver disease: differentiating fibrosis from steatohepatitis, World J. Gastroenterol. 16 (2010) 966–972.
[33] S. Bota, I. Sporea, M. Peck-Radosavljevic, R. Şirli, H. Tanaka, H. Iijima, et al., The influence of aminotransferase levels on liver stiffness assessed by acoustic radiation force impulse elastography: a retrospective multicentre study, Dig. Liver Dis. 45 (2013) 762–768.
[34] J.M. Pawlotsky, A. Aghemo, D. Back, G. Dusheiko, X. Forns, M. Puoti, C. Sarrazin, EASL recommendations on treatment of hepatitis C, J. Hepatol. 63 (2015) 199–236.
[35] J.A. Talwalkar, D.M. Kurtz, S.J. Schoenleber, C.P. West, V.M. Montori, Ultrasound-based transient elastography for the detection of hepatic fibrosis: systematic review and meta-analysis, Clin. Gastroenterol. Hepatol. 5 (2007) 1214–1220.

[36] M. Friedrich-Rust, M.F. Ong, S. Martens, C. Sarrazin, J. Bojunga, S. Zeuzem, E. Herrmann, Performance of transient elastography for the staging of liver fibrosis: a meta-analysis, Gastroenterology 134 (2008) 960–974.

[37] E.A. Tsochatzis, K.S. Gurusamy, S. Ntaoula, E. Cholangitas, B.R. Davidson, A.K. Burroughs, Elastography for the diagnosis of severity of fibrosis in chronic liver disease: a meta-analysis of diagnostic accuracy, J. Hepatol. 54 (2011) 650–659.

[38] R. Şirli, I. Sporea, A. Tudora, A. Deleanu, A. Popescu, Transient elastographic evaluation of subjects without known hepatic pathology: does age change the liver stiffness? J. Gastrointestin. Liver Dis. 18 (2009) 57–60.

[39] D. Roulot, S. Czernichow, H. Le Clésiau, J.L. Costes, A.C. Vergnaud, M. Beaugrand, Liver stiffness values in apparently healthy subjects: influence of gender and metabolic syndrome, J. Hepatol. 48 (2008) 606–613.

[40] C. Corpechot, A. El Naggar, R. Poupon, Gender and liver: is the liver stiffness weaker in weaker sex? Hepatology 44 (2006) 513–514.

[41] S.U. Kim, G.H. Choi, W.K. Han, B.K. Kim, J.Y. Park, D.Y. Kim, et al., What are "true normal" liver stiffness values using FibroScan?: a prospective study in healthy living liver and kidney donors in South Korea, Liver Int. 30 (2010) 268–274.

[42] L. Castéra, G. Sebastiani, B. Le Bail, V. de Lédinghen, P. Couzigou, A. Alberti, Prospective comparison of two algorithms combining non-invasive methods for staging liver fibrosis in chronic hepatitis C, J. Hepatol. 52 (2010) 191–198.

[43] M. Ziol, A. Handra-Luca, A. Kettaneh, C. Christidis, F. Mal, F. Kazemi, et al., Noninvasive assessment of liver fibrosis by measurement of stiffness in patients with chronic hepatitis C, Hepatology 41 (2005) 48–54.

[44] I. Sporea, R. Şirli, A. Deleanu, A. Tudora, M. Curescu, M. Cornianu, D. Lazăr, Comparison of the liver stiffness measurement by transient elastography with the liver biopsy, World J. Gastroenterol. 14 (2008) 6513–6517.

[45] M. Ziol, P. Marcellin, D. Douvin, V. de Lédinghen, R. Poupon, M. Beaugrand, Liver stiffness cut off values in HCV patients: validation and comparison in an independent population, Hepatology 44 (Suppl. 1) (2006) 269.

[46] European Association for the Study of the Liver, EASL clinical practice guidelines: management of hepatitis C virus infection, J. Hepatol. 55 (2011) 245–264.

[47] Y.E. Chon, E.H. Choi, K.J. Song, J.Y. Park, D.Y. Kim, K.H. Han, et al., Performance of transient elastography for the staging of liver fibrosis in patients with chronic hepatitis B: a meta-analysis, PLoS One 7 (2012) e44930.

[48] I. Sporea, R. Şirli, A. Deleanu, A. Tudora, A. Popescu, M. Curescu, S. Bota, Liver stiffness measurements in patients with HBV vs HCV chronic hepatitis: a comparative study, World J. Gastroenterol. 16 (2010) 4832–4837.

[49] M. Yoneda, M. Yoneda, K. Fujita, M. Inamori, M. Tamano, H. Hiriishi, A. Nakajima, Transient elastography in patients with non-alcoholic fatty liver disease (NAFLD), Gut 56 (2007) 1330–1331.

[50] R. Kumar, A. Rastogi, M.K. Sharma, V. Bhatia, P. Tyagi, P. Sharma, et al., Liver stiffness measurements in patients with different stages of nonalcoholic fatty liver disease: diagnostic performance and clinicopathological correlation, Dig. Dis. Sci. 58 (2013) 265–274.

[51] R. Kwok, Y.K. Tse, G.L. Wong, Y. Ha, A.U. Lee, M.C. Ngu, et al., Systematic review with meta-analysis: non-invasive assessment of non-alcoholic fatty liver disease—the role of transient elastography and plasma cytokeratin-18 fragments, Aliment. Pharmacol. Ther. 39 (2014) 254–269.

[52] C. Corpechot, A. El Naggar, A. Poujol-Robert, M. Ziol, D. Wendum, O. Chazouillères, et al., Assessment of biliary fibrosis by transient elastography in patients with PBC and PSC, Hepatology 43 (2006) 1118–1124.
[53] M. Friedrich-Rust, C. Müller, A. Winckler, S. Kriener, E. Herrmann, J. Holtmeier, et al., Assessment of liver fibrosis and steatosis in PBC with FibroScan, MRI, MR-spectroscopy, and serum markers, J. Clin. Gastroenterol. 44 (2010) 58–65.
[54] C. Corpechot, F. Carrat, A. Poujol-Robert, F. Gaouar, D. Wendum, O. Chazouillères, R. Poupon, Noninvasive elastography-based assessment of liver fibrosis progression and prognosis in primary biliary cirrhosis, Hepatology 56 (2012) 198–208.
[55] E. Nguyen-Khac, D. Chatelain, B. Tramier, C. Decrombecque, B. Robert, J.P. Joly, et al., Assessment of asymptomatic liver fibrosis in alcoholic patients using fibroscan: prospective comparison with seven non-invasive laboratory tests, Aliment. Pharmacol. Ther. 28 (2008) 1188–1198.
[56] C.S. Pavlov, G. Casazza, D. Nikolova, E.A. Tsochatzis, A.K. Burroughs, V.T. Ivashkin, C. Gluud, Transient elastography for diagnosis of stages of hepatic fibrosis and cirrhosis in people with alcoholic liver disease, Cochrane Database Syst. Rev. 1 (2015) CD010542.
[57] V. de Lédinghen, D. Douvin, A. Kettaneh, M. Ziol, D. Roulot, P. Marcellin, et al., Diagnosis of hepatic fibrosis and cirrhosis by transient elastography in HIV/hepatitic C virus-coinfected patients, J. Acquir. Immune Defic. Syndr. 41 (2006) 175–179.
[58] S. Vergara, J. Macías, A. Rivero, A. Gutiérrez-Valencia, M. González-Serrano, D. Merino, et al., Grupo para el Estudio de las Hepatitis Viricas de la SAEI. The use of transient elastometry for assessing liver fibrosis in patients with HIV and hepatitis C virus coinfection, Clin. Infect. Dis. 45 (2007) 969–974.
[59] C. Rigamonti, M.F. Donato, M. Fraquelli, F. Agnelli, G. Ronchi, G. Casazza, et al., Transient elastography predicts fibrosis progression in patients with recurrent hepatitis C after liver transplantation, Gut 57 (2008) 821–827.
[60] G. Crespo, S. Lens, M. Gambato, J.A. Carrión, Z. Mariño, M.C. Londoño, et al., Liver stiffness 1 year after transplantation predicts clinical outcomes in patients with recurrent hepatitis C, Am. J. Transplant. 14 (2014) 375–383.
[61] G. Crespo, G. Castro-Narro, I. García-Juárez, C. Benítez, P. Ruiz, L. Sastreet al, Usefulness of liver stiffness measurement during acute cellular rejection in liver transplantation, Liver Transplant. 22 (2016) 298–304.
[62] L. Castéra, J. Foucher, P.H. Bernard, F. Carvalho, D. Allaix, W. Merrouche, et al., Pitfalls of liver stiffness measurement: a 5-year prospective study of 13,369 examinations, Hepatology 51 (2010) 828–835.
[63] R. Şirli, I. Sporea, A. Deleanu, L. Culcea, M. Szilaski, A. Popescu, M. Dănilă, Comparison between the M and XL probes for liver fibrosis assessment by transient elastography, Med. Ultrason. 16 (2014) 119–122.
[64] L. Sandrin, B. Fourquet, J.M. Hasquenoph, S. Yon, C. Fournier, F. Mal, et al., Transient elastography: a new non-invasive method for assessment of hepatic fibrosis, Ultrasound Med. Biol. 29 (2003) 1705–1713.
[65] R.P. Myers, G. Pomier-Layrargues, R. Kirsch, A. Pollett, A. Duarte-Rojo, D. Wong, et al., Feasibility and diagnostic performance of the FibroScan® XL probe for liver stiffness measurement in overweight and obese patients, Hepatology 55 (2012) 199–208.
[66] I. Sporea, R. Şirli, R. Mare, A. Popescu, S.C. Ivaşcu, Feasibility of Transient Elastography with M and XL probes in real life, Med. Ultrason. 18 (2016) 7–10.

[67] V. de Lédinghen, B. Le Bail, L. Rebouissoux, C. Fournier, J. Foucher, V. Miette, et al., Liver stiffness measurement in children using FibroScan: feasibility study and comparison with Fibrotest, aspartate transaminase to platelets ratio index, and liver biopsy, J. Pediatr. Gastroenterol. Nutr. 45 (2007) 443–450.

[68] I. Goldschmidt, C. Streckenbach, C. Dingemann, E.D. Pfister, A. di Nanni, A. Zapf, U. Baumann, Application and limitations of transient liver elastography in children, J. Pediatr. Gastroenterol. Nutr. 57 (2013) 109–113.

[69] M. Sasso, M. Beaugrand, V. de Lédinghen, D. Douvin, P. Marcellin, R. Poupon, et al., Controlled attenuation parameter (CAP): a novel VCTE™ guided ultrasonic attenuation measurement for the evaluation of hepatic steatosis: preliminary study and validation in a cohort of patients with chronic liver disease from various causes, Ultrasound Med. Biol. 36 (2010) 1825–1835.

[70] M. Sasso, S. Audière, A. Kemgang, F. Gaouar, C. Corpechot, O. Chazouillères, et al., Liver steatosis assessed by controlled attenuation parameter (CAP) measured with the XL probe of the FibroScan®: a pilot study assessing diagnostic accuracy, Ultrasound Med. Biol. 42 (2016) 92–103.

[71] M. Lupşor-Platon, D. Feier, H. Ştefănescu, A. Tamas, E. Botan, Z. Spârchez, et al., Diagnostic accuracy of controlled attenuation parameter measured by transient elastography for the non-invasive assessment of liver steatosis: a prospective study, J. Gastrointestin. Liver Dis. 24 (2015) 35–42.

[72] K.Q. Shi, J.Z. Tang, X.L. Zhu, L. Ying, D.W. Li, J. Gao, Y.X. Fang, et al., Controlled attenuation parameter for the detection of steatosis severity in chronic liver disease: a meta-analysis of diagnostic accuracy, J. Gastroenterol. Hepatol. 29 (2014) 1149–1158.

[73] F. Vizzutti, U. Arena, R. Romanelli, L. Rega, M. Foschi, S. Colagrande, et al., Liver stiffness measurement predicts severe portal hypertension in patients with HCV-related cirrhosis, Hepatology 45 (2007) 1087–1090.

[74] T. Reiberger, A. Ferlitsch, B.A. Payer, M. Pinter, M. Homoncik, M. Peck-Radosavljevic, Vienna Hepatic Hemodynamic Lab, Non-selective β-blockers improve the correlation of liver stiffness and portal pressure in advanced cirrhosis, J. Gastroenterol. 47 (2012) 561–568.

[75] I. Sporea, I. Ratiu, R. Şirli, A. Popescu, S. Bota, Value of transient elastography for the prediction of variceal bleeding, World J. Gastroenterol. 17 (2011) 2206–2210.

[76] K.Q. Shi, Y.C. Fan, Z.Z. Pan, X.F. Lin, W.Y. Liu, Y.P. Chen, M.H. Zheng, Transient elastography: a meta-analysis of diagnostic accuracy in evaluation of portal hypertension in chronic liver disease, Liver Int. 33 (2013) 62–71.

[77] R. Masuzaki, R. Tateishi, H. Yoshida, E. Goto, T. Sato, T. Ohki, et al., Prospective risk assessment for hepatocellular carcinoma development in patients with chronic hepatitis C by transient elastography, Hepatology 49 (2009) 1954–1961.

[78] D.Y. Kim, K.J. Song, S.U. Kim, E.J. Yoo, J.Y. Park, S.H. Ahn, K.H. Han, Transient elastography-based risk estimation of hepatitis B virus-related occurrence of hepatocellular carcinoma: development and validation of a predictive model, OncoTargets Ther. 6 (2013) 1463–1469.

[79] G.L. Wong, H.L. Chan, C.K. Wong, C. Leung, C.Y. Chan, P.P. Ho, et al., Liver stiffness based optimization of hepatocellular carcinoma risk score in patients with chronic hepatitis B, J. Hepatol. 60 (2014) 339–345.

[80] F. Nascimbeni, P. Lebray, L. Fedchuk, C.P. Oliveira, M.R. Alvares-da-Silva, A. Varault, et al., Significant variations in elastometry measurements made within

short-term in patients with chronic liver diseases, Clin. Gastroenterol. Hepatol. 13 (2015) 763–771.

[81] R. D'Ambrosio, A. Aghemo, M. Fraquelli, M.G. Rumi, M.F. Donato, V. Paradis, et al., The diagnostic accuracy of Fibroscan for cirrhosis is influenced by liver morphometry in HCV patients with a sustained virological response, J. Hepatol. 59 (2013) 251–256.

[82] I. Sporea, R. Lupuşoru, R. Mare, A. Popescu, L. Gheorghe, S. Iacob, R. Şirli, Dynamics of liver stiffness values by means of transient elastography in patients with HCV liver cirrhosis undergoing interferon free treatment, J. Gastrointestin. Liver Dis. 26 (2017) 145–150.

[83] I. Sporea, R. Şirli, A. Deleanu, A. Popescu, M. Focşa, M. Dănilă, A. Tudora, Acoustic radiation force impulse elastography as compared to transient elastography and liver biopsy in patients with chronic hepatopathies, Ultraschall der Med. 32 (Suppl. 1) (2011) S46–S52.

[84] G. Ferraioli, C. Tinelli, R. Lissandrin, M. Zicchetti, B. Dal Bello, G. Filice, C. Filice, Point shear Waves elastography method for assessing liver stiffness, World J. Gastroenterol. 20 (2011) 4787–4796.

[85] C.Y. Son, S.U. Kim, W.K. Han, G.H. Choi, H. Park, S.C. Yang, et al., Normal liver elasticity values using acoustic radiation force impulse imaging: a prospective study in healthy living liver and kidney donors, J. Gastroenterol. Hepatol. 27 (2012) 130–136.

[86] A. Popescu, I. Sporea, R. Şirli, S. Bota, M. Focşa, M. Dănilă, et al., The mean values of liver stiffness assessed by acoustic radiation force impulse elastography in normal subjects, Med. Ultrason. 13 (2011) 33–37.

[87] R. Madhok, C. Tapasvi, U. Prasad, A.K. Gupta, A. Aggarwal, Acoustic radiation force impulse imaging of the liver: measurement of the normal mean values of the shearing waves velocity in a healthy liver, J. Clin. Diagn. Res. 7 (2013) 39–42.

[88] M. Friedrich-Rust, K. Wunder, S. Kriener, F. Sotoudeh, S. Richter, J. Bojunga, et al., Liver fibrosis in viral hepatitis: noninvasive assessment with acoustic radiation force impulse imaging versus transient elastography, Radiology 252 (2009) 595–604.

[89] I. Sporea, S. Bota, M. Peck-Radosavljevic, R. Şirli, H. Tanaka, H. Iijima, et al., Acoustic radiation force impulse elastography for fibrosis evaluation in patients with chronic hepatitis C: an international multicenter study, Eur. J. Radiol. 81 (2012) 4112–4118.

[90] S. Bota, H. Herkner, I. Sporea, P. Salzl, R. Şirli, A.M. Neghina, M. Peck-Radosavljevic, Meta-analysis: ARFI elastography versus transient elastography for the evaluation of liver fibrosis, Liver Int. 33 (2013) 1138–1147.

[91] J. Nierhoff, A.A. Chávez Ortiz, E. Herrmann, S. Zeuzem, M. Friedrich-Rust, The efficiency of acoustic radiation force impulse imaging for the staging of liver fibrosis: a meta-analysis, Eur. Radiol. 23 (2013) 3040–3053.

[92] M. Friedrich-Rust, M. Lupşor, R. de Knegt, V. Dries, P. Buggisch, M. Gebel, et al., Point shear wave elastography by acoustic radiation force impulse quantification in comparison to transient elastography for the noninvasive assessment of liver fibrosis in chronic hepatitis C: a prospective international multicenter study, Ultraschall der Med. 36 (2015) 239–247.

[93] G. Ferraioli, C. Tinelli, R. Lissandrin, M. Zicchetti, B. Dal Bello, C. Filice, Performance of ElastPQ® shear waves elastography technique for assessing fibrosis in chronic viral hepatitis, J. Hepatol. 58 (Suppl. 1) (2013) S7.

[94] G. Ferraioli, L. Maiocchi, R. Lissandrin, C. Tinelli, A. De Silvestri, C. Filice, Liver fibrosis study Group, Accuracy of the ElastPQ technique for the assessment of liver

fibrosis in patients with chronic hepatitis C: a "real life" single center study, J. Gastrointestin. Liver Dis. 25 (2016) 331–335.
[95] R. Mare, I. Sporea, O. Grădinaru-Taşcău, A. Popescu, R. Şirli, The performance of point shear Waves elastography using ARFI technique – ElastPQ in chronic hepatopathies, United Eur. Gastroenterol. J. 3 (Suppl. 1) (2015) A11.
[96] J.J. Ma, H. Ding, F. Mao, H.C. Sun, C. Xu, W.P. Wang, Assessment of liver fibrosis with elastography point quantification technique in chronic hepatitis B virus patients: a comparison with liver pathological results, J. Gastroenterol. Hepatol. 29 (2014) 814–819.
[97] I. Sporea, S. Bota, O. Grădinaru-Taşcău, R. Şirli, A. Popescu, Comparative study between two point shear waves electrographic techniques: acoustic radiation force impulse (ARFI) elastography and ElastPQ technique, Med. Ultrason. 16 (2014) 309–314.
[98] A. Mulabecirovic, A.B. Mjelle, O.H. Gilja, M. Vesterhus, R.F. Havre, Repeatability of shear wave elastography in liver fibrosis phantoms-Evaluation of five different systems, PLoS One 13 (2018) e0189671.
[99] G. Ferraioli, L. Maiocchi, R. Lissandrin, C. Tinelli, A. De Silvestri, C. Filice, Ruling-in and ruling-out significant fibrosis and cirrhosis in patients with chronic hepatitis C using a shear wave measurement method, J. Gastrointestin. Liver Dis. 26 (2017) 139–143.
[100] J. Boursier, G. Isselin, I. Fouchard-Hubert, F. Oberti, N. Dib, J. Lebigot, et al., Acoustic radiation force impulse: a new ultrasonographic technology for the widespread noninvasive diagnosis of liver fibrosis, Eur. J. Gastroenterol. Hepatol. 22 (2010) 1074–1084.
[101] F. Piscaglia, V. Salvatore, R. Di Donato, M. D'Onofrio, S. Gualandi, A. Gallotti, et al., Accuracy of virtual Touch acoustic radiation force impulse (ARFI) imaging for the diagnosis of cirrhosis during liver ultrasonography, Ultraschall der Med. 32 (2011) 167–175.
[102] L. Rizzo L, V. Calvaruso, B. Cacopardo, N. Alessi, M. Attanasio, S. Petta, et al., Comparison of transient elastography and acoustic radiation force impulse for non-invasive staging of liver fibrosis in patients with chronic hepatitis C, Am. J. Gastroenterol. 106 (2011) 2112–2120.
[103] H. Takahashi, N. Ono, Y. Eguchi, T. Eguchi, Y. Kitajima, Y. Kawaguchi, et al., Evaluation of acoustic radiation force impulse elastography for fibrosis staging of chronic liver disease: a pilot study, Liver Int. 30 (2010) 538–545.
[104] A. Guzmán-Aroca, M. Reus, J.D. Berná-Serna, L. Serrano, C. Serrano, A. Gilabert, A. Cepero, Reproducibility of shear Waves velocity measurements by acoustic radiation force impulse imaging of the liver: a study in healthy volunteers, J. Ultrasound Med. 30 (2011) 975–979.
[105] M. D'Onofrio, A. Gallotti, R.P. Mucelli, Tissue quantification with acoustic radiation force impulse imaging: measurement repeatability and normal values in the healthy liver, Am. J. Roentgenol. 195 (2010) 132–136.
[106] S. Bota, I. Sporea, R. Şirli, A. Popescu, M. Dănilă, D. Costăchescu, Intra and interoperator reproducibility of acoustic radiation force impulse (ARFI) elastography – preliminary results, Ultrasound Med. Biol. 38 (2012) 1103–1108.
[107] C. Cassinotto, J. Boursier, V. de Lédinghen, J. Lebigot, B. Lapuyade, P. Cales, et al., Liver stiffness in nonalcoholic fatty liver disease: a comparison of supersonic shear imaging, FibroScan® and ARFI with liver biopsy, Hepatology 63 (2016) 1817–1827.

[108] C. Cassinotto, B. Lapuyade, A. Mouries, J.B. Hiriart, J. Vergniol, D. Gaye, et al., Noninvasive assessment of liver fibrosis with impulse elastography: comparison of supersonic shear imaging with ARFI and FibroScan®, J. Hepatol. 61 (2014) 550–557.

[109] E. Herrmann, V. de Lédinghen, C. Cassinotto, W. Chu, V. Leung, G. Ferraioli, et al., Assessment of biopsy-proven liver fibrosis by two-dimensional shear wave elastography: an individual patient data-based meta-analysis, Hepatology 67 (2018) 260–272.

[110] E. Bavu, J.L. Gennisson, M. Couade, J. Bercoff, V. Mallet, M. Fink, et al., Noninvasive in vivo liver fibrosis evaluation using supersonic shear imaging: a clinical study on 113 hepatitis C virus patients, Ultrasound Med. Biol. 37 (2011) 1361–1373.

[111] G. Ferraioli, C. Tinelli, B. Dal Bello, M. Zicchetti, G. Filice, C. Filice, Liver Fibrosis Study Group, Accuracy of real-time shear waves elastography for assessing liver fibrosis in chronic hepatitis C: a pilot study, Hepatology 56 (2012) 2125–2133.

[112] G. Ferraioli, C. Tinelli, M. Zicchetti, E. Above, G. Poma, M. Di Gregorio, C. Filice, Reproducibility of real-time shear waves elastography in the evaluation of liver elasticity, Eur. J. Radiol. 81 (2012) 3102–3106.

[113] T.V. Moga, A.M. Stepan, C. Pienar, F. Bende, A. Popescu, R. Şirli, et al., Intra- and inter-observer reproducibility of a 2-D shear wave elastography technique and the impact of ultrasound experience in achieving reliable data, Ultrasound Med. Biol. (18) (2018) pii: S0301-5629 30138-8.

[114] F. Bende, A. Mulabecirovic, I. Sporea, A. Popescu, R. Sirli, O.H. Gilja, et al., Assessing liver stiffness by 2-D shear wave elastography in a healthy cohort, Ultrasound Med. Biol. 44 (2018) 332–341.

[115] F. Bende, I. Sporea, R. Şirli, A. Popescu, R. Mare, B. Miutescu, et al., Performance of 2D-SWE.GE for predicting different stages of liver fibrosis, using Transient Elastography as the reference method, Med. Ultrason. 19 (2017) 143–149.

[116] E.S. Lee, J.B. Lee, H.R. Park, J. Yoo, J.I. Choi, H.W. Lee, et al., Shear wave liver elastography with a propagation map: diagnostic performance and inter-observer correlation for hepatic fibrosis in chronic hepatitis, Ultrasound Med. Biol. 43 (2017) 1355–1363.

[117] G. Ferraioli, L. Maiocchi, R. Lissandrin, C. Tinelli, C. Filice, Accuracy of the latest release of a 2D shear wave elastography method for staging liver fibrosis in patients with chronic hepatitis C: preliminary results, Dig. Liver Dis. (Suppl. 1) (2016) e62–63.

[118] A. Popescu, I. Sporea, R. Lupusoru, M. Danila, A. Lazar, C. Foncea, R. Sirli, Diagnosis Performance with Transient Elastography of a 2D-Shear Wave Elastography Technique: Preliminary Results, ECR, 2018. Poster C-2467.

[119] I. Sporea, R. Mare, O. Grădinaru-Taşcău, A. Popescu, R. Şirli, Feasibility of four ultrasound shear wave elastographic methods for liver stiffness assessment, United Eur. Gastroenterol. J. 3 (Suppl. 1) (2015) 160–174.

[120] I. Sporea, One or more elastographic methods for liver fibrosis assessment? Med. Ultrason. 17 (2015) 137–138.

[121] A. Lyshchik, T. Higashi, R. Asato, S. Tanaka, J. Ito J, J.J. Mai, et al., Thyroid gland tumor diagnosis at US elastography, Radiology 237 (2005) 202–211.

[122] M. Friedrich-Rust, M.F. Ong, E. Herrmann, V. Dries, P. Samaras, S. Zeuzem, C. Sarrazin, Real-time elastography for noninvasive assessment of liver fibrosis in chronic viral hepatitis, Am. J. Roentgenol. 188 (2007) 758–764.

[123] M. Friedrich-Rust, A. Schwarz, M. Ong, V. Dries, P. Schirmacher, E. Herrmann, et al., Real-time tissue elastography versus FibroScan for noninvasive assessment of liver fibrosis in chronic liver disease, Ultraschall der Med. 30 (2009) 478–484.

[124] K. Fujimoto, M. Kato, A. Tonomura, et al., Evaluation of liver fibrosis using Hitachi teal-time elastography, Kanzo 51 (2010) 539–541.

[125] N. Yada, M. Kudo, H. Morikawa, K. Fujimoto, M. Kato, N. Kawada, Assessment of liver fibrosis with real-time tissue elastography in chronic viral hepatitis, Oncology 84 (2013) 13–20.

[126] K. Kobayashi, H. Nakao, T. Nishiyama, Y. Lin, S. Kikuchi, Y. Kobayashi, et al., Diagnostic accuracy of real-time tissue elastography for the staging of liver fibrosis: a meta-analysis, Eur. Radiol. 25 (2015) 230–238.

[127] M. Kudo, T. Shiina, F. Moriyasu, H. Iijima, R. Tateishi, N. Yada, et al., JSUM ultrasound elastography practice guidelines: liver, J. Med. Ultrason. 40 (2013) 325–357.

[128] D. Cosgrove, F. Piscaglia, J. Bamber, J. Bojunga, J.M. Correas, O.H. Gilja, et al., EFSUMB guidelines and recommendations on the clinical use of ultrasound elastography. Part 2: clinical applications, Ultraschall der Med. 34 (2013) 238–253.

CHAPTER

# 6 Thermal therapy monitoring using elastography

**Kullervo Hynynen**

*Physical Sciences Platform, Sunnybrook Research Institute; Department of Medical Biophysics and Institute of Biomaterials and Biomedical Engineering, University of Toronto, Toronto, ON, Canada*

## 1. Introduction

Open surgery for the removal of tissue has been slowly replaced by minimally invasive procedures over the past three decades, resulting in reduced side effects, hospital stay, and cost. These benefits are further enhanced when minimally invasive probes are used to induce high temperatures to coagulate tissues in order to ablate the tissue function. Thermal ablation has been effectively used in the clinic for multiple conditions, such as tumor ablation [1], cardiac arrhythmia elimination [2], functional brain surgery [3], and cosmetic surgery [4]. These ablative temperatures can be induced by invasive probes utilizing radio frequency (RF) [5], laser [6], microwave [7], or ultrasound energy [8], or completely noninvasively by using focused ultrasound (FUS) [9,10]. Although many of these treatments are effective, there are, however, large variations between patients, resulting in undertreatment with relatively high recurrence rates.

With the introduction of magnetic resonance imaging (MRI), which monitors the temperature elevations and thermal exposures, more consistent results have been achieved [11,12]. The cost of MRI scanners, however, limits the use of this method to more affluent countries, and even then, only with some indications. Therefore there is a need for a more affordable monitoring method for thermal ablation.

Ultrasound has been extensively studied for temperature monitoring with some success in animal tissues [13]; however, there is still inadequate evidence to assess how well these methods may work in clinical situations. Moreover, thermometry provides a metric that can only be used to determine the potential for tissue damage, but the actual thermal damage threshold is tissue-type dependent [14–16], so it would be better if thermal coagulation could be detected directly.

It is well known that tissue stiffness increases when proteins are thermally coagulated [17–19]. For this reason, ultrasound elastography has been explored for thermal coagulation monitoring and control. In this chapter, we will review the current experience and the future potential of these methods.

Tissue Elasticity Imaging. https://doi.org/10.1016/B978-0-12-809662-8.00006-1

# 2. Principles and techniques

## 2.1 Thermal effects on tissues

The effect of elevated temperature on cells and tissues has been studied extensively and is well understood. The elevated temperature-induced bioeffects are a nonlinear function of both time and temperature [15,16,20–22]. For mild temperature elevations, there could be no permanent effect on tissue. If the elevated temperature of a few degrees is maintained several minutes, then reversible increase in blood flow and perfusion can be induced [23]. Temperatures between 42 and 45°C, maintained for 10–120 min, have been used to sensitize tissues for radiation therapy [24]. However, if these exposures are maintained long enough, then irreversible thermal coagulation of proteins and tissue necrosis are induced [14].

The threshold for thermal damage varies significantly among tissues. For example, brain damage is induced with an 18-minute exposure at 43°C, whereas muscle tissue requires 240 min at the same temperature[14,15] (Fig. 6.1). For exposures of a few seconds, temperatures above approximately 55°C are needed to cause protein denaturing, resulting in irreversible tissue damage [25]. The elevated temperature can block the microvasculature [26], and even large blood vessels [27–29], stopping blood perfusion in the coagulated tissue volume. When the temperatures reach 100°C in tissue, water boils and results in gas formation and tissue fragmentation [30,31]. To compare different thermal exposures, the thermal dose that converts the temperature-time exposure to an equivalent time at 43°C has been used [15]. Although the thermal dose does not take into account factors such as induction of enhanced tissue resistance to elevated temperatures (thermal tolerance) [32,33], it is commonly used to quantify thermal treatments and is more accurate than just using temperature or energy alone.

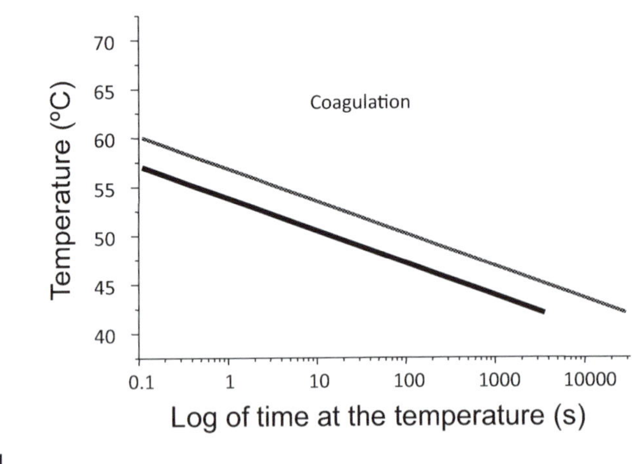


**FIGURE 6.1**

The thermal damage threshold temperature as a function of the log of the exposure duration. The two lines bracket the experimental data reviewed in [14,15].

## 2.2 Clinical use of thermal exposures for therapy

The most used thermal therapy in clinic today is thermal coagulation of the tissue, which disrupts the function of the tissue. Although low-temperature hyperthermia has been used for sensitizing tumors for radiation therapy [24] and chemotherapy, their use is not common today, and therefore, only the ablative treatments are discussed here.

### *2.2.1 Tumor ablations*

Various minimally invasive and noninvasive methods have been used to thermally ablate both benign and malignant tumors.

#### 2.2.1.1 Liver

Secondary liver cancer is a common problem with a poor prognosis. Colorectal cancer, which metastasizes to the liver in over 50% of patients [34], occurs frequently, with 157,000 new cases a year being reported in the United States [35]. Patients with liver metastases rarely survive for more than 1 year following tumor detection. The most effective treatment for these patients is surgical resection, which has a 5-year survival of 20%–30% [36,37]. However, only about 10,000 of these patients [36] are suitable for surgical resection, which itself carries a significant complication rate (about 15% of the patients suffer a major complication [38] and the surgical mortality rate being 2%–5% [37]). Percutaneous injections, cryotherapy, chemo- or radio-embolization and ablation using implantable microwave antennae, and RF or fiber-optic laser probes have been proposed as an alternative to liver resection [39–46]. In [47], 179 metachronous colorectal carcinoma hepatic metastases in 117 patients were treated with percutaneous internally cooled RF ablation (lesion diameter, 0.9–9.6 cm). The mean survival time of this patient group was 36 months, and 36% of the treated developed local recurrence, which is similar to that which has been reported in other studies by this and other groups [48–52]. In 1981, Mack et al. reported of 7148 laser applications on liver metastases in 705 patients [53] with a mean survival time of 41.8 months for colorectal liver metastases and 4.3 years in breast cancer metastases. These results are comparable to patients treated by surgical resection. However, the results are more variable with larger tumors.

#### 2.2.1.2 Breast

Breast tissue is readily accessible by ultrasound because it has no bone or air lying in intervening tissue layers along the beam path. Initial clinical trials have been conducted with both ultrasound and MRI guidance for the ablation of both benign and malignant tumors. MRI-guided FUS feasibility tests were conducted on breast fibroadenomas [54], followed by limited clinical studies on breast cancer patients, which confirmed its ability to ablate tumor tissue [55]. Subsequent work has shown the approach to be well tolerated and capable of ablating a high percentage of tumor volume [56,57]. Significant work has also been conducted for ultrasound-guided FUS [58], RF- [59], microwave-, [60] and laser-induced [61] thermal ablation of breast tumors.

#### 2.2.1.3 Prostate

Both interstitial and transrectal heating methods have been used to ablate prostate cancer. Most patients have been treated with FUS guided by diagnostic ultrasound [62,63]. Although the clinical outcomes are encouraging the rate of side effects could potentially be reduced if exposure control was implemented [64].

#### 2.2.1.4 Bone

Thermal ablations are under investigation and hold considerable promise for the treatment of osteosarcoma and metastatic bone tumors. Metastatic bone tumors arise in over 50% of cancer patients and are associated with persistent and disabling pain that is frequently refractory to radiotherapy, chemotherapy, and analgesics [65]. For FUS treatments, the high attenuation and acoustic impedance of bone makes penetration and focusing challenging; but at the same time, high absorption and low thermal conductivity permit the rapid elevation of temperature. Primary neoplasms may be associated with cortical degradation, which can facilitate ultrasound access. Results have been reported for osteosarcoma ablation under ultrasound guidance, in combination with neoadjuvant therapy, for limb salvage as well as for palliative purposes [66]. Clinical studies employing MRI-guided FUS for the palliative treatment of bone metastases—where periosteal nerve destruction is sought for metastases present in the superficial aspects of bone—have been effective [67]. Similarly, bone tumor ablation using invasive probes has been shown to be effective [65,68].

#### 2.2.1.5 Other malignancies

Minimally invasive thermal ablations of brain tumors using RF or laser have been investigated and have a good safety profile [69,70]. The use of FUS for brain treatments is in its initial stages, with the skull presenting a significant issue for tumors outside the center of the brain [71,72].

### *2.2.2 Cardiac ablation*

Cardiac intervention for the treatment of tachyarrhythmias has evolved from pharmacologic therapy to surgical elimination of arrhythmogenic foci and circuits to catheter-based ablation procedures. Ablation via percutaneously placed catheters has become the standard form of therapy for many tachyarrhythmias, surpassing pharmacologic and surgical methods [2,73,74]. Most often, arrhythmias are eliminated without the need for surgery by heating, and thus destroying, arrhythmogenic tissue using RF energy delivered from the tip of a catheter that is inserted percutaneously via an artery or vein and guided fluoroscopically to the heart. During this type of procedure, the arrhythmogenic area is mapped with a catheter and ablated using RF energy. Previously, energy delivery was guided by precise electric information recorded from the mapping and ablation electrode on the tip of the catheter. Recently, however, the role of anatomy in the genesis of arrhythmias has been recognized. More often now, ablation attempts are targeted based on anatomic considerations [75].

Similar to liver and other organs, other energy sources that have been tested for cardiac ablation [76] include laser [77], microwave [78], and ultrasound [79].

Despite significant advances, catheter ablation has major limitations affecting both efficacy and safety. The inability to accurately create transmural, continuous lesions limits its efficacy. For this reason, intraoperative or endoscopic ultrasound ablation methods have been developed [80].

### 2.2.3 Hypertension

Hypertension remains a major health problem despite the use of antihypertensive drugs. There are many patients (around 5% prevalence) [81,82] who are resistant to medical treatments and thus suffer reduced life expectancy. Radical sympathectomy was found to be effective in reducing blood pressure but was associated with complications, side effects, and a long recovery time [83]. This former surgical method was recently refined and limited to renal sympathetic denervation with a minimally invasive catheter-based approach [84]. The refined method has been shown to be relatively safe and effective [85]; however, multi-institutional studies failed—the ablation of renal nerves was not sufficiently consistent, most likely because the ablations were done without real-time exposure monitoring and control [86].

### 2.2.4 Back pain

Facet joint arthropathy affects up to 15% of patients who have chronic low-back pain [87]. Thermal ablation of the facet joint using RF currents is moderately effective for low-back pain relief [88]. MRI-guided FUS thermal ablation of bone tumors has been shown to be an effective method for pain control [67,89,90], and it has been proposed for facet joint ablation [91]. Although the initial clinical results are encouraging [92], the MRI-guided procedure is expensive and less costly alternatives are needed.

### 2.2.5 Functional brain surgery

Invasive RF probes have been used to perform functional brain surgery in which small volumes of brain tissue are thermally ablated, for example, for the elimination of essential tremor [93]. Recently, this focal ablation has also been successfully executed using MRI-guided FUS [3].

### 2.2.6 Cosmetic surgery

Thermal ablation induced by FUS is used to coagulate connective tissue in the skin for skin tightening [4].

## 2.3 Need for exposure monitoring

All the heating methods need to transfer energy from the probe or applicator to the tissue to be treated. Therefore the amount of energy deposited depends on the tissue properties related to the method, i.e., electric, optical, or ultrasound properties of the targeted tissue. As there can be spatial variations in these properties, the deposited amount can be spatially variable. For external devices, the tissue energy attenuation in the overlying tissues influences the amount of energy available at the target site, thus introducing another factor in the uncertainty of the deposited energy. The actual temperature elevation depends also on the heat transfer properties of the tissue. The local blood perfusion rate varies between locations, tumors, organs, and individuals and can be a major contributor in the uncertainty of the thermal exposure [94]. The use of short exposures (a few seconds) reduces the impact of perfusion [95] and allows even tissues close to large blood vessels to be heated [96]. MRI-thermometry-monitored FUS treatments give us an idea of the importance of the variations in the induced temperature elevations. In their study, McDannold et al. [97] measured the temperature elevation at the same depth in multiple locations in uterine fibroids and then repeated the same in other patients with the similar depth of exposure. The results are summarized in Fig. 6.2, which shows that the temperature elevation and acoustic power can vary from location to location and from patient to patient. The average temperature elevation power in the patient with best heating was approximately three times the average in the patient with the lowest heating. These results highlight the importance of temperature or thermal exposure monitoring.

## 2.4 Principles of elastography for thermal therapy monitoring

Because the mechanical properties of a tissue depend on both its macroscopic and microscopic structures, its stiffness and shear modulus decrease initially when the temperature increases (Fig. 6.3). This change is reversible if temperature increases are not producing permanent tissue damage. However, when tissue coagulation is reached,

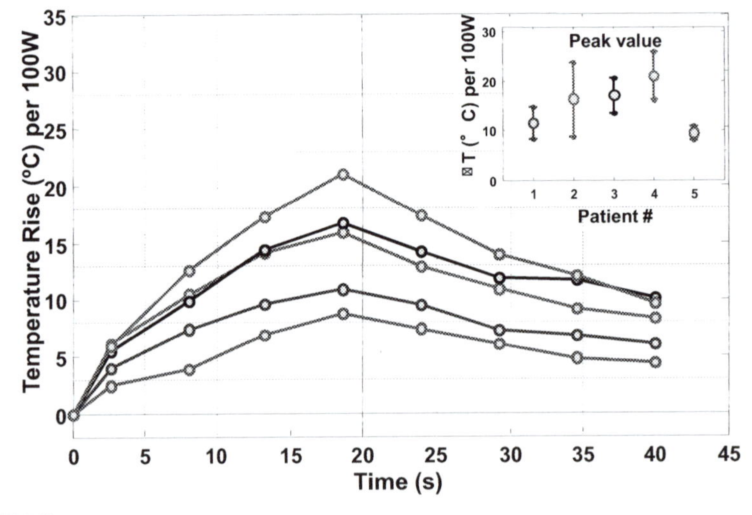


**FIGURE 6.2**

The mean temperature increase as a function of time for all sonication procedures performed in five different patients. The end of sonication is marked with a line at 20 s [97].

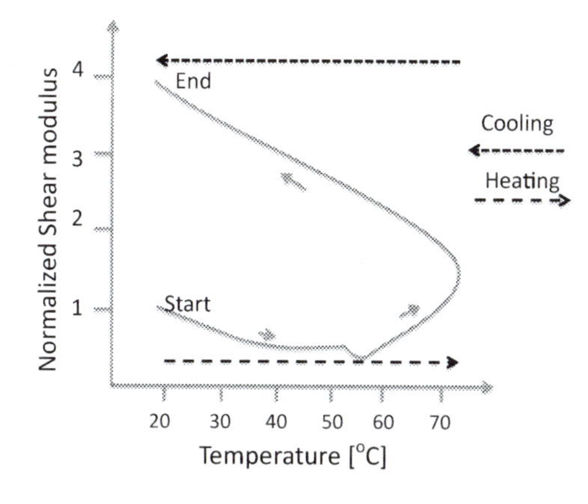


**FIGURE 6.3**

Normalized shear modulus (normalized to the start value at room temperature) in ex vivo muscle during heating with a focused ultrasound beam and then cooling after the exposure.

*Redrawn based on the data from T. Wu, J.P. Felmlee, J.F. Greenleaf, S.J. Riederer, R.L. Ehman, Assessment of thermal tissue ablation with MR elastography, Magn. Reson. Med. 45(1) (2001)80–87.*

the stiffness will quickly increase. This increase in stiffness is irreversible and stays even after the temperature has cooled back to the baseline. The increase in tissue stiffness is almost instantaneous and can be used as a marker for tissue coagulation [17].

Fig. 6.4 gives an example of a dynamic tissue motion before and after thermal ablation at the focus of an FUS therapy beam in an in vivo muscle tissue when the therapy beam is turned on and off at the frequency of 100 Hz. The tissue at the focus moves at the modulation frequency of the therapy beam because of the absorbed energy resulting in a radiation force that pushes the tissue. The amplitude of this motion reduces when tissue coagulates. The motion amplitude decreases as the modulation frequency increases. For practical purposes, frequencies of 50–100 Hz offer reasonable compromise between the motion amplitude and the detection of the local tissue stiffness changes. Fig. 6.5 shows the comparison between MRI thermometry and the tissue motion amplitude during FUS heating. When the time point at which the tissue motion amplitude drops is compared with the MRI-derived temperature, there is good correlation with the temperature and the thermal dose. This indicates that elastography changes are fast enough to be used to control thermal exposures with durations of 10s of seconds or longer that are typical in current clinical practice. It is not known if this holds true with shorter exposures.

As described in the earlier chapters, shear wave speed is proportional to the Young's modulus and thus the tissue stiffness. As a result of the thermal coagulation of tissue, the shear wave speed is increased and can be used to determine the Young's modulus of the tissue [98]. Therefore mapping to the tissue-motion-induced shear wave can be used to monitor thermal ablation.

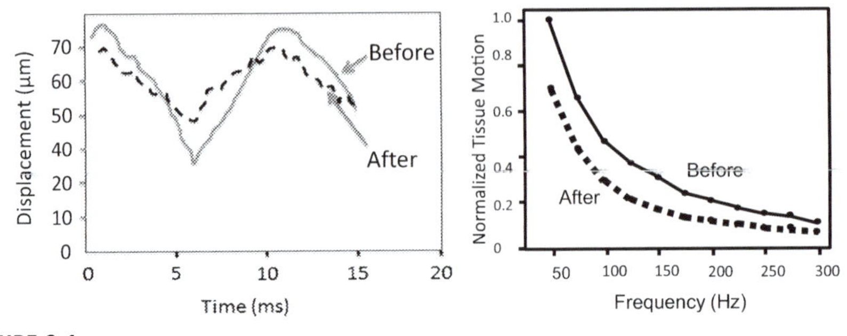

**FIGURE 6.4**

(A) Radiation-force-induced displacement of muscle tissue in vivo before and after thermal coagulation when the therapeutic focused ultrasound beam was modulated at the frequency of 100 Hz. (B) Frequency dependency of the tissue motion as a function of the radiation force modulation frequency.

*Redrawn based on the data from L. Curiel, R. Chopra, K. Hynynen, In vivo monitoring of focused ultrasound surgery using local harmonic motion, Ultrasound Med. Biol. 35(1) (2009) 65–78.*

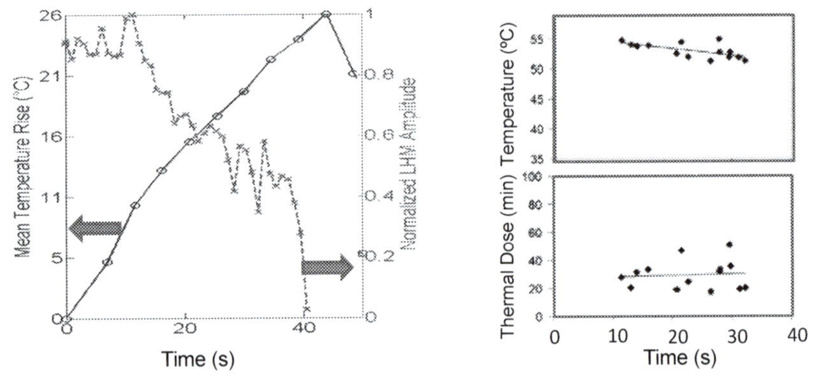


**FIGURE 6.5**

Comparison with magnetic resonance imaging (MRI) thermometry. (A) Temperature elevation measured by MRI thermometry and the focal tissue motion mapped by local harmonic motion (LHM) in vivo in rabbit muscle during sonication with a focused ultrasound beam that was modulated during sonication. As the ultrasound beam was used to induce the tissue motion, the LHM signal ends when the sonication ends. (B) (*top*) The temperature measured with MRI thermometry as a function of sonication duration that was required to induce reduction in the tissue motion. (*bottom*) The thermal dose calculated from the MRI thermometry. Each of the data points represent a separate location of tissue and separate sonication [18].

## 3. Elastographic methods for thermal therapy monitoring

Probe-based thermal therapy methods can be monitored with an external ultrasound imaging array. The main concern is to be able to obtain a good coverage of the ablated tissue volume. For the noninvasive FUS ablation methods the ultrasound imaging and therapy applicators can be combined and a single integrated device used.

The most common way, so far, to use elastography to monitor thermal therapy is to perform the measurement, both before and after the treatment, and then use the tissue stiffness change during the treatment as an indicator of coagulation. For this approach, any of the elastographic methods discussed in the other chapters would work.

To truly control the thermal exposure, the stiffness changes need to be monitored during the exposure and should be used to control the treatments. This requires online mapping of the tissue stiffness. Owing to the duration of current thermal coagulation procedures, time resolution of a few seconds is adequate. The time resolution of the MRI thermometry is approximately 3–5 s in most cases. This means that any of the elastographic methods could be used to control the treatments.

The simplest method is to use quasi-static (strain) elastography in which the tissue compression is caused by the ultrasound imaging probe [99]. The imaging system can be completely separate from the ablation device, making it applicable with any of the heating methods, provided that some means of proper alignment of the imaging plane and heating field can be accomplished. In the FUS treatments, the imaging array is often integrated in the middle of the therapy transducer, thus assuring that the heated and imaged tissue volumes are the same. For transrectal treatments, FUS applicator with a built-in imaging probe has been used [100]. These applicators are surrounded by a degassed, temperature-controlled water bolus whose volume can be controlled. By increasing the water volume, the prostate tissue can be compressed, thus providing a means to map the stiffness during thermal therapy delivery [100]. Similarly, the microwave or RF electrode used for invasive ablation can be displaced and then the tissue strain mapped [101]. This electrode displacement elastography (EDE) produces strain around the heating probe and should thus be more accurate than other methods introducing the disturbance externally.

To noninvasively excite the tissue locally at the target, the radiation-force-based methods have shown promise. In this approach, a focused beam is used to cause tissue to move by induced radiation force. The tissue motion resulting from the local radiation force is tracked typically with ultrasound imaging. In the first approach, shear wave elastography (SWE)—the radiation-force-induced disturbance—generates a shear wave in the tissue [98]. By measuring the shear wave speed, the local Young's modulus can be calculated and used as an online indicator for coagulation. Supersonic shear imaging [102] is based on measuring the shear wave speed. In this technique, multiple sonication procedures, causing tissue displacement, are performed in a row such that a shear wave front propagating in transverse direction is created. The propagation of the shear wave is tracked using ultrafast plane wave imaging [102]. The shear modulus of the tissue can then be estimated using the propagation properties of the shear wave, and the shear wave speed maps can be used to determine the volume of the coagulated tissue [103]. In the acoustic radiation force impulse (ARFI) method, an FUS beam is used to push the tissue for less than 1 ms and then the tissue relaxation back to baseline is followed using ultrasound echoes [104].

Finally, the therapy beam itself can be modulated to provide time-varying stress via radiation force and the induced tissue motion to determine relative tissue stiffness changes. These radiation-force-based elastographic techniques produce the stress directly at the focal location while utilizing the diagnostic ultrasound echoes to detect tissue displacements. Using two ultrasound beams, with slightly different frequencies and that overlap at the focal zones, can also induce the focal tissue motion. Now, a time-varying radiation force is induced at the different frequency due to the interference of the two waves. It has been proposed that the tissue motion could produce an acoustic signal that could be detected by an external hydrophone [105]. This method has been tested for monitoring thermal coagulation, in which the therapy transducer is divided into two apertures, producing mixed results [106]. However, more consistent results have been obtained by detecting the motion with a diagnostic ultrasound A-line [18,107]. A simulation study investigating the magnitude of the tissue motion determined that the maximum tissue motion was achieved when the therapy beam with a single frequency was modulated [108]. This method, called local harmonic motion (LHM) imaging, modulates the FUS such that a periodic tissue motion is induced. This method has shown good success in controlling ablation in animal models [109–118].

## 4. Diseases and applications

As described in Section 2, there are many tumors and normal tissue conditions that are treated by thermal ablation in a clinical or an experimental setting. Most of the probe-based thermal ablations of tissue are done open-loop, based on prior experience and/or on the temperature measured on the energy delivery probe. In some cases, computed tomographic (CT) imaging, MRI, or ultrasound contrast-enhanced imaging is used to determine if vascular occlusion was reached after the treatment. Noninvasive FUS treatments, when done by ultrasound guidance, most often rely on nonspecific grayscale changes in the ultrasound image intensity after sonication. This has been shown to be a poor predictor of tissue coagulation. So far, only a limited number of experimental studies have explored the possibility of using elastography to control FUS treatments.

### 4.1 Tumor treatments

Lesion sizes in liver after interstitial microwave ablation were measured from B-mode ultrasound and quasi-static elastography obtained by applying compressions with the diagnostic ultrasound probe and then compared with CT. CT areas of ablation in humans were 1.78 times the area measured with elastography and twice the area measured with B-scans [119], indicating the inability of both these methods to accurately measure thermal lesion dimensions in humans.

EDE has been evaluated in patients with hepatocellular carcinoma treated using microwave ablation. The ablation areas averaged $13 \pm 5$ $cm^2$ on EDE, compared with $8 \pm 3$ $cm^2$ on B-mode imaging [120], indicating the superiority of elastography over B-mode ultrasound imaging in lesion detection.

Shear wave imaging (SWI) prediction of thermal coagulation size after RF ablation in pig liver showed good correlation with histologic findings [103]. ARFI has been clinically tested after RF ablation of liver tumors and shown to demonstrate changes induced by the coagulation [121]. When ARFI images were compared with contrast-enhanced ultrasound images of the liver, the lesion size was found to be slightly smaller in the ARFI images [101].

LHMs have been tested for monitoring tumor ablation. In a rabbit model of VX2 tumor, sonication was executed under MRI thermometry while simultaneously modulating the therapy beam and measuring the tissue motion. The results indicated a good ability to detect tissue coagulation, provided there was no confounding motion and that a good echo signal for motion tracking was available [109]. In a simulation study, it was shown that successive sonication may result in a situation where the change in tissue motion is reduced by the prior sonication [122,123]. In another study, with mice, evidence of the ability of LHM to detect coagulated tumors was shown [115]. Similar results have been obtained in a number of other studies [110,124–126].

## 4.2 Prostate

A transrectal ultrasound applicator with ultrasound imaging guidance was used to ablate prostate cancer in 20 patients, and then ultrasound elastograms and MRI T1- and T2-weighted images were obtained to evaluate the ablated tissue volume. The ablated tissue volume was measured from the images and a statistically significant correlation was found between elastographic and MRI measurements. The elastographic measurements generally underestimated the volume measured by MRI, which has been shown to be a good estimator for tissue coagulation [100].

## 4.3 Cardiac ablation

The ARFI method has been tested for monitoring RF coagulation in animal heart [127]. In these studies, a cardiac ablation catheter was used to heat the heart muscle while another ultrasound imaging catheter was inserted into the heart to visualize the sonication. The diagnostic array used a 1-ms burst of sonication to cause the tissue to move and then the tissue relaxation was mapped. These early in vivo experiments show feasibility. SWE with intracardiac catheter was tested in vivo in sheep with RF-intracardiac catheter ablation, showing that lesion depths were predicted well with SWE [128].

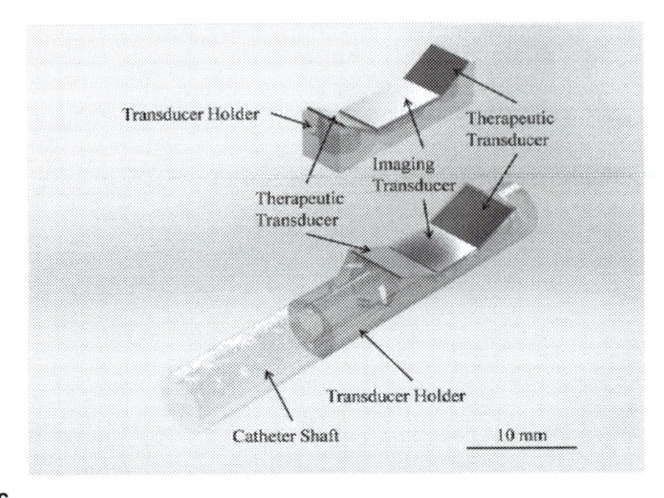


**FIGURE 6.6**

Ultrasound cardiac ablation catheter with pulse echo transducer for local harmonic motion monitoring [129].

Ideally, both the imaging and therapy devices should be integrated together to reduce complexity for catheter-based arrhythmia ablations. Recently, the feasibility of such a catheter was experimentally demonstrated. In this device, two angled ultrasound transducers were used to deliver energy for tissue heating. Modulating the acoustic power of these transducers induced tissue motion. A pulse echo transducer was located in the middle of these two transducers such that the A-line tracked the central line of the heating zone (Fig. 6.6). Both in vitro and in vivo experiments showed feasibility of detecting tissue coagulation, and potentially the lesion depth, during sonication. A similar integrated therapy and imaging device was developed for transesophageal FUS cardiac ablation and SWI and has shown promise with in vivo experiments [128].

## 5. Future opportunities

Minimally invasive thermal ablations have been shown to be safe and effective treatments with hugely reduced recovery time [130]. Currently, MRI thermometry is the dominant monitoring and control method, but it lacks widespread use because of the cost and complexity of treatment execution. Ultrasound elastography can detect thermal coagulation of tissue and can thus provide means for treatment monitoring and control at a lower cost. Therefore it is possible that ultrasound elastography could replace most of the current MRI-guided methods, with perhaps the exception of brain, where the skull bone makes it difficult to obtain adequate signals to detect tissue motion. With reduction in the cost and complexity, it is expected that the use of thermal therapy methods would increase.

One of the requirements for wider use is to be able to perform the measurements three-dimensionally and to be able to predict the tissue damage online. Fast three-dimensional elastography would be a significant boost for the method's ability to become more generally used for control of thermal treatments.

As for specific diseases, the greatest opportunities are in providing guidance and control for liver, breast, prostate, pancreas, and bone tumor ablations. Cardiac ablation and renal denervation are areas in which ultrasound elastography induced control of the ablation could significantly increase treatment consistency and have an impact on many people's lives.

## 6. Conclusion

It is well known that thermal coagulation of tissue will increase its stiffness. This change has been detected after the treatments that have used elastic imaging. Quasi-static elastography appears to detect the coagulation but not the ablated tissue volume. SWI and LHM provide a much more precise lesion boundary estimate and may be useful for after-exposure monitoring. In vivo studies have shown that these changes are fast enough to provide a useful flag for control methods to determine the thermal ablation end point. Both SWI and LHM have been integrated in experimental therapy devices and the results, so far, show promise. These radiation-force-based methods may provide the needed exposure control to allow diagnostic ultrasound-guided methods to provide adequate exposure control for reliable clinical outcomes. However, the utilization of ultrasound elastography in monitoring and control is still in its infancy. More research should be conducted before it can be widely adopted in clinical practice.

## References

[1] H. Rhim, S.N. Goldberg, G.D. Dodd, L. Solbiati, H.K. Lim, M. Tonolini, O.K. Cho, Essential techniques for successful radio-frequency thermal ablation of malignant hepatic tumors, RadioGraphics 21 (2001) S17–S35. Spec No.

[2] H. Calkins, Catheter ablation for cardiac arrhythmias, Med. Clin. N. Am. 85 (2) (2001) 473–502, xii.

[3] W.J. Elias, N. Lipsman, W.G. Ondo, P. Ghanouni, Y.G. Kim, W. Lee, M. Schwartz, K. Hynynen, A.M. Lozano, B.B. Shah, D. Huss, R.F. Dallapiazza, R. Gwinn, J. Witt, S. Ro, H.M. Eisenberg, P.S. Fishman, D. Gandhi, C.H. Halpern, R. Chuang, P.K. Butts, T.S. Tierney, M.T. Hayes, G.R. Cosgrove, T. Yamaguchi, K. Abe, T. Taira, J.W. Chang, A randomized trial of focused ultrasound thalamotomy for essential tremor, N. Engl. J. Med. 375 (8) (2016) 730–739.

[4] K. Minkis, M. Alam, Ultrasound skin tightening, Dermatol. Clin. 32 (1) (2014) 71–77.

[5] S.N. Goldberg, Radiofrequency tumor ablation: principles and techniques, Eur. J. Ultrasound 13 (2) (2001) 129–147.

[6] E. Schena, P. Saccomandi, Y. Fong, Laser ablation for cancer: past, present and future, J. Funct. Biomater. 8 (2) (2017) E19.
[7] J.I. Hernandez, M.F. Cepeda, F. Valdes, G.D. Guerrero, Microwave ablation: state-of-the-art review, Onco Targets Ther. 8 (2015) 1627–1632, https://doi.org/10.2147/OTT.S81734.
[8] J.E. Zimmer, K. Hynynen, D.S. He, F.I. Marcus, The feasibility of using ultrasound for cardiac ablation, IEEE Trans. Biomed. Eng. 42 (9) (1995) 891–897.
[9] K. Hynynen, MRIgHIFU: a tool for image-guided therapeutics, J. Magn. Reson. Imaging 34 (3) (2011) 482–493.
[10] H.G. Ter, HIFU tissue ablation: concept and devices, Adv. Exp. Med. Biol. 880 (2016) 3–20, https://doi.org/10.1007/978-3-319-22536-4_1.
[11] C. Tempany, E. Stewart, N. McDannold, B. Quade, F. Jolesz, K. Hynynen, MR Imaging-Guided Focused Ultrasound Surgery of Leiomyoma Of The Uterus: Early Experience, 2001.
[12] F.M. Fennessy, C.M. Tempany, N.J. McDannold, M.J. So, G. Hesley, B. Gostout, H.S. Kim, G.A. Holland, D.A. Sarti, K. Hynynen, F.A. Jolesz, E.A. Stewart, Uterine leiomyomas: MR imaging-guided focused ultrasound surgery–results of different treatment protocols, Radiology 243 (3) (2007) 885–893.
[13] E.S. Ebbini, H.G. Ter, Ultrasound-guided therapeutic focused ultrasound: current status and future directions, Int. J. Hyperth. 31 (2) (2015) 77–89.
[14] P.S. Yarmolenko, E.J. Moon, C. Landon, A. Manzoor, D.W. Hochman, B.L. Viglianti, M.W. Dewhirst, Thresholds for thermal damage to normal tissues: an update, Int. J. Hyperth. 27 (4) (2011) 320–343.
[15] S.A. Sapareto, W.C. Dewey, Thermal dose determination in cancer therapy, Int. J. Radiat. Oncol. Biol. Phys. 10 (6) (1984) 787–800.
[16] M.W. Dewhirst, B.L. Viglianti, M. Lora-Michiels, M. Hanson, P.J. Hoopes, Basic principles of thermal dosimetry and thermal thresholds for tissue damage from hyperthermia, Int. J. Hyperth. 19 (3) (2003) 267–294.
[17] T. Wu, J.P. Felmlee, J.F. Greenleaf, S.J. Riederer, R.L. Ehman, Assessment of thermal tissue ablation with MR elastography, Magn. Reson. Med. 45 (1) (2001) 80–87.
[18] L. Curiel, R. Chopra, K. Hynynen, In vivo monitoring of focused ultrasound surgery using local harmonic motion, Ultrasound Med. Biol. 35 (1) (2009) 65–78.
[19] E.S. Brosses, M. Pernot, M. Tanter, The link between tissue elasticity and thermal dose in vivo, Phys. Med. Biol. 56 (24) (2011) 7755–7765.
[20] A.R. Moritz, F.C. Henriques Jr., Studies of thermal injury. II. The relative importance of time and surface temperature in the causation of cutaneous burns, Am. J. Pathol. 23 (1947) 695–720.
[21] G. Crile, The effect of heat and radiation on cancers implanted on the feet of mice, Cancer Res. 23 (1963) 372–380.
[22] J. Landry, N. Marceau, Rate-Limiting events in hyperthermic cell killing, Radiat. Res. 75 (1978) 573–585.
[23] B. Emami, C.W. Song, Physiological mechanisms in hyperthermia: a review, Int. J. Radiat. Oncol. Biol. Phys. 10 (2) (1984) 289–295.
[24] J. Overgaard, The current and potential role of hyperthermia in radiotherapy, Int. J. Radiat. Oncol. Biol. Phys. 16 (1989) 535–549.
[25] P.P. Lele, Threshold and mechanisms of ultrasonic damage to organized animal tissues, in: Proceedings of a Symposium on Biological Effects and Characterization of Ultrasound Sources, Rockville, MD, June 1–3, 1977, pp. 224–239.

[26] K. Hynynen, A. Darkazanli, C. Damianou, E. Unger, J.F. Schenck, The usefulness of contrast agent and GRASS imaging sequence for MRI guided noninvasive ultrasound surgery, Investig. Radiol. 29 (1994) 897–903.
[27] C. Delon-Martin, C. Vogt, E. Chigner, C. Guers, J.Y. Chapelon, D. Cathignol, Venous thrombosis generation by means of high-intensity focused ultrasound, Ultrasound Med. Biol. 21 (1) (1995) 113–119.
[28] K. Hynynen, A. Chung, V. Culucci, F.A. Jolesz, Potential adverse effects of high intensity focused ultrasound exposure on blood vessel in vivo, Ultrasound Med. Biol. 22 (2) (1996) 193–201.
[29] K. Hynynen, V. Colucci, A. Chung, F.A. Jolesz, Noninvasive Blood Vessel Occlusion with MR Imaging-Guided Focused Ultrasound, one ninty seventh ed., 1995, p. 219.
[30] N.B. Smith, K. Hynynen, The feasibility of using focused ultrasound for transmyocardial revascularization, Ultrasound Med. Biol. 24 (7) (1998) 1045–1054.
[31] V.A. Khokhlova, J.B. Fowlkes, W.W. Roberts, G.R. Schade, Z. Xu, T.D. Khokhlova, T.L. Hall, A.D. Maxwell, Y.N. Wang, C.A. Cain, Histotripsy methods in mechanical disintegration of tissue: towards clinical applications, Int. J. Hyperth. 31 (2) (2015) 145–162.
[32] A.A. Martinez, A. Meshorer, J.L. Meyer, G.M. Hahn, L.F. Fajardo, S.D. Prionas, Thermal sensitivity and thermotolerance in normal porcine tissues, Cancer Res. 43 (5) (1983) 2072–2075.
[33] S.A. Sapareto, Thermal isoeffect dose: addressing the problem of thermotolerance, Int. J. Hyperth. 3 (4) (1987) 297–305.
[34] B.A. Ward, D.L. Miller, J.A. Frank, Prospective evaluation of hepatic imaging studies in the resection of colorectal metastases: correlation with surgical findings, Surgery 105 (1989) 180–187.
[35] C.C. Boring, T.S. Squires, T. Tong, Cancer statistics 1993, Cancer 44 (1993) 7–29.
[36] Registery of hepatic metastases, Resection of liver for colorectal casinoma metastases: a multi-institutional study of indications for resection, Surgery 103 (1988) 278–288.
[37] J. Scheele, R. Stangl, A. Altendorf-Hofmann, F.P. Gall, Indicators of prognosis after hepatic resection for colorectal secondaries, Surgery 110 (1989) 13–19.
[38] A. Masters, A.C. Steger, S.G. Bown, Role of interstitial therapy in the treatment of liver cancer, Br. J. Surg. 78 (1991) 518–523.
[39] J.L. McCall, M.W. Booth, D.L. Morris, Hepatic cryotherapy for metastatic liver tumors, Br. J. Hosp. Med. 54 (8) (1995) 378–381.
[40] T. Livraghi, S. Lazzaroni, F. Meloni, G. Torzilli, C. Vettori, Intralesional ethanol in the treatment of unresectable liver cancer, World J. Surg. 19 (6) (1995) 801–806.
[41] E. Liapi, C.C. Georgiades, K. Hong, J.F. Geschwind, Transcatheter arterial chemoembolization: current technique and future promise, Tech. Vasc. Interv. Radiol. 10 (1) (2007) 2–11.
[42] B. Sangro, R. Salem, A. Kennedy, D. Coldwell, H. Wasan, Radioembolization for hepatocellular carcinoma: a review of the evidence and treatment recommendations, Am. J. Clin. Oncol. 34 (4) (2011) 422–431.
[43] S.A. Curley, B.S. Davidson, R.Y. Fleming, F. Izzo, L.C. Stephens, P. Tinkey, D. Cromeens, Laparoscopically guided bipolar radiofrequency ablation of areas of porcine liver, Surg. Endosc. 11 (7) (1997) 729–733.
[44] H.D. Freisenhausen, Vascular arrangement and capillary density in the brain of the rabbit [in German], Acta Anat. 62 (4) (1965) 539–562.

[45] M. Sato, Y. Watanabe, S. Udeda, S. Iseki, Y. Abe, N. Sato, S. Kimura, K. Okubo, M. Onji, Microwave coagulation therapy for hepatocellular carcinoma, Gastroenterology 110 (5) (1996) 1507–1514.
[46] T.J. Vogl, P.K. Muller, R. Hammerstingl, N. Weinhold, M.G. Mack, C. Philipp, M. Deimling, J. Beuthan, W. Pegios, H. Reiss, H.-P. Lemmens, R. Felix, Malignant liver tumors treated with MR imaging-guided laser-induced thermotherapy: technique and prospective results, Radiology 196 (1995) 257–265.
[47] L. Solbiati, T. Livraghi, S.N. Goldberg, T. Ierace, F. Meloni, M. Dellanoce, L. Cova, E.F. Halpern, G.S. Gazelle, Percutaneous radio-frequency ablation of hepatic metastases from colorectal cancer: long-term results in 117 patients, Radiology 221 (1) (2001) 159–166.
[48] L. Solbiati, S.N. Goldberg, T. Ierace, T. Livraghi, F. Meloni, M. Dellanoce, S. Sironi, G.S. Gazelle, Hepatic metastases: percutaneous radio-frequency ablation with cooled-tip electrodes, Radiology 205 (2) (1997) 367–373.
[49] T. Livraghi, S.N. Goldberg, F. Monti, A. Bizzini, S. Lazzaroni, F. Meloni, S. Pellicano, L. Solbiati, G.S. Gazelle, Saline-enhanced radio-frequency tissue ablation in the treatment of liver metastases, Radiology 202 (1) (1997) 205–210.
[50] S. Rossi, M. Di Stasi, E. Buscarini, P. Quaretti, F. Garbagnati, L. Squassante, C.T. Paties, D.E. Silverman, L. Buscarini, Percutaneous RF interstitial thermal ablation in the treatment of hepatic cancer, Am. J. Roentgenol. 167 (3) (1996) 759–768.
[51] S. Rossi, E. Buscarini, F. Garbagnati, M. Di Stasi, P. Quaretti, M. Rago, A. Zangrandi, S. Andreola, D. Silverman, L. Buscarini, Percutaneous treatment of small hepatic tumors by an expandable RF needle electrode, Am. J. Roentgenol. 170 (4) (1998) 1015–1022.
[52] L. Solbiati, T. Ierace, S.N. Goldberg, S. Sironi, T. Livraghi, R. Fiocca, G. Servadio, G. Rizzatto, P.R. Mueller, A. Del Maschio, G.S. Gazelle, Percutaneous US-guided radio-frequency tissue ablation of liver metastases: treatment and follow-up in 16 patients, Radiology 202 (1) (1997) 195–203.
[53] M.G. Mack, R. Straub, K. Eichler, K. Engelmann, S. Zangos, A. Roggan, D. Woitaschek, M. Bottger, T.J. Vogl, Percutaneous MR imaging-guided laser-induced thermotherapy of hepatic metastases, Abdom. Imag. 26 (4) (2001) 369–374.
[54] K. Hynynen, O. Pomeroy, D.N. Smith, P.E. Huber, N.J. McDannold, J. Kettenbach, J. Baum, S. Singer, F.A. Jolesz, MR imaging-guided focused ultrasound surgery of fibroadenomas in the breast: a feasibility study, Radiology 219 (1) (2001) 176–185.
[55] D.C. Gianfelice, H. Mallouche, L. Lepanto, R. Poisson, E. Nassif, G. Breton, MR-guided focused ultrasound ablation of primary breast neoplasms: works in progress, Radiology 213 (1999) 106–107.
[56] D. Gianfelice, A. Khiat, M. Amara, A. Belblidia, Y. Boulanger, MR imaging-guided focused ultrasound surgery of breast cancer: correlation of dynamic contrast-enhanced MRI with histopathologic findings, Breast Cancer Res. Treat. 82 (2) (2003) 93–101.
[57] H. Furusawa, K. Namba, S. Thomsen, F. Akiyama, A. Bendet, C. Tanaka, Y. Yasuda, H. Nakahara, Magnetic resonance-guided focused ultrasound surgery of breast cancer: reliability and effectiveness, J. Am. Coll. Surg. 203 (1) (2006) 54–63.
[58] F. Wu, Z.B. Wang, H. Zhu, W.Z. Chen, J.Z. Zou, J. Bai, K.Q. Li, C.B. Jin, F.L. Xie, H.B. Su, Extracorporeal high intensity focused ultrasound treatment for patients with breast cancer, Breast Cancer Res. Treat. 92 (1) (2005) 51–60.

[59] F. Izzo, R. Thomas, P. Delrio, M. Rinaldo, P. Vallone, A. DeChiara, G. Botti, G. D'Aiuto, P. Cortino, S.A. Curley, Radiofrequency ablation in patients with primary breast carcinoma: a pilot study in 26 patients, Cancer 92 (8) (2001) 2036–2044.

[60] R.A. Gardner, H.I. Vargas, J.B. Block, C.L. Vogel, A.J. Fenn, G.V. Kuehl, M. Doval, Focused microwave phased array thermotherapy for primary breast cancer, Ann. Surg. Oncol. 9 (4) (2002) 326–332.

[61] M.C.L. Peek, M. Douek, Ablative techniques for the treatment of benign and malignant breast tumours, J. Ther. Ultrasound. 5:18 (2017) 18–0097, https://doi.org/10.1186/s40349-017-0097-8.

[62] A. Gelet, J.Y. Chapelon, R. Bouvier, R. Souchon, C. Pangaud, A.F. Abdelrahim, D. Cathignol, J.M. Dubernard, Treatment of prostate cancer with transrectal focused ultrasound: early clinical experience, Eur. Urol. 29 (1996) 174–183.

[63] N.T. Sanghvi, F.J. Fry, R. Bihrle, R.S. Foster, M.H. Phillips, J. Syrus, A.V. Zaitsev, C.W. Hennige, Noninvasive surgery of prostate tissue by high-intensity focused ultrasound, IEEE Trans. Ultrason. Ferroelectr. Freq. Control 43 (6) (1996) 1099–1110.

[64] T. Uchida, M. Nakano, S. Hongo, S. Shoji, Y. Nagata, T. Satoh, S. Baba, Y. Usui, T. Terachi, High-intensity focused ultrasound therapy for prostate cancer, Int. J. Urol. 19 (3) (2012) 187–201.

[65] R.C. Foster, J.M. Stavas, Bone and soft tissue ablation, Semin. Interv. Radiol. 31 (2) (2014) 167–179.

[66] F. Wu, Extracorporeal high intensity focused ultrasound in the treatment of patients with solid malignancy, Minim. Invasive Ther. Allied Technol. 15 (1) (2006) 26–35.

[67] M.D. Hurwitz, P. Ghanouni, S.V. Kanaev, D. Iozeffi, D. Gianfelice, F.M. Fennessy, A. Kuten, J.E. Meyer, S.D. LeBlang, A. Roberts, J. Choi, J.M. Larner, A. Napoli, V.G. Turkevich, Y. Inbar, C.M. Tempany, R.M. Pfeffer, Magnetic resonance-guided focused ultrasound for patients with painful bone metastases: phase III trial results, J. Natl. Cancer Inst. 106 (5) (2014) dju082.

[68] C.L. Brace, Radiofrequency and microwave ablation of the liver, lung, kidney, and bone: what are the differences? Curr. Probl. Diagn. Radiol. 38 (3) (2009) 135–143.

[69] K. Sugiyama, T. Sakai, I. Fujishima, H. Ryu, K. Uemura, T. Yokoyama, Stereotactic interstitial laser-hyperthermia using Nd-YAG laser, Stereotact. Funct. Neurosurg. 54–55 (1990) 501–505.

[70] R. Medvid, A. Ruiz, R.J. Komotar, J.R. Jagid, M.E. Ivan, R.M. Quencer, M.B. Desai, Current applications of MRI-guided laser interstitial thermal therapy in the treatment of brain neoplasms and epilepsy: a radiologic and neurosurgical overview, Am. J. Neuroradiol. 36 (11) (2015) 1998–2006.

[71] J. Song, A. Pulkkinen, Y. Huang, K. Hynynen, Investigation of standing-wave formation in a human skull for a clinical prototype of a large-aperture, transcranial MR-guided focused ultrasound (MRgFUS) phased array: an experimental and simulation study, IEEE Trans. Biomed. Eng. 59 (2) (2012) 435–444.

[72] A. Pulkkinen, Y. Huang, J. Song, K. Hynynen, Simulations and measurements of transcranial low-frequency ultrasound therapy: skull-base heating and effective area of treatment, Phys. Med. Biol. 56 (15) (2011) 4661–4683.

[73] D.E. Haines, S. Nath, New horizons in catheter ablation, J. Interv. Cardiol. 8 (6 Suppl. l) (1995) 845–856.

[74] D. Lin, F.E. Marchlinski, Advances in ablation therapy for complex arrhythmias: atrial fibrillation and ventricular tachycardia, Curr. Cardiol. Rep. 5 (5) (2003) 407–414.

[75] T.J. Bunch, M.J. Cutler, Is pulmonary vein isolation still the cornerstone in atrial fibrillation ablation? J. Thorac. Dis. 7 (2) (2015) 132−141.

[76] D. Keane, New catheter ablation techniques for the treatment of cardiac arrhythmias, Card. Electrophysiol. Rev. 6 (4) (2002) 341−348.

[77] G.M. Vincent, J. Fox, B.A. Benedick, J. Hunter, J.A. Dixon, Laser catheter ablation of simulated ventricular tachycardia, Lasers Surg. Med. 7 (5) (1987) 421−425.

[78] J.C. Tardif, P.W. Groeneveld, P.J. Wang, C.J. Haugh, N.A. Estes III, S.L. Schwartz, N.G. Pandian, Intracardiac echocardiographic guidance during microwave catheter ablation, J. Am. Soc. Echocardiogr. 12 (1) (1999) 41−47.

[79] M.D. Lesh, P. Guerra, F.X. Roithinger, Y. Goseki, C. Diederich, W.H. Nau, M. Maguire, K. Taylor, Novel catheter technology for ablative cure of atrial fibrillation, J. Interv. Card Electrophysiol. 4 (Suppl. 1) (2000) 127−139.

[80] S. Mitnovetski, A.A. Almeida, J. Goldstein, A.W. Pick, J.A. Smith, Epicardial high-intensity focused ultrasound cardiac ablation for surgical treatment of atrial fibrillation, Heart Lung Circ. 18 (1) (2009) 28−31.

[81] P.A. Sarafidis, G.L. Bakris, Resistant hypertension: an overview of evaluation and treatment, J. Am. Coll. Cardiol. 52 (22) (2008) 1749−1757.

[82] M. Moser, J.F. Setaro, Clinical practice. Resistant or difficult-to-control hypertension, N. Engl. J. Med. 355 (4) (2006) 385−392.

[83] Y. Huan, D.L. Cohen, Renal denervation: a potential new treatment for severe hypertension, Clin. Cardiol. 36 (1) (2013) 10−14.

[84] H. Krum, M. Schlaich, R. Whitbourn, P.A. Sobotka, J. Sadowski, K. Bartus, B. Kapelak, A. Walton, H. Sievert, S. Thambar, W.T. Abraham, M. Esler, Catheter-based renal sympathetic denervation for resistant hypertension: a multicentre safety and proof-of-principle cohort study, Lancet 373 (9671) (2009) 1275−1281.

[85] V. Papademetriou, M. Doumas, K. Tsioufis, Renal sympathetic denervation for the treatment of difficult-to-control or resistant hypertension, Int. J. Hypertens. 2011 (2011) 196518, https://doi.org/10.4061/2011/196518.

[86] D.L. Bhatt, D.E. Kandzari, W.W. O'Neill, R. D'Agostino, J.M. Flack, B.T. Katzen, M.B. Leon, M. Liu, L. Mauri, M. Negoita, S.A. Cohen, S. Oparil, K. Rocha-Singh, R.R. Townsend, G.L. Bakris, A controlled trial of renal denervation for resistant hypertension, N. Engl. J. Med. 370 (15) (2014) 1393−1401.

[87] S.P. Cohen, S.N. Raja, Pathogenesis, diagnosis, and treatment of lumbar zygapophysial (facet) joint pain, Anesthesiology 106 (3) (2007) 591−614.

[88] C.W. Slipman, A.L. Bhat, R.V. Gilchrist, Z. Issac, L. Chou, D.A. Lenrow, A critical review of the evidence for the use of zygapophysial injections and radiofrequency denervation in the treatment of low back pain, Spine J. 3 (4) (2003) 310−316.

[89] D. Gianfelice, C. Gupta, W. Kucharczyk, P. Bret, D. Havill, M. Clemons, Palliative treatment of painful bone metastases with MR imaging–guided focused ultrasound, Radiology 249 (1) (2008) 355−363.

[90] B. Liberman, D. Gianfelice, Y. Inbar, A. Beck, T. Rabin, N. Shabshin, G. Chander, S. Hengst, R. Pfeffer, A. Chechick, A. Hanannel, O. Dogadkin, R. Catane, Pain palliation in patients with bone metastases using MR-guided focused ultrasound surgery: a multicenter study, Ann. Surg. Oncol. 16 (1) (2009) 140−146.

[91] S. Harnof, Z. Zibly, L. Shay, O. Dogadkin, A. Hanannel, Y. Inbar, I. Goor-Aryeh, I. Caspi, Magnetic resonance-guided focused ultrasound treatment of facet joint pain: summary of preclinical phase, J. Ther. Ultrasound. 2 (2014) 9, https://doi.org/10.1186/2050-5736-2-9.

[92] E.M. Weeks, M.W. Platt, W. Gedroyc, MRI-guided focused ultrasound (MRgFUS) to treat facet joint osteoarthritis low back pain–case series of an innovative new technique, Eur. Radiol. 22 (12) (2012) 2822–2835.

[93] A. Niranjan, A. Jawahar, D. Kondziolka, L.D. Lunsford, A comparison of surgical approaches for the management of tremor: radiofrequency thalamotomy, gamma knife thalamotomy and thalamic stimulation, Stereotact. Funct. Neurosurg. 72 (2–4) (1999) 178–184.

[94] M.W. Dewhirst, L. Prosnitz, D. Thrall, D. Prescott, S. Clegg, C. Charles, J. Macfall, G. Rosner, T. Samulski, E. Gillette, S. LaRue, Hyperthermic treatment of malignant diseases: current status and a view toward the future, Semin. Oncol. 24 (6) (1997) 616–625.

[95] B.E. Billard, K. Hynynen, R.B. Roemer, Effects of physical parameters on high temperature ultrasound hyperthermia, Ultrasound Med. Biol. 16 (1990) 409–420.

[96] L.N. Dorr, K. Hynynen, The effect of tissue heterogeneities and large blood vessels on the thermal exposure induced by short high power ultrasound pulses, Int. J. Hyperth. 8 (1992) 45–59.

[97] N. McDannold, C.M. Tempany, F.M. Fennessy, M.J. So, F.J. Rybicki, E.A. Stewart, F.A. Jolesz, K. Hynynen, Uterine leiomyomas: MR imaging-based thermometry and thermal dosimetry during focused ultrasound thermal ablation, Radiology 240 (1) (2006) 263–272.

[98] A.P. Sarvazyan, O.V. Rudenko, S.D. Swanson, J.B. Fowlkes, S.Y. Emelianov, Shear wave elasticity imaging: a new ultrasonic technology of medical diagnostics, Ultrasound Med. Biol. 24 (9) (1998) 1419–1435.

[99] J. Ophir, S.K. Alam, B.S. Garra, F. Kallel, E.E. Konofagou, T. Krouskop, C.R. Merritt, R. Righetti, R. Souchon, S. Srinivasan, T. Varghese, Elastography: imaging the elastic properties of soft tissues with ultrasound, J. Med. Ultrason. 29 (4) (2002) 155.

[100] L. Curiel, R. Souchon, O. Rouviere, A. Gelet, J.Y. Chapelon, Elastography for the follow-up of high-intensity focused ultrasound prostate cancer treatment: initial comparison with MRI, Ultrasound Med. Biol. 31 (11) (2005) 1461–1468.

[101] P. Wiggermann, K. Brunn, J. Rennert, M. Loss, H. Wobser, A.G. Schreyer, C. Stroszczynski, E.M. Jung, Monitoring during hepatic radiofrequency ablation (RFA): comparison of real-time ultrasound elastography (RTE) and contrast-enhanced ultrasound (CEUS): first clinical results of 25 patients, Ultraschall der Med. 34 (6) (2013) 590–594.

[102] J. Bercoff, M. Pernot, M. Tanter, M. Fink, Monitoring thermally-induced lesions with supersonic shear imaging, Ultrason. Imaging 26 (2) (2004) 71–84.

[103] A. Mariani, W. Kwiecinski, M. Pernot, D. Balvay, M. Tanter, O. Clement, C.A. Cuenod, F. Zinzindohoue, Real time shear waves elastography monitoring of thermal ablation: in vivo evaluation in pig livers, J. Surg. Res. 188 (1) (2014) 37–43.

[104] K.R. Nightingale, M.L. Palmeri, On the feasibility of remote palpation using acoustic radiation force, Acoust. Soc. 110 (2001) 625. America through AIP.

[105] M. Fatemi, J.F. Greenleaf, Ultrasound-stimulated vibro-acoustic spectrography, Science 280 (3) (1998) 82–85.

[106] E. Konofagou, J. Thierman, K. Hynynen, Ultrasound Surgery Monitoring Using Ultrasound-Stimulated Acoustic Emission: Simulations and Experimental Results, 2000.

[107] E.E. Konofagou, K. Hynynen, Localized harmonic motion imaging: theory, simulations and experiments, Ultrasound Med. Biol. 29 (10) (2003) 1405–1413.

[108] J. Heikkila, K. Hynynen, Investigation of optimal method for inducing harmonic motion in tissue using a linear ultrasound phased array—a simulation study, Ultrason. Imaging 28 (2) (2006) 97—113.
[109] L. Curiel, Y. Huang, N. Vykhodtseva, K. Hynynen, Focused ultrasound treatment of VX2 tumors controlled by local harmonic motion, Phys. Med. Biol. 54 (11) (2009) 3405—3419.
[110] Y. Han, S. Wang, T. Payen, E. Konofagou, Fast lesion mapping during HIFU treatment using harmonic motion imaging guided focused ultrasound (HMIgFUS) in vitro and in vivo, Phys. Med. Biol. 62 (8) (2017) 3111—3123.
[111] T. Payen, C.F. Palermo, S.A. Sastra, H. Chen, Y. Han, K.P. Olive, E.E. Konofagou, Elasticity mapping of murine abdominal organs in vivo using harmonic motion imaging (HMI), Phys. Med. Biol. 61 (15) (2016) 5741—5754.
[112] Y. Han, S. Wang, H. Hibshoosh, B. Taback, E. Konofagou, Tumor characterization and treatment monitoring of postsurgical human breast specimens using harmonic motion imaging (HMI), Breast Cancer Res. 18 (1) (2016) 46—0707.
[113] J.-E. Damber, A. Bergh, L. Daehlin, V. Petrow, M. Landstrom, Effects of 6-methylene progesterone on growth, morphology, and blood flow of Dunning R3327 prostatic adenocarcinoma, The Prostate 20 (1992) 187—197.
[114] D.J. Coleman, F.L. Lizzi, J. Driller, A.L. Rosado, S. Chang, T. Iwamoto, D. Rosenthal, Therapeutic ultrasound in the treatment of glaucoma, Ophthalmology 92 (1985) 339—346.
[115] H. Chen, G.Y. Hou, Y. Han, T. Payen, C.F. Palermo, K.P. Olive, E.E. Konofagou, Harmonic motion imaging for abdominal tumor detection and high-intensity focused ultrasound ablation monitoring: an in vivo feasibility study in a transgenic mouse model of pancreatic cancer, IEEE Trans. Ultrason. Ferroelectr. Freq. Control 62 (9) (2015) 1662—1673.
[116] J. Chen, G.Y. Hou, F. Marquet, Y. Han, F. Camarena, E. Konofagou, Radiation-force-based estimation of acoustic attenuation using harmonic motion imaging (HMI) in phantoms and in vitro livers before and after HIFU ablation, Phys. Med. Biol. 60 (19) (2015) 7499—7512.
[117] Y. Han, G.Y. Hou, S. Wang, E. Konofagou, High intensity focused ultrasound (HIFU) focal spot localization using harmonic motion imaging (HMI), Phys. Med. Biol. 60 (15) (2015) 5911—5924.
[118] G.Y. Hou, F. Marquet, S. Wang, I.Z. Apostolakis, E.E. Konofagou, High-intensity focused ultrasound monitoring using harmonic motion imaging for focused ultrasound (HMIFU) under boiling or slow denaturation conditions, IEEE Trans. Ultrason. Ferroelectr. Freq. Control 62 (7) (2015) 1308—1319.
[119] C. Correa-Gallego, A.M. Karkar, S. Monette, P.C. Ezell, W.R. Jarnagin, T.P. Kingham, Intraoperative ultrasound and tissue elastography measurements do not predict the size of hepatic microwave ablations, Acad. Radiol. 21 (1) (2014) 72—78.
[120] W. Yang, T.J. Ziemlewicz, T. Varghese, M.L. Alexander, N. Rubert, A.N. Ingle, M.G. Lubner, J.L. Hinshaw, S.A. Wells, F.T. Lee Jr., J.A. Zagzebski, Post-procedure evaluation of microwave ablations of hepatocellular carcinomas using electrode displacement elastography, Ultrasound Med. Biol. 42 (12) (2016) 2893—2902.
[121] B.J. Fahey, R.C. Nelson, S.J. Hsu, D.P. Bradway, D.M. Dumont, G.E. Trahey, In vivo guidance and assessment of liver radio-frequency ablation with acoustic radiation force elastography, Ultrasound Med. Biol. 34 (10) (2008) 1590—1603.

[122] J. Heikkila, K. Hynynen, Simulations of lesion detection using a combined phased array LHMI-technique, Ultrasonics 48 (6–7) (2008) 568–573.

[123] J. Heikkila, L. Curiel, K. Hynynen, Local harmonic motion monitoring of focused ultrasound surgery–a simulation model, IEEE Trans. Biomed. Eng. 57 (1) (2010) 185–193.

[124] B. Baseri, J.J. Choi, Y.S. Tung, E.E. Konofagou, Multi-modality safety assessment of blood-brain barrier opening using focused ultrasound and definity microbubbles: a short-term study, Ultrasound Med. Biol. 36 (9) (2010) 1445–1459.

[125] G.Y. Hou, F. Marquet, S. Wang, E.E. Konofagou, Optimization of real-time acoustical and mechanical monitoring of high intensity focused ultrasound (HIFU) treatment using harmonic motion imaging for high focused ultrasound (HMIFU), Conf. Proc. IEEE Eng. Med. Biol. Soc. (2013) 6281–6284, https://doi.org/10.1109/EMBC.2013.6610989.

[126] V. Suomi, Y. Han, E. Konofagou, R.O. Cleveland, The effect of temperature dependent tissue parameters on acoustic radiation force induced displacements, Phys. Med. Biol. 61 (20) (2016) 7427–7447.

[127] B.J. Fahey, K.R. Nightingale, S.A. McAleavey, M.L. Palmeri, P.D. Wolf, G.E. Trahey, Acoustic radiation force impulse imaging of myocardial radiofrequency ablation: initial in vivo results, IEEE Trans. Ultrason. Ferroelectr. Freq. Control 52 (4) (2005) 631–641.

[128] W. Kwiecinski, F. Bessiere, E.C. Colas, W.A. N'djin, M. Tanter, C. Lafon, M. Pernot, Cardiac shear-wave elastography using a transesophageal transducer: application to the mapping of thermal lesions in ultrasound transesophageal cardiac ablation, Phys. Med. Biol. 60 (20) (2015) 7829–7846.

[129] M. Carias, K. Hynynen, Combined therapeutic and monitoring ultrasonic catheter for cardiac ablation therapies, Ultrasound Med. Biol. 42 (1) (2016) 196–207.

[130] E.A. Stewart, B. Gostout, J. Rabinovici, H.S. Kim, L. Regan, C.M. Tempany, Sustained relief of leiomyoma symptoms by using focused ultrasound surgery, Obstet. Gynecol. 110 (2 Pt 1) (2007) 279–287.

CHAPTER

# Thyroid elastography

7

**Manjiri Dighe**

*Department of Radiology, Abdominal imaging section, University of Washington, Seattle, WA, United States*

## 1. Thyroid pathology

Thyroid nodules are common and their detection is increasing with the increasing use of imaging studies such as ultrasonography (US) and computed tomography. Thyroid US is able to confirm the presence of a thyroid nodule when physical examination is equivocal. However, US has low accuracy to differentiate between benign and malignant lesions. Presently fine-needle aspiration (FNA) is recommended as an additional diagnostic method in the evaluation of thyroid nodules with a size of >10 mm and with suspicious US findings or concerning history [1–5]. A classic criterion for malignancy upon palpation is a hard or firm consistency [6,7]. Ultrasound elastography is a new technique that can evaluate stiffness in a tissue and help differentiate between benign and malignant nodules. In addition to nodular disease affecting the thyroid, a wide spectrum of diffuse diseases affect the thyroid gland and noninvasive evaluation of these diseases by US will be helpful.

## 2. Strain elastography

### 2.1 Introduction

Strain elastography evaluates the stiffness in tissues by applying external compression. It is defined as a change in length during compression divided by the length of the tissue after compression using Young's modulus. Young's modulus evaluates the relationship between the compression (stress) and strain and is defined as $E =$ stress/strain [8]. Commercial US elastography equipment cannot measure the applied stress and hence direct quantification of the strain is not possible and only relative stiffness is displayed. The stress in strain elastography (SE) is usually applied by external compression either manually with a transducer by acoustic radiation force impulse (ARFI) imaging or using physiologic shifts in the patient, such as carotid artery pulsations. When manual stress is used, compression with the transducer is applied by continuously and uniformly compressing and decompressing the skin of the patient by a few millimeters at a time. The elastogram is calculated by

Tissue Elasticity Imaging. https://doi.org/10.1016/B978-0-12-809662-8.00007-3

evaluating the change in signals from before- to after-compression images and displayed either as a color overlay on the B-mode image or in a side-by-side split-screen format. The elastographic display of stiffness provides qualitative information of stiffness in a tissue and is called as strain histogram. Color image is displayed in a continuum of colors from red to green to blue, depicting soft (high strain), intermediate (equal strain), and hard (no strain), respectively. Tissue stiffness can also be displayed in a grayscale format. At present, there is no color standard and the elastogram display varies based on the manufacturer [8,9]. Strain information can also be displayed in a semiquantitative way as a strain value and compared with that of normal tissue to calculate a strain ratio.

## 2.2 Strain histograms

Strain histograms are a way to display the elastographic information of stress in a tissue in a color-based format or in gray scale. These histograms can then be further scored subjectively with a visual scoring system. The Tsukuba scoring system was first devised for breast US using a 5-point scale [10], which was then adapted and used in thyroid elastography. Utilizing the Tsukuba scoring system, prior investigators found that scores 4–5 were highly predictive of malignancy ($P < .0001$), with a sensitivity of 97%, specificity of 100%, positive predictive value (PPV) of 100%, and negative predictive value (NPV) of 98% [11,12]. A modified 4-point system was then created from the Tsukuba scoring system for thyroid nodules by Asteria et al. [13], as shown in Table 7.1. A 2-pattern scoring system has also been used by combining Tsukuba 1 + 2 + 3 score in score 1 and Tsukuba 4 + 5 in score 2 with excellent sensitivity and specificity [12]. Score 1 was defined as <50% blue and hence soft nodules and score 2 was defined as >50% blue and hence hard nodules [14]. Examples of nodules classified according to the scoring system are shown in Fig. 7.1, and an example of difference between benign and malignant nodule is shown in Fig. 7.2.

Most systems use a color-based scoring system, but some systems have used gray scoring systems. Lyshchik et al. used a 4-point grayscale scoring system, as shown in Table 7.2. Ding et al. [15] used image texture analysis by changing the original color

**Table 7.1** The Tsukuba 5-point scoring system description used in breast ultrasonography [10].

| Score | Description |
|---|---|
| 1 | Deformability of the entire lesion |
| 2 | Deformability of most of the lesion with some stiff areas |
| 3 | Deformability of the peripheral portion of the lesion with stiff tissue in the center |
| 4 | Entire lesion is stiff |
| 5 | Entire lesion and surrounding tissue are stiff |

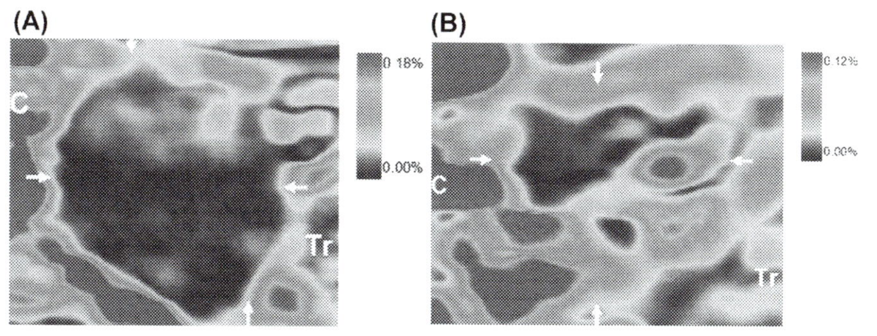


**FIGURE 7.1**

The classification of strain histogram in 2 nodules using the 5-point classification (blue [gray in print version], hard; red [black in print version], soft). (A) A nodule (*arrows*) with uniform blue (gray in print version) color in the strain histogram image, suggesting a type 5 lesion. (B) A nodule (*arrows*) with a combination of blue (gray in print version) and red (black in print version) in the nodule, suggesting a type 3 lesion.

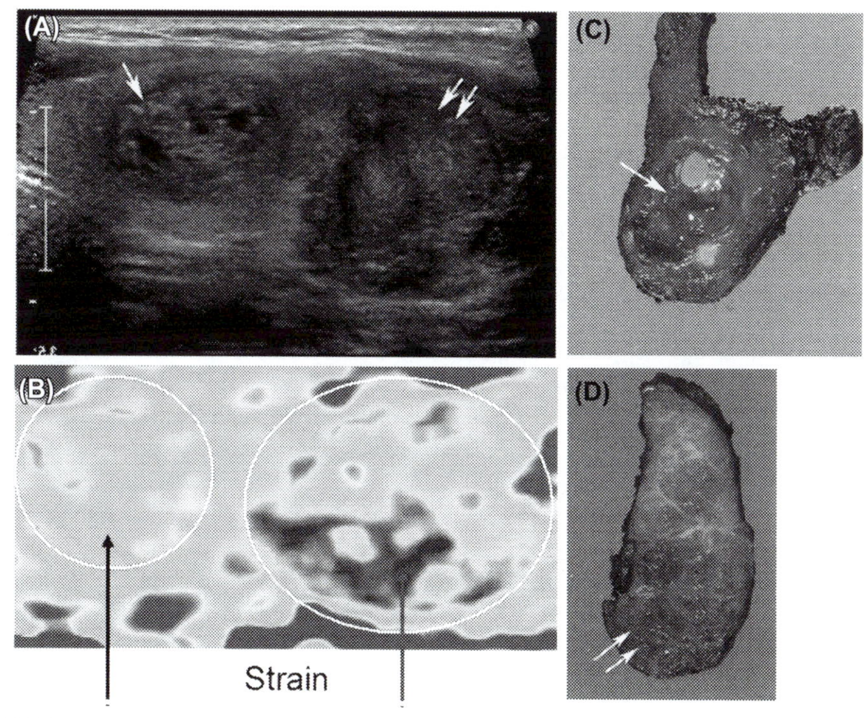


**FIGURE 7.2**

A 44-year-old woman with multiple nodules in the thyroid gland presented for routine surveillance. She had 2 nodules in the right lobe (A), the lower pole nodule (*double arrows*) had mildly increased in size, whereas the mid-pole to upper pole nodule (*arrow*) was stable. (B) Strain elastography showed uniform low stiffness in the upper pole nodule (*black arrow* [left arrow in print version]) while the lower pole nodule (*red arrow* [right arrow in print version]) appeared to have some stiff areas in it. Fine-needle aspiration of the lower pole nodule was performed, which was suspicious for papillary carcinoma, and histopathologic examination confirmed a colloid nodule in the mid-pole to upper pole nodule (*arrow*) (C), while the lower pole nodule (*double arrows*) was consistent with papillary carcinoma (D).

**Table 7.2** Lyshchik et al. 4-point scoring system for grayscale-based scoring [49].

| Score | Description |
|---|---|
| 1 | Very dark lesion |
| 2 | Markedly darker than the surrounding parenchyma |
| 3 | Slightly darker than the surrounding parenchyma |
| 4 | Equally as bright or brighter than the surrounding parenchyma |

thyroid elastograms from a red-green-blue color space to the hue saturation value color space and extracted the texture features. They had a classification accuracy of 93.6% to differentiate between benign and malignant nodules. Histogram analysis of color elastograms has also been used for the diagnosis and management of patients with diffuse thyroid disease [16,17].

## 2.3 Strain ratio

Strain ratio is a semiquantitative method of analyzing tissue stiffness and is the ratio between the strain in adjacent thyroid tissue and the nodule. Two types of strain ratio have been defined: parenchyma to nodule strain ratio (PNSR), which is the mean strain in the normal thyroid parenchyma divided by the mean strain within the thyroid nodule, and the muscle to nodule strain ratio (MNSR), which is the mean strain in the adjacent strap muscle divided by the strain in the nodule (Fig. 7.3) [18].

Several studies have proposed various numbers for the strain ratio to distinguish between benign and malignant nodules, as shown in Table 7.3; however, an agreement on the critical cutoff point has not been reached [15,19–23].

Another approach exploits the pulsation of the adjacent carotid artery to provide stress on the nodule, called as elasticity contrast index [24]. In this method, the transducer is held still on the skin over the nodule/thyroid with light contact and after asking the patient to hold his/her breath for a few seconds, elastography data is acquired over 3–4 s. Two regions of interest (ROIs) are then drawn, one close to the area of highest strain adjacent to the carotid artery in the surrounding muscle and another within the thyroid nodule at the region of the lowest strain (highest stiffness) to calculate the thyroid stiffness index [25]. The elasticity contrast index is calculated by co-concurrence matrix comparison of benign and malignant features in a nodule by using complex calculations [24].

## 2.4 Examination technique

The patient should be placed supine with a pillow under the neck as with routine thyroid US. The transducer is placed over the nodule with gel interposed. If manual compression is used, light manual compression is applied in a rhythmic fashion to create uniform compression over the nodule. Each nodule should have at least 2 acquisitions of SE.

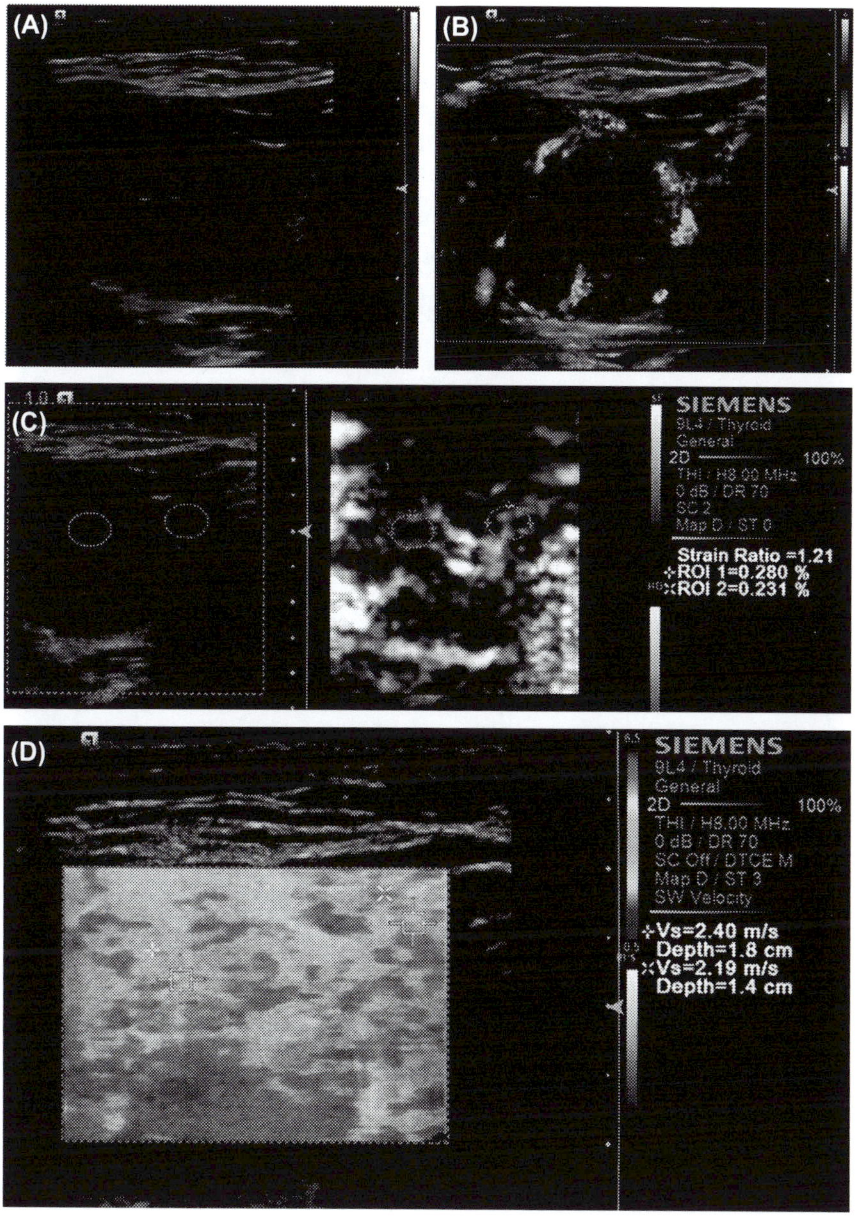

**FIGURE 7.3**

Images from a 36-year-old woman with (A) a heterogeneous 2.2-cm mass (B) with small cystic areas and moderate internal blood flow. (C) Strain elastography was performed that showed a parenchyma to nodule strain ratio of 1.21 and (D) a stiffness of 2.19 m/s on two-dimensional shear wave elastography. The lesion was negative for malignancy on biopsy.

*Case courtesy Dr. Richard Barr.*

**Table 7.3** The specificity, sensitivity, NPV, and PPV in nodules in various studies, with cutoff values defined as shown.

| | Strain type | | Specificity (%) | Sensitivity (%) | PPV (%) | NPV (%) | Cutoff |
|---|---|---|---|---|---|---|---|
| Lyschik et al. [49] | SE | Strain index | 96 | 82 | – | – | Strain index value > 4 |
| | | Margin ratio | 88 | 36 | – | – | Margin regularity score > 3 |
| | | Tumor ratio | 92 | 46 | – | – | Tumor area ratio > 1 |
| Kagoya et al. [19] | SE | Scores 3 and 4 | 64 | 73 | – | – | |
| | | SR > 1.5 | 90 | 50 | – | – | |
| Xing et al. [20] | SE | SR | 85.7 | 97.8 | 88 | 97.8 | SR = 3.79 |
| | | 4-Point scoring system | 81.1 | 88.8 | 80 | 89.5 | |
| Ding et al.[15] | SE | Hard area ratio | 81.2 | 98.2 | – | – | Hard area ratio = 0.45 |
| | | SR | 73.2 | 89.3 | – | – | SR = 2.73 |
| | | Color score | 68.1 | 78.6 | – | – | Color score = 2.5 |
| Cantisani et al.[21] | SE | SR | 91.7 | 97.3 | 87.8 | 98.2 | SSR > 2 |
| Ning et al. [35] | SE | ES | 72 | 82 | – | – | ES = 3.5 |
| Zhang et al. [69] | ARFI and SE | EI | 66.7 | 65.9 | 40.3 | 85.1 | SR = 4.225 |
| | | SR | 83 | 82 | – | – | SWV = 2.87 |
| Gu et al. [70] | ARFI | SWV | 93.42 | 86.36 | 79.17 | 95.95 | SWV = 2.55 |
| Ciledag et al. [18] | SE | SR | 82.1 | 85.7 | 33.3 | 98.2 | SR = 2.31 |
| Wang et al. [71] | SE | EI | 80 | 78 | – | – | SR ≥ 2.9 |
| | | SR | 92 | 87 | – | – | |
| Friedrich-Rust et al. [42] | ARFI | SWV | 91–95 | | | | No cutoff values given |
| | | ES | 91 | | | | |
| Zhang et al. [72] | ARFI | SWV | 95.7 | 96.8 | 93.75 | 97.8 | SWV = 2.84 |
| | | SWR | 91.9 | 81.7 | 77.03 | 93.83 | |
| Hou et al. [73] | ARFI | SWV | 89.23 | 80 | 69.56 | 93.54 | SWV = 2.42 |
| Wang et al.[74] | SE | SR | 91.38 | 80.77 | – | – | SR = 3.855. |
| | | 5-Point score | 78.45 | 84.62 | | | ES = 3.5 |
| Zhan et al. [75] | ARFI | SWV | 85.3 | 94.4 | 77.2 | 96.6 | SWV = 2.85 |
| Cantisani et al. [76] | SE | SR | 89 | 93 | 82 | 94 | SR = 2 |
| Cantisani et al. [77] | SE | ES | 88 | 95 | 97 | 91 | SR = 2 |

| | | | | | | | |
|---|---|---|---|---|---|---|---|
| Cantisani et al. [78] | SE | SR, operator 1 | 92 | 93 | – | – | SR = 2.02 |
| | | SR, operator 2 | 79 | 84 | – | – | SR = 1.86 |
| Grazhdani et al. [79] | ARFI | SWV, operator 1 | 75 | 90 | 90.91 | 96.55 | SWV = 2.365 |
| | | SWV, operator 2 | 72 | 90 | 90 | 96.9 | |
| Cakir et al. [80] | SE | SI | 96.30 | 98.77 | 91.95 | 99.45 | SI = 6.66 |
| Guazzaroni et al. [81] | SE | ES | 68 | 91 | 27 | 98 | SS > 2 |
| | | SR | 83 | 81.8 | 39 | 97 | SR ≥ 3.28 |
| Aydin et al. [82] | SE | PNSR | 50.6 | 95.6 | 35 | 97.6 | PNSR = 3.14 |
| | | MNSR | 92.8 | 95.6 | 78.6 | 98.7 | MNSR = 1.85 |
| Xu et al. [83] | ARFI | SWV | 83.4 | 71.6 | – | – | SWV = 2.87 |
| Calvete et al. [84] | ARFI | SWV | 96 | 85.7 | – | – | SWV = 2.5 |
| Zhuo et al. [85] | ARFI | SWV | 96.2 | 96.3 | – | – | SWV = 2.545 |
| Xu et al. [86] | ARFI | SWV | 76.9 | 68.2 | 62.5 | 81.1 | SWV = 2.87 |
| Zhang et al. [87] | ARFI | VTI classification (4 grades) | 89.6 | 95.7 | 86.27 | 96.77 | Cutoff > grade III |
| | | VTI area ratio | 86.6 | 91.3 | 82.35 | 93.55 | Area ratio = 1.08 |
| | | SWV | 85.1 | 91.3 | 80.77 | 93.44 | SWV = 2.90 |
| Huang et al. [88] | ARFI | VTI | 90.4 | 73.8 | 83.3 | 84.2 | SE score, 4 or above |
| | p-SWE | SWV | 76.6 | 82 | 69.4 | 86.7 | SWV > 2.64 |
| Hamidi et al. [89] | ARFI | SWV | 82.3 | 100 | – | – | SWV = 2.66 |
| Liu et al. [90] | ARFI | SE | 89.7 | 80 | – | – | SE > 4 |
| | p-SWE | SWV | 57.9 | 80 | – | – | SWV > 2.15 |
| Wang et al. [91] | SE | ES | 71.1 | 91.7 | | | SR = 5.03 |
| | | SR | 92.1 | 75 | | | |
| Friedrich-Rust et al. [92] | SE | SV | 56 | 81 | 92 | 32 | Cutoff = 0.17 |
| | | SR | 58 | 78 | 92 | 30 | Cutoff = 2.66 |
| Cantisani et al. [23] | SE | SR | 93 | 90.6 | 82.8 | 96.4 | SR = 2.09 |

ARFI, *acoustic radiation force impulse;* EI, *elasticity index;* ES, *elasticity score;* MNSR, *muscle to nodule strain ratio;* NPV, *negative predictive value;* p-SWE, *point shear wave elastography;* PNSR, *parenchyma to nodule strain ratio;* PPV, *positive predictive value;* SE, *strain elastography;* SR, *strain ratio;* SWV, *shear wave velocity;* VTI, *Virtual Touch imaging.*

As SE uses US, an adequate B-mode image should be acquired to get a diagnostic elastogram. However, if the nodule is deep, higher stress needs to be applied, which could result in poor-quality elastograms in the near field. The focal zone should be placed to optimize the B-mode image in the ROI to generate a good SE image.

As SE displays relative elasticity within an ROI, the ROI is set as large as possible to acquire the whole nodule and adjacent normal thyroid tissue, avoiding including large vessels, bones, and nonthyroid tissue [26].

### 2.5 Interobserver and intraobserver variabilities

As manual compression is used in SE and the accuracy of SE depends on the operator's skill and experience, training and experience in acquiring the elastograms is essential. Park et al. [27] reported first on the interobserver agreement for SE for thyroid cancers and found that the agreement was very poor for SE in contrast to conventional US, which had much better agreement. Newer machines now have a quality display that gives real-time feedback to the operator about the stress applied to the tissue. Recent reports show substantial or almost perfect agreement within multiple operators as shown in Table 7.4. No in vivo studies of intraobserver variability have been reported in the literature.

### 2.6 Practical advice, tips, and limitations

It is important to keep in mind that SE displays the relative strain of the structures in the ROI because the stress applied by the transducer is not quantifiable, hence strain values do not represent the elasticity modulus directly [28]. As strain changes with applied stress, its numeric value cannot be compared between 2 lesions or 2 individuals [25,28]. At present, there is no color standard and strain histogram display varies based on the manufacturer [8,9], so care must be taken to note the settings of the machine before interpreting the histograms. Strain values obtained during elastography cover a very large range of values; however, the range of values in color-coded images is low and hence can lead to mismatch in the values and display

**Table 7.4** The interobserver variability in strain elastography in different studies.

| Reference | Statistical method | Statistical value |
|---|---|---|
| Merino et al. [93] | Cohen's kappa statistic | 0.82 |
| Ragazzoni et al. [94] | Cohen's kappa statistic | 0.64 |
| Kim et al. [95] | Cohen's kappa statistic | 0.738 |
| Calvete et al. [33] | Cohen's kappa statistic | 0.838 |
| Cantisani et al. [78] | Cohen's kappa statistic | 0.95 |
| Cantisani et al. [96] | Cohen's kappa statistic | 0.71–0.79 |

provided. Optimization of the elastogram for soft material will incorrectly display the difference in stiff tissues similar to aliasing seen in color Doppler images. Hence, optimization of the displayed scale according to the stiffness of the material scanned is imperative to obtain quality elastograms.

Nodules that are in the isthmus are more difficult for SE because of their superficial location and lack of normal reference tissue [29]. In addition, deep nodules may also be difficult for SE because of the stress decay due to their distance from the transducer [30]. Nodule characteristics such as calcifications and fibrosis are associated with increased stiffness, irrespective of the underlying pathologic condition and can produce unreliable results (Fig. 7.4) [16,31].

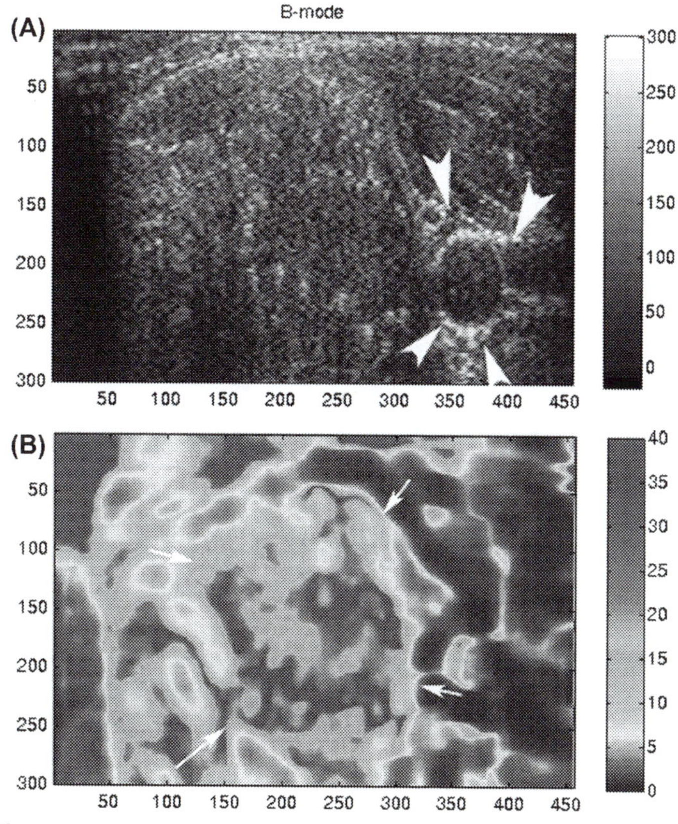


**FIGURE 7.4**

A 37-year-old woman presented with (A) a large nodule (*white arrows*) in the left lobe of the thyroid with small calcifications in it. Red dots mark the position of the carotid for strain evaluation. (B) Strain elastography showed multiple areas of high stiffness within the nodule (in between the arrows) seen as areas of blue (gray in print version). Fine-needle aspiration was performed that showed an indeterminate nodule with suspicious areas of malignancy in it. Hemithyroidectomy was performed that showed a nodular goiter with areas of hemorrhage and fibrosis in it.

A strain quality indicator is helpful to provide real-time feedback about the quality of compression. Various manufacturers have different recommendations; however, a quality factor of 3–4 (Hitachi) or above 50 (Siemens) has been recommended [32,33]. For a very stiff nodule, the scale can be increased to enable the display of the upper limit to allow differentiation between relatively stiff and soft areas; similarly, in a very soft nodule the scale should be lowered to enable the display of the relatively hard regions in a nodule.

Several factors can affect the results of elastography, including nodule characteristics, experience of the operator, and motion artifacts from respiration and carotid pulsation. Transverse scans can be more susceptible to interference from carotid artery pulsations than longitudinal scans, and hence, longitudinal scans are preferable for elastography with external compression [34,35].

## 3. Shear wave elastography

### 3.1 Introduction

In shear wave elastography (SWE), an ultrasound push pulse (shear wave) is sent into the tissue and the minute displacement of tissues produced by the acoustic radiation force it generates sets up transverse shear waves that travel away from the push pulse lines. The induced shear wave speed (SWS) is dependent on the stiffness of the tissue, with stiffer tissue conducting shear waves faster. The US machine is able to measure the SWS in meters per second (m/s) from which Young's modulus can be calculated in kilopascals (kPa) after making some assumptions regarding the tissue properties [8,9].

### 3.2 Different methods of shear wave elastographic imaging of the thyroid

SWE can be performed using the shear wave technique point shear wave elastography (p-SWE) or ARFI imaging in which a small ROI of fixed size is placed at the desired site while performing real-time B-mode imaging. The tissue in the ROI is mechanically excited using acoustic pulses to generate localized tissue displacements, which result in shear-wave propagation away from the region of excitation and are tracked using ultrasonic correlation-based methods [36]. The second method, two-dimensional shear wave elastography (2D-SWE) captures the waves that propagate from the stimulated tissue in question with an ultrafast US tracking method and then displays real-time information in terms of velocity or estimated tissue stiffness expressed in kilopascals. In this method, a larger ROI is placed and when activated, a color-coded map of the SWS is displayed in the field of view (Fig. 7.5). One or more measurement ROIs can then be placed within the larger ROI to get SWS measurement, which correlates with the stiffness in tissues.

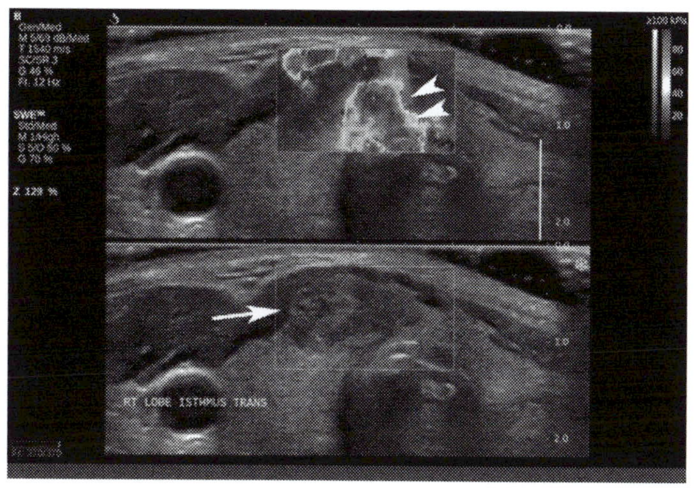


**FIGURE 7.5**

A 27-year-old man presented with a palpable nodule in the midline. (A) Ultrasonography showed a small hypoechoic nodule (*arrow*) in the isthmus, which was difficult to evaluate with (B) ultrasound elastography because of its superficial location. A fingerlike artifact (*arrowheads*) was seen extending from the trachea into the isthmus, which suggested a compression artifact.

## 3.3 Review of literature

Multiple papers have been published using the two methods of shear wave imaging described previously. The sensitivities of these studies range from 0.35 to 1.00 with the majority between 0.68 and 0.95 and specificities range from 0.71 to 0.97. To date, 4 meta-analyses have been performed including 6000 nodules [37–40]. All the meta-analyses concluded that SWE imaging is a useful complement to B-mode US in differentiating benign from malignant nodules, and hence, they concluded that SWE can be useful to select patients with thyroid nodules for surgery [37].

In the study by Zhang et al., 174 pathologically proven thyroid nodules (139 benign, 35 malignant) in 154 patients were included. US, Virtual Touch imaging (VTI), and Virtual Touch quantification (VTQ) were performed and interpreted by two readers with different experience who were blinded to the results and independently scored the likelihood of malignancy using a 5-point scale in three different image-reading sets [41]. The specificity of both readers improved significantly after viewing the VTI/VTI and VTQ images (all $P < .05$). The confidence in characterizing a nodule as benign or malignant increased after reviewing VTI and VTQ images versus conventional US images for the senior reader ($P < .05$). The authors concluded that adding p-SWE improved the specificity in diagnosing malignant thyroid nodules compared with conventional US on its own. Results of the other studies using SWE for the evaluation of thyroid nodules are listed in Table 7.5.

**Table 7.5** Sensitivity and specificity of shear wave elastography in the evaluation of thyroid nodules.

| Author | Method(s) | Cutoff, m/s (kPa) | Sensitivity | Specificity |
|---|---|---|---|---|
| Hamidi et al.[89] | p-SWE | 2.66 (21) | 1.00 | 0.82 |
| Sebag et al.[60] | | 4.65 (65) | 0.82 | 0.97 |
| Bhatia et al.[97] | | 3.45 (34.5) | 0.77 | 0.71 |
| Veyrieres et al.[50] | 2D-SWE | 4.70 (66) | 0.80 | 0.91 |
| Slapa et al.[98] | | 4.65 (65) | | |
| Liu et al.[99] | 2D-SWE | 3.65 (39.3) | 0.66 | 0.84 |
| Park et al.[100] | | 5.3 Mean (85)<br>5.6 Max (94) | 0.95 | |
| Zhang et al. [72] | p-SWE | 2.84 | 0.97 | 0.96 |
| Hou et al. [73] | p-SWE | 2.42 | 0.80 | 0.89 |
| Zhang et al. [69] | p-SWE | 2.87 | 0.66 | 0.67 |
| Bojunga et al. [44] | p-SWE | 2.57 | 0.35 | 0.79 |
| Gu et al. [70] | p-SWE | 2.56 | 0.86 | 0.93 |
| Han et al. [101] | p-SWE | 2.75 | | |
| Xu et al. [83] | p-SWE + ARFI SE | | 0.68 | 0.77 |
| Liu et al. [102] | 2D-SWE | 3.60 (38.3) | 0.68 | 0.87 |

2D-SWE, *two-dimensional shear wave elastography;* ARFI, *acoustic radiation force impulse;* p-SWE, *point shear wave elastography;* SE, *strain elastography.*

## 3.4 Interobserver and intraobserver variabilities

Results of 4 studies evaluating the interobserver variability of SWE are listed in Table 7.6. Intraobserver variability was evaluated in a single study by Friedrich-Rust et al. [42] and they concluded that even with a low value of concordance rate of $k = 0.35$, the variability did not affect the differentiation between benign and malignant nodules.

**Table 7.6** The results of interobserver variability of shear wave elastography.

| Study | Statistic used | Result |
|---|---|---|
| Grazhdani et al. [79] | Concordance rate | 0.75 |
| | Interclass correlation | 0.97 |
| Veyrieres et al. [50] | Interclass correlation | 0.97 |
| Zhang et al. [69] | Intraobserver variability | 0.90 |
| | Interobserver variability | 0.86 |
| Brezak et al. [103] | Concordance correlation coefficient | 0.97–0.93 |
| | Intrareader coefficient of variation | 13%–14% |
| | Interreader coefficient of variation | 21% |

## 3.5 Examination technique

SWE images are acquired in a similar fashion as SE images, with the main difference being that no external compression is applied during the acquisition of images. Any excess pressure from the transducer produces superficial hardening artifacts and increases the stiffness. After acquisition of the images, measurement ROIs can be placed in the stiffest part of the lesion and on the adjacent thyroid or muscle to obtain quantitative readouts as well as ratios.

## 3.6 Practical advice and tips

ROIs in p-SWE are not modifiable and are available in 2 sizes: 5 × 6 mm and 20 × 20 mm; hence, with p-SWE the ROI may include the surrounding normal thyroid, cystic areas, or calcification and may affect the measurements. The VTQ measurement box is also not modifiable and can include surrounding tissue. SWS values can be measured up to 9 m/s and higher speeds are displayed as "x.xx m/s". This should be taken into consideration in very stiff nodules. The penetration depth in ARFI imaging is limited to about 5.5 cm, and hence, large thyroid and very large and deep nodules cannot be properly assessed using ARFI quantification [43]. Similar to SE, nodules located in the isthmus are difficult due to the superficial location (Fig. 7.6).

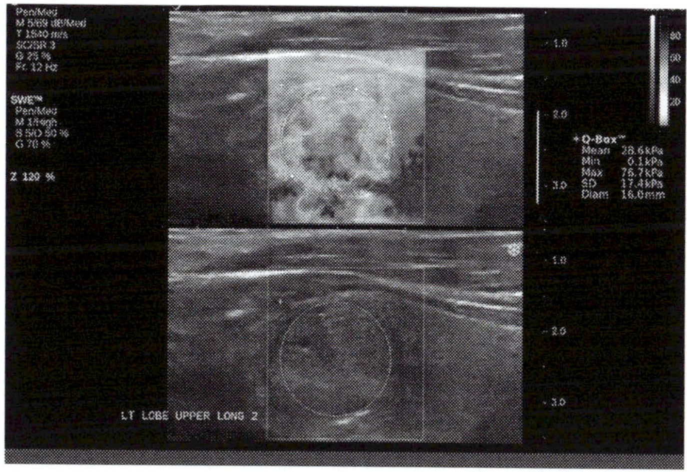


**FIGURE 7.6**

A 34-year-old man presented with a nodule in the right lobe, which was iso- to hypoechoic on B-mode ultrasonography (A) (within the circular region of interest) and showed small areas of stiffness on ultrasound elastography (B). This lesion had a maximum stiffness of 76.7 kPa in the stiff areas and was shown to be a papillary carcinoma on fine-needle aspiration.

As stiffness ratios can be calculated from SWE, it is important to remember that stiffness ratios obtained from SWE are not the same as the elasticity index (EI). It is important to acquire 5–10 measurements to obtain a median to yield a reliable accuracy [26,44,45].

Multiple guidelines have been published concerning ultrasound elastography and its usefulness in routine practice. Some of these include the World Federation for Ultrasound in Medicine and Biology guidelines [46], the European Thyroid Association guidelines [47], and the European Federation of Societies for Ultrasound in Medicine and Biology guidelines [48]. These guidelines should be reviewed to understand how to perform thyroid elastography correctly and also how to use it appropriately in clinical practice.

## 3.7 Interpretation of results

SE appearance of normal thyroid has been reported by 2 studies. First, the study by Lyshchik et al. [49] mentioned that the normal thyroid was easily identifiable; however, pulsation artifacts from the adjacent carotid artery compromised some of the imaging. Second, Bae et al. [28] in their study described that the strain in the thyroid gland was quite homogeneous with the adjacent trachea appearing stiffer and the neck muscles appearing softer than the thyroid (Fig. 7.7). The thyroid stiffness index values calculated by their technique, which are similar to the EI except the carotid artery is used as a reference, were 4.3–5.2 for normal volunteers, which was much softer than the nodules seen in malignant cases [28]. Three studies reported the stiffness values of the normal thyroid gland using SWE imaging [42,50,51] and these values range from 1.60 ± 0.18 m/s for p-SWE to 2.6 ± 1.8 m/s for 2D-SWE.

Zhan et al. [37] in their meta-analysis of 16 studies that included a total of 2436 nodules in 2147 patients for ARFI imaging studies found that the overall mean sensitivity and specificity of ARFI imaging for the differentiation of thyroid nodules were 80% and 85%, respectively, and hence concluded that ARFI imaging had a high sensitivity and specificity for the identification of thyroid nodules. Ghajarzadeh

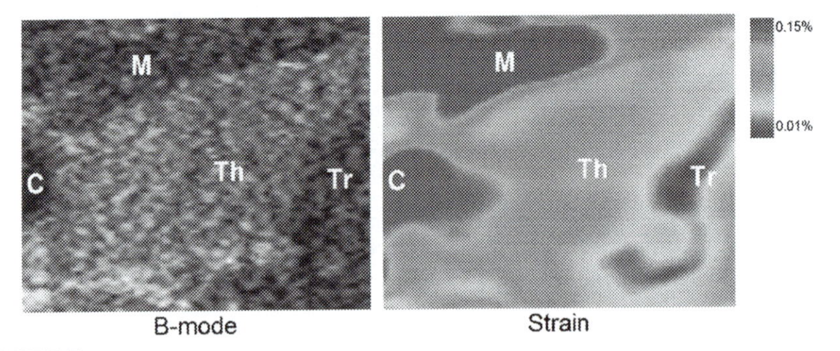


**FIGURE 7.7**

Normal thyroid in a 24-year-old man. Note the uniform appearance on strain elastography. C, carotid artery; M, muscle; *Th*, thyroid; *Tr*, trachea.

et al. [52] in their meta-analysis found that the elasticity scores between 2 and 3 showed a sensitivity of 86% and a specificity of 66.7%; however, the highest sensitivity of the test was achieved by an elasticity score threshold between 1 and 2, which increased the sensitivity to 98.3%.

Sun et al. in their study involving 5481 nodules in 4468 patients for elasticity score studies and 1063 nodules in 983 patients for strain ratio studies found that strain ratio had better performance than elasticity score assessment. The areas under the curve for the elasticity score and strain ratio were 0.8941 and 0.9285, respectively [53].

Fukuhara et al. [51] evaluated the utility of ARFI SWE for diagnosing chronic autoimmune thyroiditis (CAT) and to verify the effect of fibrotic thyroid tissue on shear wave velocity (SWV) and found that the SWV for CAT (2.47 ± 0.57 m/s) was significantly higher than that for control healthy subjects (1.59 ± 0.41 m/s) ($P < .001$). The area under the receiver operating characteristic (AUROC) curve value for CAT was 0.899, and the SWV cutoff value was 1.96 m/s. The sensitivity, specificity, and diagnostic accuracy were 87.4%, 78.7%, and 85.1%, respectively [51]. Kim et al. [54] found the cutoff mean value of the EI using carotid artery as the internal compression source, in cases of diffuse thyroid disease, to be 27.6 kPa and a maximum value of 41.3 kPa, with a sensitivity of 40.9% and a specificity of 82.9%.

Sporea et al. [55] found that thyroid stiffness values assessed by means of ARFI SWE in healthy subjects (2 ± 0.40 m/s) were significantly lower than those in Graves disease (GD) (2.67 ± 0.53 m/s) ($P < .0001$) and in patients with CAT (2.43 ± 0.58 m/s) ($P = .0002$), but the differences were not significant between patients with GD and CAT ($P = .053$). The optimal cutoff value for the prediction of diffuse thyroid pathology was 2.36 m/s with sensitivity, specificity, PPV, and accuracy of 62.5%, 79.5%, 87.6%, and 72.7%, respectively, for the presence of diffuse thyroid gland pathology (AUROC curve value = 0.804).

Stiff areas are seen in the thyroid gland in subacute thyroiditis and these stiff areas can be stiffer than the ones seen in chronic thyroiditis as well. They can resemble malignant nodules on elastography [56,57]. Xie et al. [56] found that real-time strain ultrasound elastography does not provide conclusive information in the diagnosis and differential diagnosis of subacute thyroiditis due to its inability to distinguish between subacute thyroiditis and thyroid cancer.

Yang et al. found that the MNSRs of patients with hyperthyroidism, Hashimoto thyroiditis, and subacute thyroiditis were 2.30 ± 1.08, 7.04 ± 7.74, and 24.09 ± 13.56, respectively. The strain ratio of the control group was 1.76 ± 0.54 [58].

Studies using SWE have reported cutoff values ranging from 3.65 to 4.70 m/s (34.5–66 kPa). Increasing the cutoff values from 1.8 to 6.7 m/s (10–132 kPa) increased the specificity from 8.9% to 100% in a study by Bhatia et al. [34]. Zhang et al. [59] in their meta-analysis found that out of the 15 studies included in their study, 15 reported an SWS cutoff in the range of 2–3 m/s, which they defined as the "Gray-Zone," with 11 studies reporting the best SWS cutoff of >2.5 m/s.

Combining US and SWE information has provided increased sensitivity and specificity in differentiating benign and malignant nodules. Sebag et al. [60] in their study found that the sensitivity and specificity for malignancy were 51.9% and 97.0% for grayscale US and 81.5% and 97.0% for the combination of grayscale US and SWE, respectively.

Owing to the presence of a diffuse fibrotic structure in CAT, Fukuhara et al. [51] in their study found that the SWS of CAT was higher than that of normal or benign thyroid nodules and closer to that of papillary thyroid carcinoma. The cutoff values for VTI (ARFI SE) and VTQ (p-SWE) are higher in patients with CAT than those in normal patients [61].

It is essential to remember that not all thyroid carcinomas are hard and some specific ones maybe soft or heterogeneous. For example, follicular carcinomas can be soft and are difficult to differentiate from benign nodules [62]. In patients with a known background of chronic thyroiditis, an MNSR will give better results than PNSR because of the stiffer appearance of the chronic thyroiditis parenchyma than the normal thyroid.

Tian et al. [63] performed a meta-analysis evaluating the diagnostic performance of real-time elastography (RTE) and SWE and found that the pooled sensitivity and specificity of RTE was 83% and 81.2%, which were higher than those of SWE alone, i.e., 78.7% and 80.5%, respectively. They concluded that RTE had higher diagnostic performance than SWE in differentiating malignant from benign nodules.

## 4. Artifacts in thyroid elastography

One of the well-known artifact in elastography is the precompression artifact from applying too much pressure on the skin during elastography image acquisition. It is also know that fluid in cystic nodules can produce two types of artifacts: lack of information in fluid and aliasing artifact distal to cystic areas. Cho et al. in their study including 13 patients with cystic breast lesions found an aliasing artifact in the tissues posterior to the cystic mass, which was seen as a blue-green-red area. This artifact suggested that the lesion is cystic, which may not be readily apparent on just the US images [64]. A similar artifact can be seen in cystic thyroid nodules as well, which can manifest as an artifactual increase in the stiffness in the solid areas posterior to the cystic portions (Fig. 7.8). In such cases, stiffness in the tissue should be measured superficial or lateral to the cystic areas and not posterior to the cystic areas, as shown in Fig. 7.8.

Bhatia et al. [34] in their paper have described a similar artifact in predominantly cystic and partially cystic thyroid nodules. They found an area of increased stiffness posterior to the cystic portion in the nodule, which they attributed to either a genuine increase in stiffness in that part of the solid tissue or an artifact from uneven stress distribution at the boundary of tissues of differing stiffness (liquid vs. solid). They also postulated that as elastograms are spatial maps of relative strain within the elastography window, low signal within fluid in cystic nodules may have resulted in an artifactual signal increase in other tissues including the solid portion of cystic nodules [34].

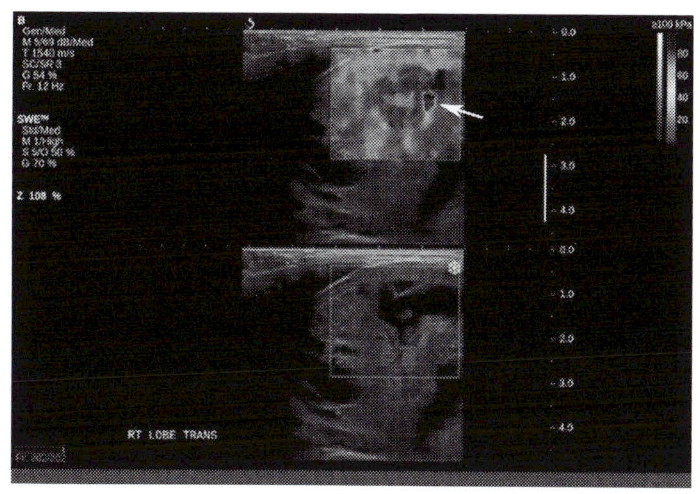


**FIGURE 7.8**

Cystic areas causing artifactual lack of information in the nodule seen on the shear wave elastographic image with a small area of artifactual stiffness (*arrow*), which was seen in the tissue beyond the cystic area.

Calcifications within nodules increase the stiffness in the nodule and can make benign nodules with macrocalcifications appear "stiff" (Fig. 7.9) [13]. Szczepanek-Parulska et al. [65] in their study found that benign lesions with micro- or macrocalcifications were stiffer than benign lesions without calcifications. Vorländer et al. [66] reported that thyroid lesions with coarse calcifications are not a suitable group for elastographic evaluation because they yield false-positive results.

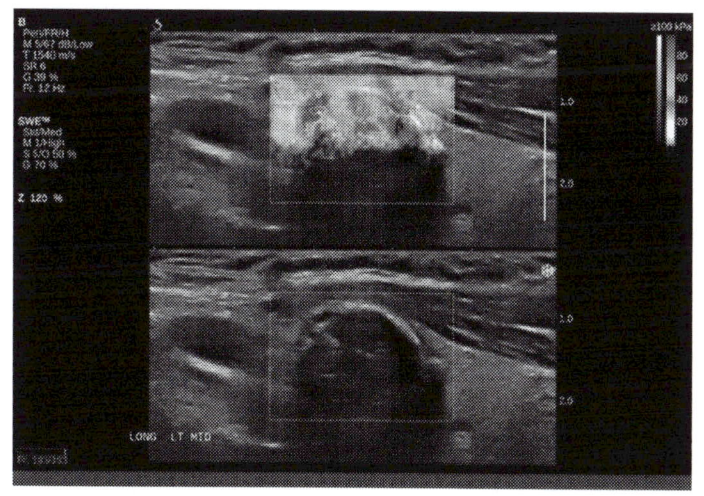


**FIGURE 7.9**

Eggshell calcification resulting in lack of information beyond the calcification on shear wave elastographic image due to the lack of shear waves passing through the calcified area.

Even rim calcifications can give incorrect ultrasound elastography results because of the lack of penetration of ultrasound shear wave beam through the rim calcification [11,49,67].

## 5. Conclusion

Ultrasound elastography is able to help noninvasively differentiate between benign and malignant nodules. Both SE and SWE have shown good sensitivity and specificity for the same. Ultrasound elastography has also shown to have an excellent NPV, which is helpful because the prevalence of thyroid malignancies is very low, i.e., when a nodule is predominantly soft or entirely soft it is almost certainly benign [68]. As mentioned by Russ et al. [47], elastography should not replace grayscale study but it should be used as a complementary tool for assessing nodules for FNA, especially due to its high NPV. Further study of artifacts seen on ultrasound elastography, as it pertains to thyroid imaging, is required.

## References

[1] D.S. Cooper, G.M. Doherty, B.R. Haugen, et al., Revised American Thyroid Association management guidelines for patients with thyroid nodules and differentiated thyroid cancer, Thyroid 19 (11) (2009) 1167–1214.

[2] M. Dietlein, H. Schicha, Lifetime follow-up care is necessary for all patients with treated thyroid nodules, Eur. J. Endocrinol. 148 (3) (2003) 377–379, author reply 81-2.

[3] H. Gharib, E. Papini, R. Paschke, et al., American association of clinical endocrinologists, associazione medici endocrinologi, and european thyroid association medical guidelines for clinical practice for the diagnosis and management of thyroid nodules, J. Endocrinol. Investig. 33 (5 Suppl. 1) (2010) 1–50.

[4] L. Hegedüs, Clinical practice. The thyroid nodule, N. Engl. J. Med. 351 (17) (2004) 1764–1771.

[5] B.R. Haugen, E.K. Alexander, K.C. Bible, et al., 2015 American thyroid association management guidelines for adult patients with thyroid nodules and differentiated thyroid cancer: the American thyroid association guidelines task force on thyroid nodules and differentiated thyroid cancer, Thyroid 26 (1) (2016) 1–133.

[6] D.S. Cooper, G.M. Doherty, B.R. Haugen, et al., Management guidelines for patients with thyroid nodules and differentiated thyroid cancer, Thyroid 16 (2) (2006) 109–142.

[7] F. Pacini, M. Schlumberger, H. Dralle, et al., European consensus for the management of patients with differentiated thyroid carcinoma of the follicular epithelium, Eur. J. Endocrinol. 154 (6) (2006) 787–803.

[8] J. Bamber, D. Cosgrove, C.F. Dietrich, et al., EFSUMB guidelines and recommendations on the clinical use of ultrasound elastography. Part 1: basic principles and technology, Ultraschall der Med. 34 (2) (2013) 169–184.

[9] T. Shiina, K.R. Nightingale, M.L. Palmeri, et al., WFUMB guidelines and recommendations for clinical use of ultrasound elastography: part 1: basic principles and terminology, Ultrasound Med. Biol. 41 (5) (2015) 1126–1147.

[10] A. Itoh, E. Ueno, E. Tohno, et al., Breast disease: clinical application of US elastography for diagnosis, Radiology 239 (2) (2006) 341–350.

[11] T. Rago, F. Santini, M. Scutari, A. Pinchera, P. Vitti, Elastography: new developments in ultrasound for predicting malignancy in thyroid nodules, J. Clin. Endocrinol. Metab. 92 (8) (2007) 2917–2922.

[12] T. Rago, P. Vitti, Role of thyroid ultrasound in the diagnostic evaluation of thyroid nodules, Best Pract. Res. Clin. Endocrinol. Metabol. 22 (6) (2008) 913–928.

[13] C. Asteria, A. Giovanardi, A. Pizzocaro, et al., US-elastography in the differential diagnosis of benign and malignant thyroid nodules, Thyroid 18 (5) (2008) 523–531.

[14] Y. Chong, J.H. Shin, E.S. Ko, B.K. Han, Ultrasonographic elastography of thyroid nodules: is adding strain ratio to colour mapping better? Clin. Radiol. 68 (12) (2013) 1241–1246.

[15] J. Ding, H. Cheng, C. Ning, J. Huang, Y. Zhang, Quantitative measurement for thyroid cancer characterization based on elastography, J. Ultrasound Med. 30 (9) (2011) 1259–1266.

[16] V. Cantisani, P. Lodise, H. Grazhdani, et al., Ultrasound elastography in the evaluation of thyroid pathology. Current status, Eur. J. Radiol. 83 (3) (2014) 420–428.

[17] J.H. Yoon, J. Yoo, E.K. Kim, et al., Real-time elastography in the evaluation of diffuse thyroid disease: a study based on elastography histogram parameters, Ultrasound Med. Biol. 40 (9) (2014) 2012–2019.

[18] N. Ciledag, K. Arda, B.K. Aribas, E. Aktas, S.K. Köse, The utility of ultrasound elastography and MicroPure imaging in the differentiation of benign and malignant thyroid nodules, Am. J. Roentgenol. 198 (3) (2012) W244–W249.

[19] R. Kagoya, H. Monobe, H. Tojima, Utility of elastography for differential diagnosis of benign and malignant thyroid nodules, Otolaryngol. Head Neck Surg. 143 (2) (2010) 230–234.

[20] P. Xing, L. Wu, C. Zhang, S. Li, C. Liu, C. Wu, Differentiation of benign from malignant thyroid lesions: calculation of the strain ratio on thyroid sonoelastography, J. Ultrasound Med. 30 (5) (2011) 663–669.

[21] V. Cantisani, V. D'Andrea, F. Biancari, et al., Prospective evaluation of multiparametric ultrasound and quantitative elastosonography in the differential diagnosis of benign and malignant thyroid nodules: preliminary experience, Eur. J. Radiol. 81 (10) (2012) 2678–2683.

[22] V. Cantisani, S. Ulisse, E. Guaitoli, et al., Q-elastography in the presurgical diagnosis of thyroid nodules with indeterminate cytology, PLoS One 7 (11) (2012) e50725.

[23] V. Cantisani, P. Maceroni, V. D'Andrea, et al., Strain ratio ultrasound elastography increases the accuracy of colour-Doppler ultrasound in the evaluation of Thy-3 nodules. A bi-centre university experience, Eur. Radiol. 26 (2016) 1441–1449.

[24] D.J. Lim, S. Luo, M.H. Kim, S.H. Ko, Y. Kim, Interobserver agreement and intraobserver reproducibility in thyroid ultrasound elastography, Am. J. Roentgenol. 198 (4) (2012) 896–901.

[25] M. Dighe, U. Bae, M.L. Richardson, T.J. Dubinsky, S. Minoshima, Y. Kim, Differential diagnosis of thyroid nodules with US elastography using carotid artery pulsation, Radiology 248 (2) (2008) 662–669.

[26] S.M. Dudea, C. Botar-Jid, Ultrasound elastography in thyroid disease, Med. Ultrason. 17 (1) (2015) 74–96.

[27] S.H. Park, S.J. Kim, E.K. Kim, M.J. Kim, E.J. Son, J.Y. Kwak, Interobserver agreement in assessing the sonographic and elastographic features of malignant thyroid nodules, Am. J. Roentgenol. 193 (5) (2009) W416–W423.

[28] U. Bae, M. Dighe, T. Dubinsky, S. Minoshima, V. Shamdasani, Y.M. Kim, Ultrasound thyroid elastography using carotid artery pulsation – preliminary study, J. Ultrasound Med. 26 (6) (2007) 797–805.

[29] Y. Wang, H.J. Dan, H.Y. Dan, T. Li, B. Hu, Differential diagnosis of small single solid thyroid nodules using real-time ultrasound elastography, J. Int. Med. Res. 38 (2) (2010) 466–472.

[30] F. Tranquart, A. Bleuzen, P. Pierre-Renoult, C. Chabrolle, M. Sam Giao, P. Lecomte, Elastosonography of thyroid lesions, J. Radiol. 89 (1 Pt 1) (2008) 35–39.

[31] C. Shuzhen, Comparison analysis between conventional ultrasonography and ultrasound elastography of thyroid nodules, Eur. J. Radiol. 81 (8) (2012) 1806–1811.

[32] C. Cappelli, I. Pirola, E. Gandossi, et al., Real-time elastography: a useful tool for predicting malignancy in thyroid nodules with nondiagnostic cytologic findings, J. Ultrasound Med. 31 (11) (2012) 1777–1782.

[33] A.C. Calvete, J.M. Rodríguez, J. de Dios Berná-Mestre, A. Ríos, D. Abellán-Rivero, M. Reus, Interobserver agreement for thyroid elastography: value of the quality factor, J. Ultrasound Med. 32 (3) (2013) 495–504.

[34] K.S. Bhatia, D.P. Rasalkar, Y.P. Lee, et al., Cystic change in thyroid nodules: a confounding factor for real-time qualitative thyroid ultrasound elastography, Clin. Radiol. 66 (9) (2011) 799–807.

[35] C.P. Ning, S.Q. Jiang, T. Zhang, L.T. Sun, Y.J. Liu, J.W. Tian, The value of strain ratio in differential diagnosis of thyroid solid nodules, Eur. J. Radiol. 81 (2) (2012) 286–291.

[36] K. Nightingale, R. Nightingale, D. Stutz, G. Trahey, Acoustic radiation force impulse imaging of in vivo vastus medialis muscle under varying isometric load, Ultrason. Imaging 24 (2) (2002) 100–108.

[37] J. Zhan, J.M. Jin, X.H. Diao, Y. Chen, Acoustic radiation force impulse imaging (ARFI) for differentiation of benign and malignant thyroid nodules–A meta-analysis, Eur. J. Radiol. 84 (11) (2015) 2181–2186.

[38] F.J. Dong, M. Li, Y. Jiao, et al., Acoustic radiation force impulse imaging for detecting thyroid nodules: a systematic review and pooled meta-analysis, Med. Ultrason. 17 (2) (2015) 192–199.

[39] B.J. Liu, D.D. Li, H.X. Xu, et al., Quantitative shear wave velocity measurement on acoustic radiation force impulse elastography for differential diagnosis between benign and malignant thyroid nodules: a meta-analysis, Ultrasound Med. Biol. 41 (12) (2015) 3035–3043.

[40] P. Lin, M. Chen, B. Liu, S. Wang, X. Li, Diagnostic performance of shear wave elastography in the identification of malignant thyroid nodules: a meta-analysis, Eur. Radiol. 24 (11) (2014) 2729–2738.

[41] Y.F. Zhang, Y. He, H.X. Xu, et al., Virtual touch tissue imaging of acoustic radiation force impulse: a new technique in the differential diagnosis between benign and malignant thyroid nodules, J. Ultrasound Med. 33 (2014) 585–595.

[42] M. Friedrich-Rust, O. Romenski, G. Meyer, et al., Acoustic Radiation Force Impulse-Imaging for the evaluation of the thyroid gland: a limited patient feasibility study, Ultrasonics 52 (1) (2012) 69–74.

[43] R.S. Goertz, K. Amann, R. Heide, T. Bernatik, M.F. Neurath, D. Strobel, An abdominal and thyroid status with acoustic radiation force impulse elastometry–a feasibility study: acoustic radiation force impulse elastometry of human organs, Eur. J. Radiol. 80 (3) (2011) e226–e230.

[44] J. Bojunga, N. Dauth, C. Berner, et al., Acoustic radiation force impulse imaging for differentiation of thyroid nodules, PLoS One 7 (8) (2012) e42735.
[45] I. Sporea, S. Bota, M. Peck-Radosavljevic, et al., Acoustic radiation force impulse elastography for fibrosis evaluation in patients with chronic hepatitis C: an international multicenter study, Eur. J. Radiol. 81 (12) (2012) 4112–4118.
[46] D. Cosgrove, R. Barr, J. Bojunga, et al., WFUMB guidelines and recommendations on the clinical use of ultrasound elastography: part 4. Thyroid, Ultrasound Med. Biol. 43 (1) (2017) 4–26, https://doi.org/10.1016/j.ultrasmedbio.2016.06.022.
[47] G. Russ, S.J. Bonnema, M.F. Erdogan, C. Durante, R. Ngu, L. Leenhardt, European thyroid association guidelines for ultrasound malignancy risk stratification of thyroid nodules in adults: the EU-TIRADS, Eur. Thyroid J. 6 (5) (2017) 225–237.
[48] D. Cosgrove, F. Piscaglia, J. Bamber, et al., EFSUMB guidelines and recommendations on the clinical use of ultrasound elastography. Part 2: clinical applications, Ultraschall der Med. 34 (3) (2013) 238–253.
[49] A. Lyshchik, T. Higashi, R. Asato, et al., Thyroid gland tumor diagnosis at US elastography, Radiology 237 (1) (2005) 202–211.
[50] J.B. Veyrieres, F. Albarel, J.V. Lombard, et al., A threshold value in Shear Wave elastography to rule out malignant thyroid nodules: a reality? Eur. J. Radiol. 81 (12) (2012) 3965–3972.
[51] T. Fukuhara, E. Matsuda, S. Izawa, K. Fujiwara, H. Kitano, Utility of shear wave elastography for diagnosing chronic autoimmune thyroiditis, J. Thyroid Res. 2015 (2015) 164548.
[52] M. Ghajarzadeh, F. Sodagari, M. Shakiba, Diagnostic accuracy of sonoelastography in detecting malignant thyroid nodules: a systematic review and meta-analysis, Am. J. Roentgenol. 202 (4) (2014) W379–W389.
[53] J. Sun, J. Cai, X. Wang, Real-time ultrasound elastography for differentiation of benign and malignant thyroid nodules: a meta-analysis, J. Ultrasound Med. 33 (3) (2014) 495–502.
[54] I. Kim, E.K. Kim, J.H. Yoon, et al., Diagnostic role of conventional ultrasonography and shearwave elastography in asymptomatic patients with diffuse thyroid disease: initial experience with 57 patients, Yonsei Med. J. 55 (1) (2014) 247–253.
[55] I. Sporea, M. Vlad, S. Bota, et al., Thyroid stiffness assessment by acoustic radiation force impulse elastography (ARFI), Ultraschall der Med. 32 (3) (2011) 281–285.
[56] P. Xie, Y. Xiao, F. Liu, Real-time ultrasound elastography in the diagnosis and differential diagnosis of subacute thyroiditis, J. Clin. Ultrasound 39 (8) (2011) 435–440.
[57] M. Ruchala, E. Szczepanek, J. Sowinski, Sonoelastography in de quervain thyroiditis, J. Clin. Endocrinol. Metab. 96 (2) (2011) 289–290.
[58] Z. Yang, H. Zhang, K. Wang, G. Cui, F. Fu, Assessment of diffuse thyroid disease by strain ratio in ultrasound elastography, Ultrasound Med. Biol. 41 (11) (2015) 2884–2889.
[59] B. Zhang, X. Ma, N. Wu, et al., Shear wave elastography for differentiation of benign and malignant thyroid nodules: a meta-analysis, J. Ultrasound Med. 32 (12) (2013) 2163–2169.
[60] F. Sebag, J. Vaillant-Lombard, J. Berbis, et al., Shear wave elastography: a new ultrasound imaging mode for the differential diagnosis of benign and malignant thyroid nodules, J. Clin. Endocrinol. Metab. 95 (12) (2010) 5281–5288.
[61] B.J. Liu, H.X. Xu, Y.F. Zhang, et al., Acoustic radiation force impulse elastography for differentiation of benign and malignant thyroid nodules with concurrent Hashimoto's thyroiditis, Med. Oncol. 32 (3) (2015) 50.

[62] A.E. Samir, M. Dhyani, A. Anvari, et al., Shear-wave elastography for the preoperative risk stratification of follicular-patterned lesions of the thyroid: diagnostic accuracy and optimal measurement plane, Radiology 277 (2) (2015) 565–573.

[63] W. Tian, S. Hao, B. Gao, et al., Comparison of diagnostic accuracy of real-time elastography and shear wave elastography in differentiation malignant from benign thyroid nodules, Medicine 94 (52) (2015) e2312.

[64] N. Cho, W.K. Moon, J.M. Chang, S.J. Kim, C.Y. Lyou, H.Y. Choi, Aliasing artifact depicted on ultrasound (US)-elastography for breast cystic lesions mimicking solid masses, Acta Radiol. 52 (1) (2011) 3–7.

[65] E. Szczepanek-Parulska, K. Woliński, A. Stangierski, E. Gurgul, M. Ruchała, Biochemical and ultrasonographic parameters influencing thyroid nodules elasticity, Endocrine 47 (2) (2014) 519–527.

[66] C. Vorländer, J. Wolff, S. Saalabian, R.H. Lienenlüke, R.A. Wahl, Real-time ultrasound elastography—a noninvasive diagnostic procedure for evaluating dominant thyroid nodules, Langenbeck's Arch. Surg. 395 (7) (2010) 865–871.

[67] Y. Hong, X. Liu, Z. Li, X. Zhang, M. Chen, Z. Luo, Real-time ultrasound elastography in the differential diagnosis of benign and malignant thyroid nodules, J. Ultrasound Med. 28 (7) (2009) 861–867.

[68] S. Nell, J.W. Kist, T.P. Debray, et al., Qualitative elastography can replace thyroid nodule fine-needle aspiration in patients with soft thyroid nodules. A systematic review and meta-analysis, Eur. J. Radiol. 84 (4) (2015) 652–661.

[69] Y.F. Zhang, H.X. Xu, Y. He, et al., Virtual touch tissue quantification of acoustic radiation force impulse: a new ultrasound elastic imaging in the diagnosis of thyroid nodules, PLoS One 7 (11) (2012) e49094.

[70] J. Gu, L. Du, M. Bai, et al., Preliminary study on the diagnostic value of acoustic radiation force impulse technology for differentiating between benign and malignant thyroid nodules, J. Ultrasound Med. 31 (5) (2012) 763–771.

[71] H.L. Wang, S. Zhang, X.J. Xin, et al., Application of real-time ultrasound elastography in diagnosing benign and malignant thyroid solid nodules, Cancer Biol. Med. 9 (2) (2012) 124–127.

[72] F.J. Zhang, R.L. Han, The value of acoustic radiation force impulse (ARFI) in the differential diagnosis of thyroid nodules, Eur. J. Radiol. 82 (11) (2013) e686–e690.

[73] X.J. Hou, A.X. Sun, X.L. Zhou, et al., The application of virtual touch tissue quantification (VTQ) in diagnosis of thyroid lesions: a preliminary study, Eur. J. Radiol. 82 (5) (2013) 797–801.

[74] H. Wang, D. Brylka, L.N. Sun, Y.Q. Lin, G.Q. Sui, J. Gao, Comparison of strain ratio with elastography score system in differentiating malignant from benign thyroid nodules, Clin. Imaging 37 (1) (2013) 50–55.

[75] J. Zhan, X.H. Diao, Q.L. Chai, Y. Chen, Comparative study of acoustic radiation force impulse imaging with real-time elastography in differential diagnosis of thyroid nodules, Ultrasound Med. Biol. 39 (12) (2013) 2217–2225.

[76] V. Cantisani, V. D'Andrea, E. Mancuso, et al., Prospective evaluation in 123 patients of strain ratio as provided by quantitative elastosonography and multiparametric ultrasound evaluation (ultrasound score) for the characterisation of thyroid nodules, Radiol. Med. 118 (6) (2013) 1011–1021.

[77] V. Cantisani, F. Consorti, A. Guerrisi, et al., Prospective comparative evaluation of quantitative-elastosonography (Q-elastography) and contrast-enhanced ultrasound for the evaluation of thyroid nodules: preliminary experience, Eur. J. Radiol. 82 (11) (2013) 1892–1898.

[78] V. Cantisani, H. Grazhdani, P. Ricci, et al., Q-elastosonography of solid thyroid nodules: assessment of diagnostic efficacy and interobserver variability in a large patient cohort, Eur. Radiol. 24 (1) (2014) 143–150.

[79] H. Grazhdani, V. Cantisani, P. Lodise, et al., Prospective evaluation of acoustic radiation force impulse technology in the differentiation of thyroid nodules: accuracy and interobserver variability assessment, J. Ultrasound 17 (1) (2014) 13–20.

[80] B. Cakir, R. Ersoy, F.N. Cuhaci, et al., Elastosonographic strain index in thyroid nodules with atypia of undetermined significance, J. Endocrinol. Investig. 37 (2) (2014) 127–133.

[81] M. Guazzaroni, A. Spinelli, I. Coco, C. Del Giudice, V. Girardi, G. Simonetti, Value of strain-ratio on thyroid real-time sonoelastography, Radiol. Med. 119 (3) (2014) 149–155.

[82] R. Aydin, M. Elmali, A.V. Polat, M. Danaci, I. Akpolat, Comparison of muscle-to-nodule and parenchyma-to-nodule strain ratios in the differentiation of benign and malignant thyroid nodules: which one should we use? Eur. J. Radiol. 83 (3) (2014) e131–e136.

[83] J.M. Xu, X.H. Xu, H.X. Xu, et al., Conventional US, US elasticity imaging, and acoustic radiation force impulse imaging for prediction of malignancy in thyroid nodules, Radiology 272 (2) (2014) 577–586.

[84] A.C. Calvete, J.D. Mestre, J.M. Gonzalez, E.S. Martinez, B.T. Sala, A.R. Zambudio, Acoustic radiation force impulse imaging for evaluation of the thyroid gland, J. Ultrasound Med. 33 (6) (2014) 1031–1040.

[85] J. Zhuo, Z. Ma, W.J. Fu, S.P. Liu, Differentiation of benign from malignant thyroid nodules with acoustic radiation force impulse technique, Br. J. Radiol. 87 (1035) (2014), 20130263.

[86] J.M. Xu, H.X. Xu, X.H. Xu, et al., Solid hypo-echoic thyroid nodules on ultrasound: the diagnostic value of acoustic radiation force impulse elastography, Ultrasound Med. Biol. 40 (9) (2014) 2020–2030.

[87] F.J. Zhang, R.L. Han, X.M. Zhao, The value of virtual touch tissue image (VTI) and virtual touch tissue quantification (VTQ) in the differential diagnosis of thyroid nodules, Eur. J. Radiol. 83 (11) (2014) 2033–2040.

[88] X. Huang, L.H. Guo, H.X. Xu, et al., Acoustic radiation force impulse induced strain elastography and point shear wave elastography for evaluation of thyroid nodules, Int. J. Clin. Exp. Med. 8 (7) (2015) 10956–10963.

[89] C. Hamidi, C. Göya, S. Hattapoğlu, et al., Acoustic radiation force impulse (ARFI) imaging for the distinction between benign and malignant thyroid nodules, Radiol. Med. 120 (6) (2015) 579–583.

[90] B.J. Liu, F. Lu, H.X. Xu, et al., The diagnosis value of acoustic radiation force impulse (ARFI) elastography for thyroid malignancy without highly suspicious features on conventional ultrasound, Int. J. Clin. Exp. Med. 8 (9) (2015) 15362–15372.

[91] J. Wang, P. Li, L. Sun, Y. Sun, S. Fang, X. Liu, Diagnostic value of strain ratio measurement in differential diagnosis of thyroid nodules coexisted with Hashimoto thyroiditis, Int. J. Clin. Exp. Med. 8 (4) (2015) 6420–6426.

[92] M. Friedrich-Rust, C. Vorlaender, C.F. Dietrich, et al., Evaluation of strain elastography for differentiation of thyroid nodules: results of a prospective DEGUM multicenter study, Ultraschall der Med. 37 (3) (2016) 262–270.

[93] S. Merino, J. Arrazola, A. Cárdenas, et al., Utility and interobserver agreement of ultrasound elastography in the detection of malignant thyroid nodules in clinical care, Am. J. Neuroradiol. 32 (11) (2011) 2142–2148.

[94] F. Ragazzoni, M. Deandrea, A. Mormile, et al., High diagnostic accuracy and interobserver reliability of real-time elastography in the evaluation of thyroid nodules, Ultrasound Med. Biol. 38 (7) (2012) 1154–1162.

[95] J.K. Kim, J.H. Baek, J.H. Lee, et al., Ultrasound elastography for thyroid nodules: a reliable study? Ultrasound Med. Biol. 38 (9) (2012) 1508–1513.

[96] V. Cantisani, P. Lodise, G. Di Rocco, et al., Diagnostic accuracy and interobserver agreement of quasistatic ultrasound elastography in the diagnosis of thyroid nodules, Ultraschall der Med. 36 (2) (2015) 162–167.

[97] K.S. Bhatia, C.S. Tong, C.C. Cho, E.H. Yuen, Y.Y. Lee, A.T. Ahuja, Shear wave elastography of thyroid nodules in routine clinical practice: preliminary observations and utility for detecting malignancy, Eur. Radiol. 22 (11) (2012) 2397–2406.

[98] R.Z. Slapa, A. Piwowonski, W.S. Jakubowski, et al., Shear wave elastography may add a new dimension to ultrasound evaluation of thyroid nodules: case series with comparative evaluation, J. Thyroid Res. 2012 (2012) 657147.

[99] B. Liu, J. Liang, Y. Zheng, et al., Two-dimensional shear wave elastography as promising diagnostic tool for predicting malignant thyroid nodules: a prospective single-centre experience, Eur. Radiol. 25 (3) (2015) 624–634.

[100] A.Y. Park, E.J. Son, K. Han, J.H. Youk, J.A. Kim, C.S. Park, Shear wave elastography of thyroid nodules for the prediction of malignancy in a large scale study, Eur. J. Radiol. 84 (3) (2015) 407–412.

[101] R. Han, F. Li, Y. Wang, Z. Ying, Y. Zhang, Virtual touch tissue quantification (VTQ) in the diagnosis of thyroid nodules with coexistent chronic autoimmune Hashimoto's thyroiditis: a preliminary study, Eur. J. Radiol. 84 (2) (2015) 327–331.

[102] B.X. Liu, X.Y. Xie, J.Y. Liang, et al., Shear wave elastography versus real-time elastography on evaluation thyroid nodules: a preliminary study, Eur. J. Radiol. 83 (7) (2014) 1135–1143.

[103] R. Brezak, D. Hippe, J. Thiel, M.K. Dighe, Variability in stiffness assessment in a thyroid nodule using shear wave imaging, Ultrasound Q. 31 (4) (2015) 243–249.

CHAPTER

# Elastography applications in pregnancy

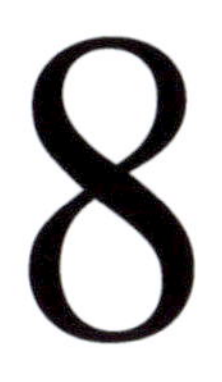

**Helen Feltovich**[1,2]

[1]*Maternal-Fetal Medicine, Intermountain Healthcare, Provo, UT, United States;* [2]*Quantitative Ultrasound Laboratory, Department of Medical Physics, University of Wisconsin, Madison, WI, United States*

## 1. Introduction

Elasticity imaging in pregnancy is not part of routine clinical practice, primarily because pregnancy tissues are not as straightforward as tissues such as breast or the liver, for which elastography applications are well established [1,2], but the field is rapidly developing and possibilities are exciting. This chapter reviews potential applications of elasticity techniques to two important pregnancy tissues: the cervix (for predicting preterm birth or success of labor induction) and the placenta (for evaluating function).

## 2. The cervix

The cervix, a cylindrical structure that protrudes into the vagina distally and is contiguous with the uterus proximally, is the gateway between intrauterine and extrauterine life. In normal pregnancy, it remains long and closed while the fetus develops, then shortens, softens, and completely dilates to allow delivery of the neonate, after which it resumes its original configuration. The cervical extracellular matrix (ECM) contains proteins (mostly collagen) and proteoglycans (e.g., decorin and hyaluronic acid) that form a supportive scaffolding and govern tissue properties such as strength and elasticity [3]. Changes within the ECM, facilitated by precisely timed molecular events, appear responsible for the remarkable cervical remodeling process [3].

While cervical modeling throughout pregnancy is necessary for normal delivery, premature changes within the cervical ECM may lead to premature dilation and shortening, and subsequent spontaneous preterm birth (sPTB) [3,4]. The precise mechanisms that initiate the changes in pregnancy tissues that lead to delivery are unclear, but it is clear that all pathways ultimately culminate in the cervix [3–5]. As such, transvaginal ultrasound measurement of cervical length is currently the best quantitative indicator of risk for sPTB. Unfortunately, cervical shortening has relatively poor predictive value; most women with a short cervix in midpregnancy

Tissue Elasticity Imaging. https://doi.org/10.1016/B978-0-12-809662-8.00008-5

will deliver at or near term, even most of those with a very short cervix [6]. Cervical softening is probably more informative; in normal pregnancy, softening initiates well before the cervix shortens [5]. Conversely, a cervix that has failed to soften by term is associated with postdates pregnancy, with its attendant risks of cesarean delivery and stillbirth [5]. However, assessment of softening is entirely subjective, based upon an individual provider's digital examination. Because accurate and quantitative assessment of cervical softening is promising for evaluating timing of delivery, elasticity imaging methods have gained recent attention.

### 2.1 Strain elastography

Strain elastography (SE) and shear wave elasticity imaging (SWEI) techniques have been applied to the cervix for assessment of stiffness/softness to investigate both risk of preterm birth and success of labor induction. SE is appealing because it is easy to perform with a standard, commercially available endovaginal transducer. This technique is suitable for assessing relative differences between target and reference tissue, such as a tumor in the breast [7]. However, in the small and heterogeneous cervix, with its complicated boundary conditions and its ubiquitous softening during pregnancy, SE is limited by the lack of surrounding tissue for reference and difficulty in standardizing the applied force [8].

Thomas et al. demonstrated feasibility of SE for measuring softness of the pregnant cervix more than a decade ago, in 2006 [9]. They applied freehand compression to the cervix with a transvaginal transducer. The elastographic images that appeared on the ultrasound screen were then analyzed by color coding the strain (red, soft; green, medium; and blue, hard). An "elasticity tissue quotient," calculated as a percentage of red to green, did not vary with gestational age. Yamaguchi et al., using the same technique and similar analysis, claimed differences in deformability with gestational age, although results are difficult to interpret because the study population and data were not shown [10]. Khalil et al. examined a small number of women ($n = 12$) at 15–33 weeks of gestation [11]. They reported that the cervix is generally softer proximally as compared with distally, although there were no gestational age-based trends. Importantly, they noted larger deformations closer to the transducer and concluded that it was difficult to determine whether deformations were due to actual compression, or simply to movement of the cervix.

Swiatkowska-Freund and Preis, in an attempt to make elastography less subjective through standardizing the force applied by the transducer, introduced a semi-quantitative evaluation metric called the "elasticity index (EI)" [12]. They held the transducer just in contact with the cervix, to minimize the applied force, and deformation was created as the patients' arterial pulsations or breathing caused the tissue to move against the transducer. The EI was based on points assigned for different colors; purple (hardest) was assigned 0 points while red (softest) was given 4, and colors in between were given intermediate points (e.g., blue = 1, green = 2, and yellow = 3 points). In a pilot study, 29 patients at term gestation were evaluated prior to induction of labor to determine whether elastography was predictive of success of vaginal delivery. EI was calculated at the internal os, external os, and the endocervical canal. They found a statistically significant difference in EI at the internal os (but not the canal or external os) between patients with successful, as opposed to failed, induction (as defined by vaginal delivery). However, they noted that technique standardization would be necessary to make meaningful comparisons between women because intersubject variability was high.

Molina et al. attempted to standardize the applied force by deforming the anterior lip of the cervix by a consistent amount (1 cm) [13]. In a cross-sectional study, they evaluated women in all trimesters of pregnancy (n = 112 at 12–40 weeks) and found that tissue closest to the probe was softest, regardless of gestational age. They concluded that elastography measurements "may be merely a reflection of the force applied by the transducer to different parts of the cervix." Fruscalzo and Schmitz agreed that "the arbitrary guideline of compressing the tissue by 1 cm is not associated in any predictable way with the actual force applied." [14].

Hernandez-Andrade et al. conducted a large cross-sectional study of women in all trimesters of pregnancy (n = 262 at 8–40 weeks) [15]. Strain was evaluated at the canal, the external os, and the internal os. They semiquantitated the approach by maintaining a consistent value on the "pressure bar" displayed on the ultrasound system's monitor. They reported that the cervix was significantly softer at external/superior locations, as compared with internal/inferior, and that strain varied by previous preterm delivery and cervical length. Using an approach similar to Hernandez-Andrade et al., Fuchs et al. reported a weak correlation between cervical strain and cervical length in the third trimester (28–39 weeks) [16]. They concluded that SE may be useful for tracking cervical softening, but Mazza et al. [17] and Maurer et al. [18] demonstrated via theoretical argument and phantom studies that no information about actual transducer force is provided by the "pressure bar" because the bar only indicates quality of the strain image itself.

Some investigators have combined SE with other imaging biomarkers in an attempt to improve its predictive value. For example, Pereira et al. evaluated the potential of elastographic score at the internal os, plus cervical length and angle of progression to predict success of labor induction (n = 66 at 35–42 weeks) [19]. They found that cervical length was a significant predictor of vaginal delivery, with no additional contribution from angle of progression or elastographic score. However, it is difficult to interpret these results because the deformation technique is not discussed and results may be simply a reflection of high variance in applied force. Two small studies using the technique proposed by Swiatkowska-Freund and Preis found significant differences in EI in women with failed, as opposed to successful, induction [20,21].

Wozniak et al. used this same basic approach for prediction of preterm birth [22]. They evaluated the internal os of low-risk women in the second trimester (n = 333). Instead of scoring the elastograms like other groups, they assigned each image to a color group: red (soft), yellow (medium soft), blue (medium hard), and purple (hard). They reported significantly higher rates of preterm births in the red and yellow groups with a sensitivity, specificity, PPV, and NPV of 85.7%, 97.6%, 98.3%, and 81.1%, respectively. Sabiani et al. found that preterm birth was more common in women (n = 72) whose first trimester cervices demonstrated greater local deformation than those with lesser deformation, although interpretation of these results is compromised by lack of discussion of their compression technique [23].

Hernandez-Andrade et al. also evaluated women in the midtrimester (n = 189 at 16–24 weeks) as well as in the first, second, and early third trimester (n = 545 at 11–28 weeks), 21 of whom delivered preterm [24,25]. They found strain measurements at the internal os to be significantly associated with preterm birth. There was no correlation between strain values and gestational age.

Hee et al. addressed the lack of a reference tissue with a reference cap that fits on the end of a transducer [26]. Young's modulus was calculated based on the cap's known elastic properties compared with strain in the cervix in response to cyclic pressure applied to the cervix with the transducer. In a study of women presenting for induction of labor (n = 49), they found a weak but statistically significant correlation between Young's modulus and cervical dilation (AUC, sensitivity, and specificity were 0.71%, 74%, and 69%, respectively). The technique performed better than Bishop score or cervical length, although results may be biased because calculation of the cervical elastic modulus assumes a homogenous stress distribution, and simulations of Maurer et al. show that the reference cap introduces a nonlinear, inhomogeneous stress distribution [18].

Parra-Saavedra et al. approached force standardization by applying pressure to a point of maximum deformation [27]. In a cross-sectional study of women in all trimesters of pregnancy (n = 1115 at 5–36 weeks), they calculated a "cervical consistency index" (CCI) by comparing the anteroposterior cervical diameter before and after maximal deformation. A significant correlation was found between CCI and gestational age ($r^2 = 0.66$, $P < .001$) but not between CCI and cervical length. They also found a significant difference in CCI between women who delivered at term versus those who delivered at <32 weeks, <34 weeks, and <37 weeks (AUC = 0.947, 0.943, and 0.907, respectively), although sensitivities were low (67%, 64%, and 45%). In a similar approach, Fruscalzo et al. applied pressure perpendicular to the anteroposterior axis until maximal deformation of the anterior cervical lip was attained, then recorded strain values during the subsequent relaxation phase [28]. Feasibility of both raw data acquisition and strain calculation was demonstrated in a study of women in the first and second trimesters of pregnancy (n = 10), and they subsequently demonstrated significant correlations of strain with both gestational age ($\rho = 0.82$, 95% CI 0.73–0.88) and cervical length ($\rho = -0.59$, 95% CI −0.72 to −0.42) in women at all trimesters of pregnancy (n = 74 at 12–42 weeks) [29]. When Fruscalzo et al. [30] applied this technique to prediction of labor induction success in women at term (n = 77), failure of induction was predicted in four of the women, and when it was used by Kobbing et al. for prediction of preterm birth in women at all trimesters (n = 182 at 11–36 weeks), significantly higher tissue strain ratios were observed in women who delivered preterm (n = 17), with sensitivity and specificity in the second trimester of 57% and 50%, respectively [31].

In summary, SE shows only moderate success for evaluating the cervix. Since SE is intended for comparison of *relative* strain, and there exists no adequate standard for comparison in either the cervix itself or a reference phantom, it is not surprising that cervical SE has not found a place in clinical practice.

## 2.2 Shear wave elasticity imaging

SWEI, or shear wave elastography (SWE), is less dependent on the transducer force (i.e., the operator) because force is generated acoustically. Therefore this approach has potential to overcome some of the challenges of SE. Typically, stiffness/softness of the tissue is measured with shear wave speed (SWS), or with conversion of the SWS into the elastic modulus (Young's modulus). Fig. 8.1 shows an example of SWS estimates (superimposed over the B-mode image) in a first trimester cervix [32].

Producing adequate shear waves in the cervix, however, is not trivial because thin, heterogeneous tissues like cervix can disrupt shear wave propagation; cervical tissue is highly attenuating because of its microstructural complexity, and current commercially available software packages cannot access quality of underlying raw data. In other words, many assumptions needed to quantify elastic moduli based on shear wave imaging are easily violated in complex tissues such as cervix. Further, a typical endovaginal transducer is not ideal because its large acoustic elements limit how close to the transducer reliable shear waves can be produced, and there is currently no commercially available alternative [33].

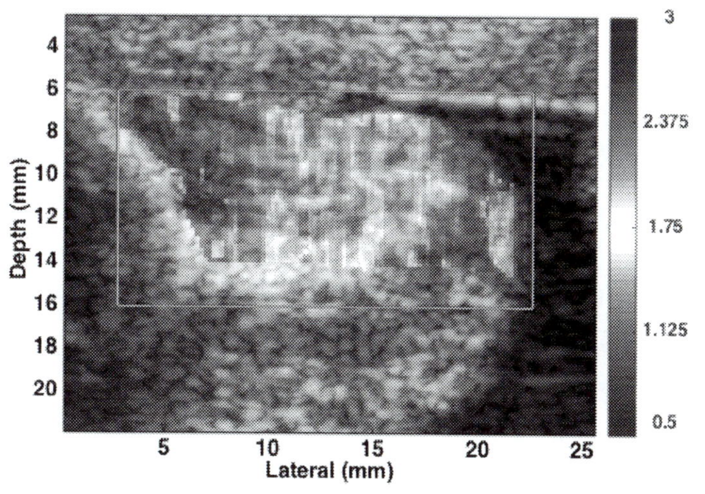


**FIGURE 8.1**

An example of shear wave speed (SWS) evaluation in a first trimester cervix. The gray box indicates region of interest. The bar on the right indicates SWS in m/s. SWS is higher near the proximal end (left) and slower near the distal end (right). For statistical analysis, SWS is reported in m/s (for example, SWS near the proximal end is around 3 m/s and on the distal end is less than 1 m/s).

Muller et al. reported a “slight” decrease in SWS in a cross-sectional study of women hospitalized for preterm labor (n = 81) as compared with controls (n = 27) at 24–35 weeks [34]. While there was an expected statistically significant difference in cervical stiffness between the first and third trimesters, they found no difference between the first and second trimesters, which is inconsistent with both clinical observation and previous studies. The authors suggested that this unexpected finding could be explained by the high variance noted in their first trimester estimates. It is possible that their use of a standard endovaginal transducer produced the high variability because shear wave generation can be unreliable close to a transducer with large acoustic elements; Fig. 8.2 shows how close to the transducer (tissue boundary) is the region of interest (ROI) (white circle).

Another possible reason Muller et al. results did not agree with the findings of other investigators is that they assessed the external (as opposed to the internal) os. Hernandez-Andrade et al. evaluated SWS in multiple areas of the cervix (including the internal and external ostia) at all trimesters [35]. They found that SWS decreased with increasing gestational age only at the internal os. Similarly, in a cross-sectional study (n = 280, low risk women at 10–40 weeks), Ono et al. found both that cervical stiffness decreases significantly with increasing gestational age, and that the internal os area seems most relevant for assessment of cervical softening [36].

Our findings corroborate those of Hernandez-Andrade et al. and Ono et al. The primary difference between our group and others is the use of a linear, instead of a typical transvaginal (curvilinear), transducer. An advantage of a curvilinear transducer is its commercial availability. However, as noted above, its larger elements, as compared with those of a linear transducer, may compromise adequate production

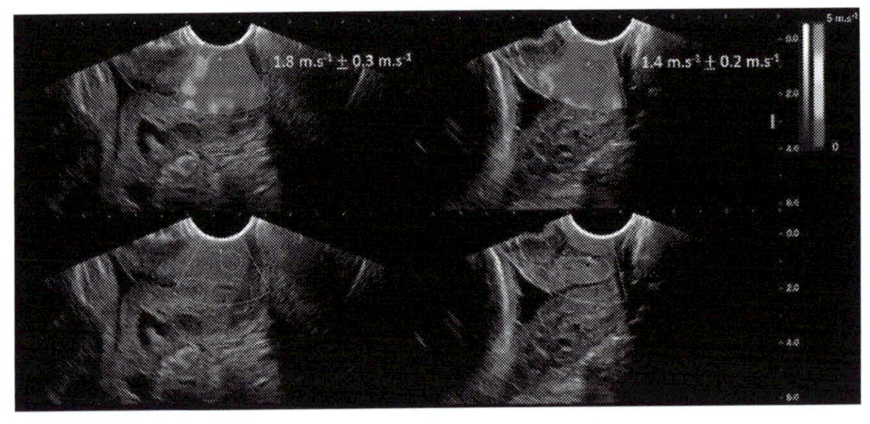


**FIGURE 8.2**

An example of how close to the transducer lies the region of interest.

*Reprinted with permission from M. Muller, D. Ait-Belkacem, M. Hessabi, J. Gennisson, et al., Assessment of the cervix in pregnant women using shear wave elastography: a feasibility study, Ultrasound Med. Biol. 41 (11) (2015) 2789–2797.*

of shear waves close to the aperture. This is not a problem for relatively large, homogeneous tissues such as liver, but can be for a small structure such as the cervix. To explore reliability of SWEI for the cervix, we studied sources of measurement variance as well as potential sources of biological variability in cervical tissue. Specifically, we scanned the cervix from hysterectomy specimens of nonpregnant women (n = 22) over its entire length and circumference [37,38]. As expected, we found that shear wave fronts were relatively planar despite transducer proximity to the ROI. Although our study was ex vivo, and those of Hernandez-Andrade et al. and Ono et al., findings were similar. For example, SWS changed progressively from the external to the internal os and estimates vary significantly in different regions of the cervix, with the proximal cervix (the area of the internal os) demonstrating the least variability between specimens. In addition, we demonstrated that the ripened cervix (with prostaglandins used clinically to soften the cervix) is significantly softer than the unripened [37,38].

In a cross-sectional study in pregnant women, we confirmed that palpable cervical softening in the third, as compared with the first, trimester cervix is quantifiable; average SWS estimates were 4.42 ± 0.32 m/s (n = 12) for the first trimester and 2.13 ± 0.66 m/s (n = 18) for the third trimester ($P < .0001$) [39]. The area under the receiver–operator curve was 0.95 (95% CI 0.82–0.99), with a sensitivity and specificity of 83% for distinguishing the early from the late pregnant cervix. In a subsequent longitudinal study of pregnant rhesus macaques, we established that SWS decreases at a rate of 6% per week (95% CI 5%–7%) throughout gestation [40]. This study also established that factors such as scanning method and specific SWEI approach significantly affect results, and therefore technique is critical.

Two studies have recently evaluated SWEI and cervical length as predictors of sPTB. The first of these, Hernandez-Andrade et al., was a prospective cohort study of SWS and cervical length at 18–24 weeks (n = 628) [41]. They found that the risk of sPTB was increased 4.5-fold (RR 4.5 (95% CI 2.1–9.8, $P = .0002$)) in women with a soft cervix as compared with those with a nonsoft cervix, and the risk was increased 18-fold (RR 18.0 (95% CI 6.6–43.9, $P < .0001$)) in women with both a soft and a short cervix as compared with women with a normal cervical length. The other, Agarwal et al., was a cross-sectional study of women (n = 34) at 28–37 weeks with symptoms of preterm labor [42]. Women were evaluated with SWEI, cervical length ultrasound, and Bishop score. Per ROC curves, a cutoff value of 2.83 m/s distinguished women who delivered preterm (n = 14) from those who delivered at term (n = 20), and they concluded that SWEI is a promising technique for preterm birth prediction.

SWEI has also been applied to the study of labor induction at term gestation; our feasibility study of women presenting for induction of labor (n = 20) showed significant differences in SWS before and after cervical ripening (mean SWS estimates of 2.53 + 0.75 and 1.54 + 0.31 m/s, respectively $P < .001$) [43]. SWS estimates were significantly correlated with digitally assessed cervical softness (as a component of the Bishop score) and marginally correlated with Bishop score.

Together, these studies suggest that, with careful technique and attention to measurement location, SWEI appears to be a useful approach to assessing cervical tissue stiffness/softness and a promising clinical tool.

## 3. The placenta

The role of the placenta is to facilitate nutrient and gas exchange between the fetus and mother; it functions as the fetus' gastrointestinal tract, lungs, and kidneys in that it supplies nutrients to, and transfers metabolic waste away from, the fetus [44]. The dysfunctional placenta, which leads to progressive fetal growth restriction (FGR) and oligohydramnios as gas and nutrient exchange becomes compromised, manifests in progressively abnormal vascular resistance detectable with Doppler ultrasound. For this reason, a dysfunctional placenta is often referred to as "stiff." It is a reasonable hypothesis that an imaging technique which addresses elasticity could provide information about placental function. Potential applications include evaluation of fetal well-being and characterization of disease state in the case of maternal disorders that are associated with FGR, such as preeclampsia, diabetes, and autoimmune diseases.

### 3.1 Strain elastography

Evaluation of the placenta with SE has shown limited success. Durhan et al. compared placentas from neonates who were normally grown (n = 30) and growth restricted (n = 25) at 36–42 weeks [45]. Their goal was to identify differences in elasticity between the normal and abnormal placenta, and correlate elasticity with histopathology. Similar to the elasticity index (EI) of Swiatkowska-Freund and Preis in the cervix, Durhan et al.'s EI is a numerical presentation of strain on a color-coded strain image. They found that placentas from neonates with FGR were stiffer than those from normally grown neonates, and that histopathological abnormalities were associated with EI values, although differences between the groups did not reach statistical significance.

Cimsit et al. used SE to evaluate pregnant women at 20–23 weeks (n = 101 normal, n = 28 with early onset preeclampsia and n = 15 with history of preeclampsia) [46]. They chose a 5 × 5 mm ROI in an area of the elastogram that was mostly green (to represent intermediate/average strain in an image where red represented high, and blue, low strain), and calculated the ratio of values in this area to values in fat. They reported a significantly different mean strain ratio in the normal group (0.9 (95% CI 0.82–0.97)) as compared with the preeclampsia group (1.56 (95% CI 1.12–2.16)) ($P < .001$). An important limitation of their study, however, was that a posterior placenta was an exclusion criterion due to inability to obtain adequate data. This resulted in ineligibility of approximately one-third of women screened (63 of 219), which is a substantial issue for clinical translation.

## 3.2 Shear wave methods

Shear wave methods to evaluate the placenta generally involve placing a 5 × 5 mm or 5 × 10 mm ROI on various regions in the placenta, usually the center and the edge. The ultrasound system displays an elastogram together with the corresponding B-mode images so that the operator can choose a relatively homogeneous area. Within this area, they place the ROI at a location which shows the greatest stiffness as determined by the colors on the elastogram. An example of shear wave imaging in an ex vivo, perfused placenta is shown in Fig. 8.3.

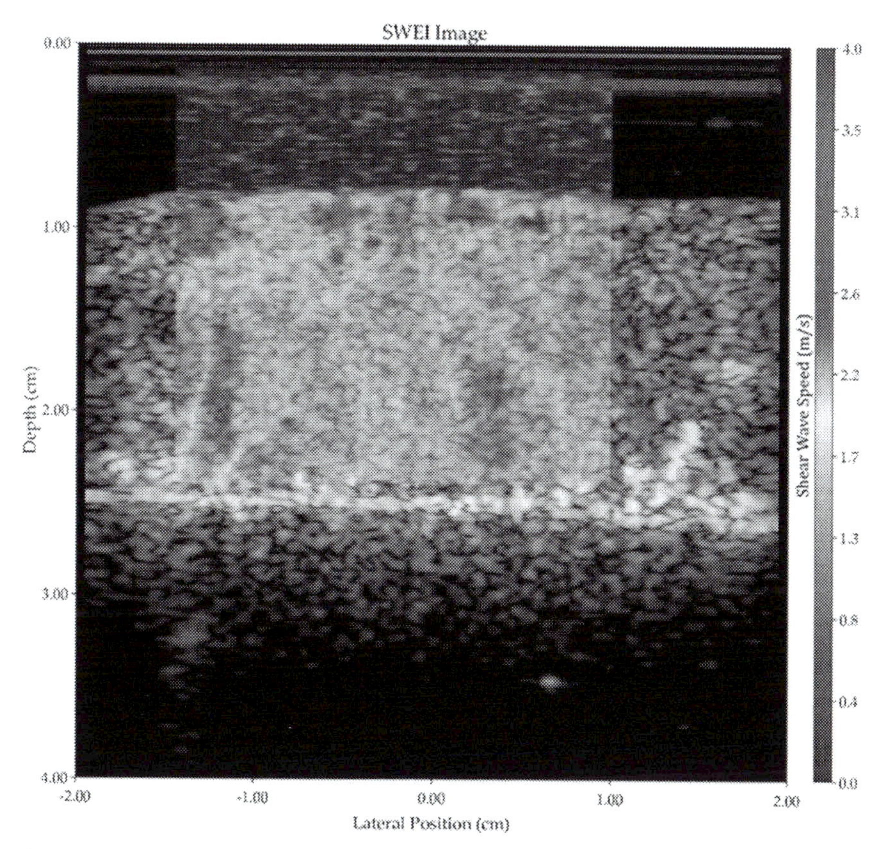


**FIGURE 8.3**

Color overlay representing shear wave speed (SWS) (0–4 m/s) on a B-mode image of an ex vivo, perfused placenta. The placenta is oriented with the chorionic plate on the top of the image. In this placenta, the typical SWS color image shows some variations, and most values are around 2 m/s.

*Reprinted with permission from S. McAleavey, K. Parker, J. Ormachea, R. Wood, et al., Shear wave elastography in the living, perfused, post-delivery placenta, Ultrasound Med. Biol. 42 (6) (2016) 1282–1288.*

The logistical issues that plague SE compromise clinical translation of shear wave methods as well because it is currently not possible to evaluate a placenta that is more than 8 cm away from the transducer (e.g., due to posterior location or maternal obesity). In addition, as explained by Calle et al. [48], even in the case of an anterior placenta, the shear wave must cross several different boundaries (e.g., skin, fat, muscle, fluid) before reaching the placenta. They also note that, when multiple interfaces are present, compression waves may be induced by the shear waves, and vice versa, which can change shear wave directionality and thus complicate interpretation of results. In addition, the presence of multiple boundaries means that the propagation distance of the wave is longer, and therefore more vulnerable to attenuation by diffraction and viscosity [48]. Another issue that complicates interpretation of results is that maternal and fetal movements are unpredictable; Ohmaru et al. [49] found that SWS values are affected by maternal deep, as compared with shallow, breathing ($P < .01$), or an active, as compared with a quiet, fetus ($P < .01$).

Studies on postdelivery placentas have utilized SWE, which depends upon acoustic radiation force to generate shear waves, or transient elastography (TE), which relies on a vibrating plate. Results are reported as shear wave speed or velocity (SWS, m/s) or average elastic modulus (kPa). One issue with the latter is that the conversion from SWS to elastic modulus is only valid under the assumption of homogenous tissue. Investigators who use this method describe careful choice of an ROI that appears homogeneous on B-mode imaging and the color elastogram, in fact the entire placenta is heterogeneous, even upon gross visual inspection.

That said, shear wave methods appear encouraging for distinguishing the abnormal from normal ex vivo placenta. For example, using an SWE method, Sugitani et al. studied placentas from women who delivered at 26–41 weeks (n = 115) and divided into three groups: normal, FGR, and pregnancy-induced hypertension [50]. They found that shear wave velocities were significantly higher in the FGR group (1.94 + 0.74 vs. 1.31 + 0.35 m/s, $P < .05$).

In a study comparing SWE to TE methods with a vibrating plate applied directly to the placenta, Simon et al. found no significant difference in SWS between the approaches (mean SWS 1.80 + 0.28 m/s at 50 Hz for TE and 1.82 + 0.13 m/s for SWE, $P = .912$) in their study of 10 placentas [51]. Another TE approach, shear wave absolute vibro-elastography (SWAVE), is described by Abeysekera et al. [52] This method analyzes the dispersion of shear wave velocity as a function of frequency. Advantages are that it also allows estimates of viscosity and, as compared with the TE method of Simon et al., generates a higher signal and therefore has better signal-to-noise ratio. In postdelivery placentas (n = 61), they reported good penetration with this approach at all frequencies, with SWS values similar to previous studies (1.23 + 0.44, 1.67 + 0.76, 1.74 + 0.72, 1.80 + 0.78, and 2.25 + 0.80 m/s at 60, 80, 90, 100, and 120 Hz, respectively). A disadvantage to clinical use is that the measurement must be repeated several times at each frequency. Also, applying a vibrating plate to a placenta in vivo is impractical. Calle et al. [48] used a TE approach involving an impulse method. This appears more suitable to the clinic because energy is emitted in a frequency band (20–80 Hz), which obviates the need for repeated experiments, and shear waves are generated by an external vibration (two vibrating rods on either side of the transducer). They evaluated (ex vivo) placentas (n = 20) with the impulse TE method and SWE, comparing results with their previous TE experiments (Simon et al.) in which the vibrator was applied directly to the placenta. Standard deviations were higher with the impulse method, which they attributed to measurement bias in part because the system is handheld instead of fixed, which means that the prestress cannot be controlled and therefore reproducibility is lower. They note that if these limitations cannot be overcome in vivo, a different method that does not need such precise shear wave directivity will be necessary.

McAleavey et al. investigated SWS variation in a "living placenta" model in which placentas were perfused immediately following delivery to replicate the in vivo situation [47]. In normal placentas (n = 11), they noted inhomogeneity and nonuniformity in their SWS color elastograms, as well as local variations, which they attributed to presence of vessels. They reported SWS of 1.5–2.5 m/s, which agreed with previous studies. They also noted localized changes in SWS in response to introduction of vasodilators and vasoconstrictors to the perfused placentas.

In pregnant women, shear wave elasticity techniques have been used to assess maternal medical disorders that directly affect the placenta, such as preeclampsia and gestational diabetes. Li et al. investigated elastic modulus and variance in placentas of women near term in normal pregnancy (n = 30) [53]. They found no difference between average elastic modulus near the edge of the placenta (7.60 + 1.71 kPa) as compared with the center of the placenta (7.84 + 1.68 Pa). Unfortunately, they also found that elasticity values did not correlate with uterine or umbilical artery resistance (measured with Doppler ultrasound), and they noted that variance was high between placentas. Cimsit et al. found that placental elasticity was significantly higher at 20–23 weeks (the time of routine fetal anomaly scan) in women with preeclampsia (n = 28) as compared with women who ultimately delivered normally grown neonates (n = 101) [54]. Mean elasticity values for the normal placenta were 2.53 kPa. An important note is that approximately one-third (63 of 204) of the women screened for the study were excluded because of posterior placenta. Another study of preeclampsia prediction via placental elasticity imaging was undertaken by Kilic et al. [55]. They studied women in the second or third trimester (n = 50), 23 of whom were preeclamptic and the other 27 normotensive. The median elasticity modulus in the central part of the placenta was 21 kPa (range, 3–71 kPa) in preeclamptic women, as compared with 4 kPa (range 1.5–14 kPa) in controls ($P < .001$). Yuksel et al. used the same approach in pregnant women with gestational diabetes (GDM, n = 33) compared with women without diabetes (n = 43) [56]. The mean values in the central part of the placenta were 10.63 + 5.97 kPa for women with GDM and 5.47 + 1.74 kPa for controls ($P < .001$).

Ohmaru et al. [49] and Alan et al. [56] studied various groups of pregnant women with SWEI. Ohmaru et al. found statistically significant differences in SWS between women with normal pregnancy (n = 143), FGR (n = 21), and hypertension (n = 15), with SWS 0.98 + 0.21, 1.28 + 0.39, and 1.60 + 0.45 m/s, respectively [48]. Similarly, Alan et al. reported higher SWS in women with preeclampsia (n = 42) as compared with those with normal pregnancy (n = 44). Mean SWS values were 1.39 (1.32–1.53) versus 1.07 (1.00–1.14), ($P < .001$) [57].

In summary, elasticity imaging techniques appear to detect differences in placental elasticity between women with and without medical disorders that can affect the placenta (preeclampsia, diabetes). At this moment, it is difficult to imagine how any of these methods could become broadly useful given their constraints with respect to depth of placenta, as well as maternal/fetal movement, but several groups are invested in this area, and the future will be interesting.

## 4. Conclusions

SE techniques can theoretically evaluate elasticity of pregnancy tissues, although standardizing the transducer force is a large obstacle to clinical use. SWEI techniques are appealing because they are less user dependent, and several groups are working on promising techniques. Meaningful clinical tools to evaluate elasticity of pregnancy tissues, based on comprehensive understanding of sources of biological and system variability, as well as objective measure of data quality and robust validation of underlying assumptions, are likely to be introduced into the clinic in the not-too-distant future.

## References

[1] R.G. Barr, K. Nakashima, D. Amy, D. Cosgrove, et al., WFUMB guidelines and recommendations for clinical use of ultrasound elastography: Part 2: Breast, Ultrasound Med. Biol. 41 (5) (2015) 1148–1160.

[2] G. Ferraioli, C. Filice, L. Castera, B. Choi, et al., WFUMB guidelines and recommendations for clinical use of ultrasound elastography: Part 3: Liver, Ultrasound Med. Biol. 41 (5) (2015) 1161–1179.

[3] J. Vink, H. Feltovich, Cervical etiology of spontaneous preterm birth, Semin. Fetal Neonat. Med. 21 (2) (2016) 106–112.

[4] R. Menon, E.A. Bonney, J. Condon, S. Mesiano, et al., Novel concepts on pregnancy clocks and alarms; redundancy and synergy in human parturition, Human Reprod. Update 22 (5) (2016) 535–560.

[5] H. Feltovich, Cervical assessment: from ancient to precision medicine, Obstet. Gynecol. 130 (1) (2017) 51–63.

[6] S. Hassan, R. Romero, S. Berry, K. Dang, et al., Patients with an ultrasonographic cervical <= 15mm have nearly a 50% risk of early spontaneous preterm delivery, Am. J. Obstet. Gynecol. 182 (2000) 1458–1467.
[7] J. Ophir, I. Cespedes, H. Ponnekanti, Y. Yazdi, X. Li, Elastography: a quantitative method for imaging the elasticity of biological tissues, Ultrason. Imaging 13 (2) (1991) 111–134.
[8] H. Feltovich, T.J. Hall, Quantitative imaging of the cervix: setting the bar, Ultrasound Obstet. Gynecol. 41 (2) (2013) 121–128.
[9] A. Thomas, Imaging of the cervix using sonoelastography, Ultrasound Obstet. Gynecol. 28 (3) (2006) 356–357.
[10] S. Yamaguchi, Y. Kamei, S. Kozuma, Y. Taketani, Tissue elastography imaging of the uterine cervix during pregnancy, J. Med. Ultrason. 34 (4) (2007) 209–210.
[11] M.R. Khalil, P. Thorsen, N. Uldbjerg, Cervical ultrasound elastography may hold potential to predict risk of preterm birth, Danish Med. J. 60 (1) (2013) A4570.
[12] M. Swiatkowska-Freund, K. Preis, Elastography of the uterine cervix: implications for success of induction of labor, Ultrasound Obstet. Gynecol. 38 (1) (2011) 52–56.
[13] F.S. Molina, L.F. Gomez, J. Florido, M.C. Padilla, K.H. Nicolaides, Quantification of cervical elastography: a reproducibility study, Ultrasound Obstet. Gynecol. 39 (6) (2012) 685–698.
[14] A. Fruscalzo, R. Schmitz, Quantitative cervical elastography in pregnancy, Ultrasound Obstet. Gynecol. 40 (5) (2012) 612.
[15] E. Hernandez-Andrade, S.S. Hassan, H. Ahn, S.J. Korzeniewski, et al., Evaluation of cervical stiffness during pregnancy using semiquantitative ultrasound elastograph, Ultrasound Obstet. Gynecol. 41 (2) (2013) 152–161.
[16] T. Fuchs, R. Woyton, M. Pomorski, A. Wiatrowski, et al., Sonoelastography of the uterine cervix as a new diagnostic tool of cervical assessment in pregnant women - preliminary report, Ginekol. Pol. 84 (1) (2013) 12–16.
[17] E. Mazza, M. Parra-Saavedra, M. Bajka, E. Gratacos, et al., In vivo assessment of the biomechanical properties of the uterine cervix in pregnancy, Prenat. Diagn. 34 (1) (2014) 33–41.
[18] M. Maurer, S. Badir, M. Pensalfini, M. Bajka, et al., Challenging the in-vivo assessment of biomechanical properties of the uterine cervix: a critical analysis of ultrasound quasi-static procedures, J. Biomech. 48 (9) (2015) 1541–1548.
[19] S. Pereira, A.P. Frick, L. Poon, A. Zamprakou, et al., Successful induction of labor: prediction by preinduction cervical length, angle of progression and cervical elastography, Ultrasound Obstet. Gynecol. 44 (4) (2014) 468–475.
[20] H. Hwang, I. Sohn, H. Kwon, Imaging analysis of cervical elastography for prediction of successful induction of labor at term, J. Ultrasound Med. 32 (6) (2013) 937–946.
[21] A. Muscatello, M.A.V. Di Nicola, N. MAstrocola, et al., Sonoelastography as method for preliminary evaluation of the uterine cervix to predict success of induction of labor, Fetal Diagn. Ther. 35 (1) (2014) 57–61.
[22] S. Wozniak, P. Czuczwar, P. Szkodziak, P. Milart, et al., Elastography in predicting preterm delivery in asymptomatic, low-risk women: a prospective observational study, BMC Pregnancy Childbirth 14 (1) (2014) 238.
[23] L. Sabiani, J. Haumonte, A. Loundou, A. Caro, et al., Cervical HI-RTE elastography and pregnancy outcome: a prospective study, Eur. J. Obstet. Gynaecol. Reprod. Biol. 186 (2015) 80–84.

[24] E. Hernandez-Andrade, R. Romero, S. Korzeniewski, A. Hyunyoung, et al., Cervical strain determined by ultrasound elastography and its association with spontaneous preterm delivery, J. Perinat. Med. 42 (2) (2014) 159–169.

[25] E. Hernandez-Andrade, M. Garcia, H. Ahn, S. Korzeniewski, et al., Strain at the internal cervical os assessed with quasi-static elastography is associated with the risk of spontaneous preterm delivery at <=34 weeks of gestation, J. Perinat. Med. 43 (6) (2015) 657–666.

[26] L. Hee, C.K. Rasmussen, J.M. Schlutter, P. Sandager, N. Uldbjerg, Quantitative sonoelastography of the uterine cervix prior to induction of labor as a predictor of cervical dilation time, Acta Obstet. Gynecol. Scand. 93 (7) (2014) 684–690.

[27] M. Parra-Saavedra, L. Gomez, A. Barrero, G. Parra, et al., Prediction of preterm birth using the cervical consistency index, Ultrasound Obstet. Gynecol. 38 (1) (2011) 44–51.

[28] A. Fruscalzo, R. Schmitz, W. Klockenbusch, J. Steinhard, Reliability of cervix elastography in the late first and second trimester of pregnancy, Ultraschall der Med. 33 (7) (2012) E101–E107.

[29] A. Fruscalzo, A. Londero, C. Frohlich, U. Mollmann, R. Schmitz, Quantitative elastography for cervical stiffness assessment during pregnancy, BioMed Res. Int. 2014 (2014), 826535.

[30] A. Fruscalzo, A.P. Londero, C. Frohlich, M. Meyer-Wittkopf, R. Schmitz, Quantitative elastography of the cervix for predicting labor induction success, Ultraschall Med. 36 (1) (2015) 65–73.

[31] K. Kobbing, A. Fruscalzo, K. Hammer, M. Mollers, et al., Quantitative elastography of the uterine cervix as a predictor of preterm delivery, J. Perinatol. 34 (1) (2014) 774–780.

[32] H. Feltovich, New techniques in evaluation of the cervix, Semin. Perinatol. 41 (8) (2017) 477–484.

[33] M. Palmeri, H. Feltovich, A. Homyk, L. Carlson, T. Hall, Evaluating the feasibility of acoustic radiation force impulse shear wave elasticity imaging of the uterine cervix with an intracavity array: a simulation study, IEEE Trans. Ultrason. Ferrelectr. Freq. Control 60 (10) (2013) 2053–2064.

[34] M. Muller, D. Ait-Belkacem, M. Hessabi, J. Gennisson, et al., Assessment of the cervix in pregnant women using shear wave elastography: a feasibility study, Ultrasound Med. Biol. 41 (11) (2015) 2789–2797.

[35] E. Hernandez-Andrade, A. Aurioles-Garibay, M. Garcia, S.J. Korzeniewski, et al., Effect of depth on shear-wave elastography estimated in the internal and external cervical os during pregnancy, J. Perinat. Med. 42 (4) (2014) 549–557.

[36] T. Ono, D. Katsura, K. Yamada, K. Hayashi, et al., Use of ultrasound shear-wave elastography to evaluate change in cervical stiffness during pregnancy, J. Obstet. Gynaecol. Res. 43 (9) (2017) 1405–1410.

[37] L. Carlson, H. Feltovich, M. Palmeri, J. Dahl, et al., Estimation of shear wave speed in the human uterine cervix, Ultrasound Obstet. Gynecol. 43 (4) (2014) 452–458.

[38] L. Carlson, P.M. Feltovich, H.,A. Munoz Del Rio, T. Hall, Statistical analysis of shear wave speed in the uterine cervix, IEEE Trans. Ultrason. Ferroelectr. Freq. Control 61 (10) (2014) 1651–1660.

[39] L.C. Carlson, T.J. Hall, M.L. Palmeri, H. Feltovich, Detection of changes in cervical softness using shear wave speed in early versus late pregnancy: an in vivo cross-sectional study, Ultrasound Med. Biol. 44 (3) (2018) 515–521.

[40] I. Rosado-Mendez, L.C. Carlson, K. Woo, A.P. Santoso, et al., Quantitative assessment of cervical softening during pregnancy in the Rhesus macaque with shear wave elasticity imaging, Phys. Med. Biol. 63 (8) (2018) 085016.

[41] E. Hernandez-Andrade, E. Maymon, S. Luewan, G. Bhatti, et al., A soft cervix, categorized by shear wave elastography, in women with short or with normal cervical length at 18-24 weeks is associated with a higher prevalence of preterm delivery, J. Perinat. Med. 46 (5) (2018) 489–501.

[42] A. Agarwal, S. Agarwal, S. Chandak, Role of acoustic radiation force impulse and shear wave velocity in prediction of preterm birth: a prospective study, Acta Radiol. 59 (6) (2018) 755–762.

[43] L. Carlson, S. Romero, M. Palmeri, A. Munoz Del Rio, et al., Changes in shear wave speed pre- and post-induction of labor: a feasibility study, Ultrasound Obstet. Gynecol. 46 (1) (2015) 93–98.

[44] A.A. Baschat, Planning management and delivery of the growth-restricted fetus, Best Pract. Res. Clin. Obstet. Gynaecol. 49 (2018) 53–65.

[45] G. Durhan, H. Unverdi, C. Deveci, M. Ireci, et al., Placental elasticity and histopathological findings in normal and intrauterine growth restriction pregnancies assessed with strain elastography in ex vivo placenta, Ultrasound Med. Biol. 43 (1) (2017) 111–118.

[46] C. Cimsit, T. Yoldemir, I. Akpinar, Strain elastography in placental dysfunction: placental elasticity differences in normal and preeclamptic pregnancies in the second trimester, Arch. Gynecol. Obstet. 291 (4) (2015) 811–817.

[47] S. McAleavey, K. Parker, J. Ormachea, R. Wood, et al., Shear wave elastography in the living, perfused, post-delivery placenta, Ultrasound Med. Biol. 42 (6) (2016) 1282–1288.

[48] S. Calle, E. Simon, M.-D. Dumoux, F. Perrotin, et al., Shear wave velocity dispersion analysis in placenta using 2-D transient elastography, J. Appl. Phys. 123 (23) (2018) 234902.

[49] T. Ohmaru, Y. Fujita, M. Sugitani, M. Shimokawa, et al., Placental elasticity evaluation using virtual touch tissue quantification during pregnancy, Placenta 36 (2015) 915–920.

[50] M. Sugitani, Y. Fujita, Y. Yumoto, K. Fukushima, et al., A new method for measurement of placental elasticity: acoustic radiation force impulse imaging, Placenta 34 (2013) 1009–1013.

[51] E.G. Simon, S. Calle, F. Perrotin, J.-P. Remenieres, Measurement of shear-wave speed dispersion in the placenta by transient elastography: a preliminary ex-vivo study, PLoS One 13 (4) (2018) e0194309.

[52] J.M. Abeysekera, M. Ma, M. Pesteie, J. Terry, et al., SWAVE imaging of placental elasticity and viscosity: proof of concept, Ultrasound Med. Biol. 43 (6) (2017) 1112–1124.

[53] W. Li, Z. Wei, R. Yan, Detection of placenta elasticity modulus by quantitative real-time shear wave imaging, Clin. Exp. Obstet. Gynecol. 39 (4) (2012) 470–473.

[54] C. Cimsit, T. Yoldemir, I. Akpinar, Shear wave elastography in placental dysfunction: comparison of elasticity values in normal and preeclamptic pregnancies in the second trimester, J. Ultrasoud Med. 34 (1) (2015) 151–159.

[55] F. Kilic, Y. Kayadibi, M.A.I. Yuksel, et al., Shear wave elastography of placenta: in vivo quantitation of placental elasticity in preeclampsia, Diagn. Interv. Radiol. 21 (3) (2015) 202–207.

[56] M. Yuksel, F. Kilic, Y. Kayadibi, E.A. Davutoglu, et al., J. Obstet. Gynecol. 36 (2016) 585–588.

[57] B. Alan, S. Tunc, E. Agacayak, A. Bilici, Diagnosis of pre-eclampsia and assessment of severity through examination of the placenta with acoustic radiation force impulse elastography, Int. J. Obstet. Gynecol. 135 (1) (2016) 43–46.

CHAPTER

# Musculoskeletal elastography

9

**M. Abd Ellah[1,2,3], M. Taljanovic[4], A.S. Klauser[1]**

*[1]Department of Radiology, Medical University of Innsbruck, Innsbruck, Tyrol, Austria; [2]Department of Diagnostic Radiology, South Egypt Cancer Institute, Assiut University, Assiut, Egypt; [3]Radiology/Neuroradiology Department Rehabilitationskliniken Ulm, Germany; [4]Department of Medical Imaging, University of Arizona, College of Medicine, Banner- University Medical Center, Tucson, AZ, United States*

## 1. Introduction

Sonoelastography (SEL) is a relatively new imaging modality used to examine tissue elasticity. It is considered as the second breakthrough in the field of ultrasonography (US) after Doppler. The principle of strain imaging was first described in 1991 by Ophir et al. [1] and later on developed by Pesavento et al. in 1999 [2] allowing for real-time SEL-based cross-sectional techniques to be applied in clinical practice.

Different tissues vary in their composition, which may affect their elasticity (deformability). Tissue elasticity is characterized by Young's modulus, which represents the ratio of stress to strain (in pascals).

SEL can display differences in tissue elasticity in the form of a color-coded image that may be superimposed on a B-mode US image. The resulting color-coded elastograms express the difference in tissue elasticity (stiffness) in different color grades, which may vary depending on the manufacturer of the ultrasound unit. Commonly, red indicates the softest tissues, blue indicates hardest tissues, and green and yellow indicate tissues of intermediate elasticity. SEL enables not only a qualitative assessment of tissue elasticity but also a quantitative assessment.

SEL has successfully been used in the diagnosis of different disease conditions, especially in the differentiation of benign and malignant lesions in the liver, thyroid, breast, and prostate [3–7]. In general, malignant soft tissue tends to be harder in SEL than normal soft tissues [8].

Different types of SEL are available, varying mainly by the way stress is applied to the tissue. The most common types are compression (strain) elastography, shear wave elastography (SWE), and transient elastography.

In the recent literature, progressive achievements of SEL in the evaluation of musculoskeletal (MSK) pathologies have been achieved [9–15].

Compression elastography is currently the most commonly described in the MSK field. However, there is a recent emergence of reports describing the feasibility and applicability of SWE in MSK imaging [10,15–18].

**Tissue Elasticity Imaging. https://doi.org/10.1016/B978-0-12-809662-8.00009-7**

## 2. Compression (strain) elastography

This technique requires the application of pressure on the target structure by the ultrasound probe. The probe should be held perpendicular to the target structure while multiple compression-relaxation cycles are applied to induce displacement. The changes in tissue displacement are converted to a color-coded strain distribution map (elastogram), which can be superimposed on a B-mode US image [19]. To avoid interindividual variability in pressure application, most of the SEL machines are supported with a strain indicator. This indicator may be seen at the side or the bottom of the screen in the form of numbers or a sinusoidal curve. The provided elastograms can display the difference in tissue elasticity in the form of four main different colors: red (indicating the softest structure in the tissue), blue (the hardest structure), and green and yellow (indicating tissues with intermediate elasticity) (Fig. 9.1).

Compression SEL mainly provides qualitative evaluation of tissues; however, semiquantitative evaluation can still be obtained by the application of a strain ratio (SR). The SR is an index of the relative elasticity between a targeted object in the elastogram and a reference tissue (usually the subcutaneous fat layer). It is determined from the elastograms by drawing a region of interest (ROI) in the target object and another ROI in the reference tissue, with calculation of an approximated semiquantitative value of tissue elasticity. This may be, to some extent, one of the limitations in the MSK system, in which a lack of the surrounding nearby reference tissue (fat) is the case in many of the examined target structures.

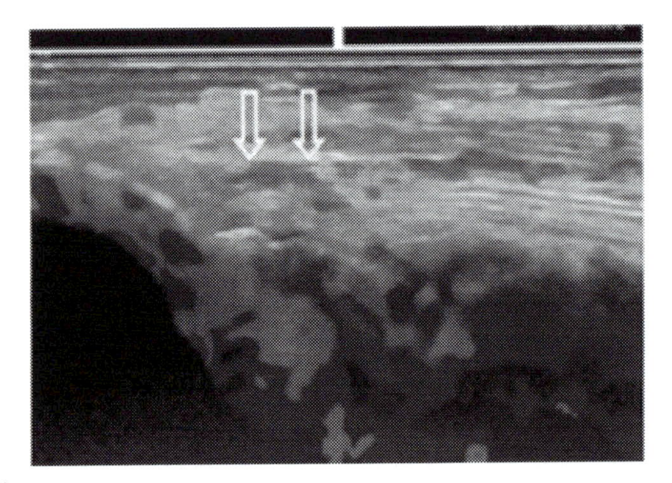

**FIGURE 9.1**

Example of an image of compression sonoelastography of the common extensor tendon of the elbow in a patient with lateral epicondylitis, showing diffuse green (light gray in print version) and blue (dark gray in print version) coloration of the tendon with red (gray in print version) and yellow (white in print version) areas within the tendon substance (*arrows*) denoting soft areas. Red indicates very soft structures, blue indicates hard structures and yellow and green are tissues of intermediate elasticity.

## 3. Shear wave elastography

SWE, also termed dynamic elastography, has a progressively growing role in the evaluation of MSK tissues. It has already been applied in several clinical and research applications [17–20].

The generated ARF results in mere micrometers of tissue displacement, which in turn induces a 50–400 Hz shear wave (SW). These SWs propagate roughly perpendicular to the axial tissue displacement, and the speed of the SW is recorded as a measure of tissue stiffness. SWs travel faster in harder materials, producing faster shear wave velocities (SWVs) than in softer tissues [21].

SWE allows for both qualitative and quantitative evaluation of tissue elasticity. As with compression SEL, color elastograms are usually generated in SWE. These elastograms allow for the measurement of tissue stiffness by drawing an ROI in the target structure, with the results presented quantitatively in kilopascals (kPa) (Fig. 9.2).

## 4. Transient elastography

This is also referred to as pulsed elastography. It is considered a type of SWE that uses a mechanical vibration of the central piston [22]. This technique provides quantitative measures only expressed in kilopascals without color elastograms [21,23]. It currently has no MSK applications, so further discussion of this technique falls outside the scope of our chapter.

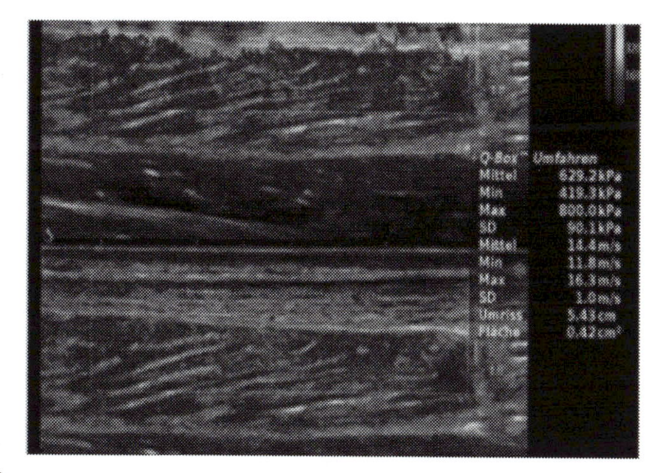


**FIGURE 9.2**

Shear wave elastographic (SWE) image of the middle and proximal thirds of Achilles tendon. The upper image is the SWE image and the corresponding B-mode image is shown below it. A region of interest was drawn within the tendon substance. Both quantitative and qualitative evaluations were presented.

# 5. Applications of sonoelastography in the musculoskeletal system

## 5.1 Tendon disorders

Tendons are the anatomic structures that connect muscles to bones and are responsible for transferring the mechanical load produced by muscles to bones. Tendon pathologies can be classified into two main categories: inflammatory enthesitis and tendinitis, denoting inflammation due to an inflammatory disease, and tendinopathy and enthesopathy, denoting mechanical, degenerative, and overuse disease [24].

Although it has been shown that US diagnosis of tendon pathology is comparable in sensitivity to that of magnetic resonance imaging (MRI), US may be limited in its ability to differentiate between the causes of pathology [24,25].

## 5.2 Achilles tendon

The Achilles tendon is a complex structure formed by the confluence of the gastrocnemius and soleus muscles, which combine together to insert onto the calcaneus. The middle third is the most common site of pathology, followed by the insertion site and finally the proximal third [24,26–28].

The incidence of acute and chronic Achilles tendinopathy has increased tremendously in the past few decades, with overuse (related to sports and occupational activities) being the main cause [29,30].

The main histopathologic changes involving the Achilles tendon are degenerative changes with collagen fiber separation, increased cellularity, neoangiogenesis, and fatty infiltration [13,31]. These factors may result in the softening of the tendon structure leading to spontaneous rupture [32,33].

Clinical differentiation between different Achilles tendon pathologic conditions (tendinopathy, partial tears, and paratendonitis) is difficult to achieve; therefore imaging is considered crucial for accurate diagnosis [12].

Achilles tendon is the first and most common tendon to be studied by SEL [9,10,13].

Achilles tendon is described generally as a hard structure in healthy subjects, appearing mainly as a homogeneous blue color in most of the cases, whereas soft Achilles tendon appearing as a mix of red and yellow colors is referred to as pathologic tendons as in the cases of Achilles tendinopathy and achillodynia [9,12,34].

In a study conducted on 80 asymptomatic Achilles tendons, De Zordo et al. [10], have shown a hard structure in 86.7% of the examined tendon thirds, with mild softening in 12.1% of tendons. This is in line with the findings of Drakonaki et al. [34] describing the Achilles tendon as a homogeneously hard structure, with inhomogeneity in some cases.

In another study by the same group [9], they demonstrated that Achilles tendon is a hard structure in about 93% of healthy tendons, while softening was encountered in half of the tendons with Achilles tendinopathy [9,35]. See the examples in Fig. 9.3.

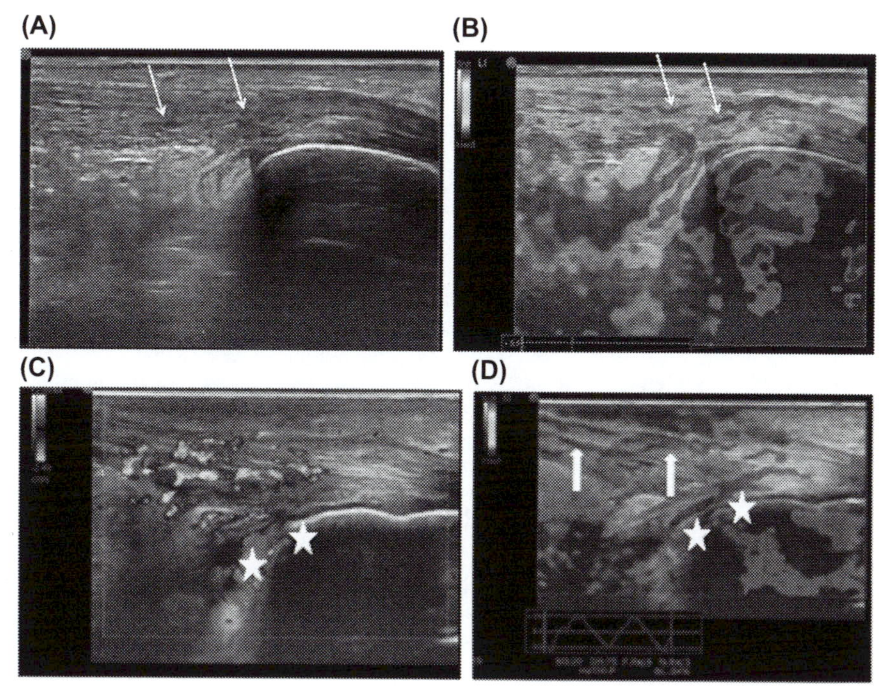


**FIGURE 9.3**

(A,B and C,D) Two different examples of pathologic Achilles tendons. (A) Longitudinal B-mode ultrasonography showing mild diffuse thickening of the Achilles tendon in the distal and middle thirds with disruption of the parallel fibrillar echo texture and multiple hypoechoic areas within the tendon substance (*arrows*). (B) Corresponding compression sonoelastographic image of the Achilles tendon of the same patient showing mild diffuse thickening of the distal and middle thirds with diffuse green (white in print version) coloration of the tendon as well as multiple red (black in print version) linear streaks (*arrows*) within the tendon substance denoting soft pathologic areas. (C) Color Doppler ultrasonographic image of the middle and lower thirds of Achilles tendon showing diffuse thickening with multiple hypoechoic areas inside, and multiple serpiginous vascular structures are seen within the tendon substance and at its lower anterior aspect. Abnormal vasculature occupies the tendon substance suggesting marked tendinopathy. Hypoechoic thin rim of fluid seen in the retrocalcaneal bursa (*stars*). (D) The corresponding compression sonoelastographic image of the same patient shows multiple linear red (black in print version) streaky irregular lines (*arrows*) within the tendon substance on a green (white in print version) background of the tendon denoting soft pathologic areas representing marked tendinopathy. Thin linear rim of red coloration seen in the retrocalcaneal bursa (*stars*) corresponding to the previously described hypoechoic thin fluid rim in (C).

In a study by Klauser et al. [13], the comparison of both B-mode US and compression SEL in reference to histopathologic analysis as the gold standard demonstrated better correlation of SEL with histologic findings. SEL could successfully detect all cases of histologic degeneration (100%), while B-mode US failed to detect it in two cases (14%).

The difference between both healthy Achilles tendons and surgically repaired Achilles ruptures was shown by Tan et al. [36] by using color grading. A homogeneous hard structure for healthy tendons was described, but a more heterogeneous appearance in surgically repaired tendons was reported 38 months after the procedure.

Current literature supports that SWs propagate much faster in stiffer (healthy) tendons than in softened (pathologic) tendons and faster in contracted tendons than in relaxed tendons. Age was found also to be an important factor in determining tendon elasticity [17,18,37–40].

DeWall et al. [37] evaluated SWE of healthy Achilles tendons from the calcaneal insertion to the medial and lateral heads of gastrocnemius aponeurosis by using different ankle positions, namely, resting (neutral) position, plantar flexion, and dorsiflexion. They found higher SWVs in the Achilles free tendon in the resting position in comparison to the passive plantar flexion. At a fixed posture, SWVs decreased significantly from the free tendon to the gastrocnemius myotendinous junction.

Aubry et al. [17] evaluated Achilles tendons at midportions in 180 asymptomatic Achilles tendons and 30 symptomatic individuals with Achilles tendinopathy. Examination was done in both longitudinal and axial scan planes in both relaxed (full passive plantar flexion) and stretched (0-degree flexion) positions. In both longitudinal and axial scans, SWVs showed lower values in cases with tendinopathy than in healthy tendons.

Chen et al. [18] also evaluated ruptured Achilles tendons in comparison to healthy tendons. They found a significant difference in elasticity between both categories.

Age effect on the stiffness of Achilles tendons was shown in a study by Ruan et al. [38] as lower values in relaxed and tension states with increasing age, which was supported by another study by Slane et al. [39] as seen in the example in Fig. 9.4.

## 5.3 Lateral epicondylitis

Lateral epicondylitis, also known as tennis elbow, represents a common source of elbow pain with a prevalence rate of 1%–3% [41]. The most common presentation is pain, hyperalgesia, and weak grip strength. Overuse is usually the most common cause of this pathology [41,42]. Diagnosis is mainly done clinically through patient history and examination, while imaging is important for excluding other causes of elbow pain and determining disease severity [43,44].

SEL has been used for the evaluation of lateral epicondylitis in several prior studies [11,45]. Despite the multiple US findings that might help diagnosing lateral

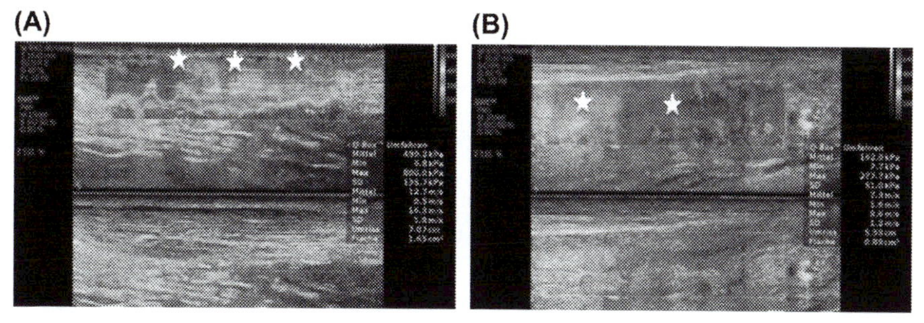


**FIGURE 9.4**

Example of shear wave elastography (SWE) of normal and ruptured Achilles tendons. (A) Longitudinal SWE image of the middle third of normal Achilles tendon with the region of interest (ROI) drawn as an interrupted line (*stars*). The tendon shows a mean stiffness value of about 490 kPa denoting hard structure. (B) Longitudinal SWE image of the contralateral ruptured Achilles tendon in the same patient with the ROI drawn as an interrupted line over the pathologic segment (*stars*). The tendon shows a mean stiffness value of about 162 kPa denoting softening.

epicondylitis, there is lack of consensus on the best B-mode US finding. Those findings include tendon thickness, hypoechogenicity, calcifications, and partial or full thickness tears, as well as bony changes such as cortical irregularities and cortical spurs [46]. On the other hand, SEL showed normal tendons as hard structures (blue to green elastograms) and detected pathologic changes as softening (yellow and red elastograms) [47].

The normal common extensor tendon origins have been described as a homogeneous blue to green elastograms, whereas typical findings of lateral epicondylitis are irregular softening of the tendon origins represented by elastograms with red and yellow in the tendon substance (Fig. 9.5). De Zordo et al. [11] evaluated 38 patients with a clinical diagnosis of lateral epicondylitis by both B-mode US and compression SEL. SEL has shown better sensitivity and accuracy in diagnosis than B-mode US, while both showed the same specificity. For SEL, sensitivity was 100%, specificity was 89%, and accuracy was about 94%, while B-mode US showed 95%, 89%, and 91%, respectively. Ahn et al. [45] examined common extensor tendons in 79 patients with lateral epicondylosis and 14 healthy subjects by using B-mode US, color Doppler ultrasonography (CDUS), and compression SEL with the calculation of the SR. Softness in elastograms was detected in 73 out of 97 tendons (75.3%). The mean SR showed significantly lower values for symptomatic tendons ($1.45 \pm 0.45$) than for asymptomatic tendons ($2.07 \pm 0.7$, $P < .001$), which indicates softer consistency of the symptomatic tendons. In another study by Kocyigit et al. [48], both qualitative and quantitative evaluation of color elastograms were used for the evaluation of 34 common extensor tendons in 17 patients with unilateral epicondylitis, with the contralateral common extensor tendons used as controls. They concluded the ability of compression SEL to differentiate between healthy

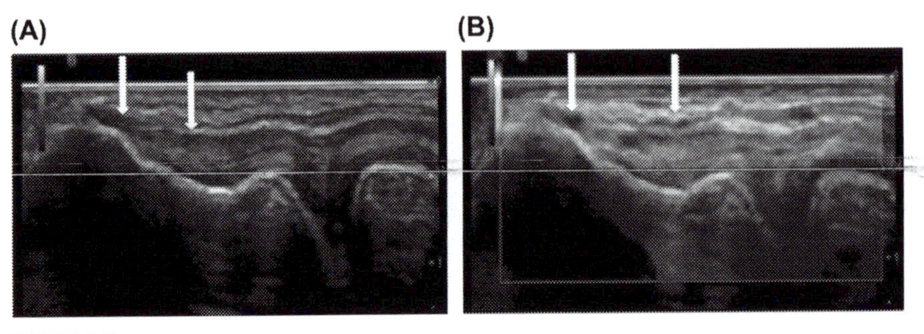

**FIGURE 9.5**

Example of lateral epicondylitis. (A) Longitudinal B-mode ultrasonographic image showing diffuse thinning (loss of convexity) of the common extensor tendon and loss of echogenicity (*arrows*), representing areas of mucoid degeneration. (B) Corresponding compression sonoelastographic image shows in the thinned-out tendon multiple variable-sized red (gray in print version) areas in a mainly yellow (white in print version) background of the tendon substance (*arrows*) with still seen green (light gray in print version) areas. Red areas denote softening of the tendon substance and both red and yellow occupy most of the tendon substance. SEL image shows higher degree of tendon pathology than the B-mode ultrasonographic image in (A). (Red indicates very soft structures, blue (dark gray in print version) hard structures and yellow and green are tissues of intremediate elasticity).

and affected elbows with good interobserver agreement, being superior to B-mode US and CDUS in discriminating lateral epicondylitis. Both color elastograms and SRs of the affected tendons were significantly different from the normal tendons.

Klauser et al. [47] studied the role of B-mode US and SEL in detecting the changes of common extensor tendons when compared to histologic examination on cadavers. Both modalities showed nearly the same ability to detect histologic changes with no significant difference ($P > .3$); however, combining both the modalities has resulted in significant increase in sensitivity ($P < .02$) without a significant change in specificity ($P > .3$), yielding an improvement in overall accuracy. Such a finding may emphasize the advantage of combining the use of both modalities in daily routine.

## 5.4 Medial epicondylitis

The data on SEL of the common flexor tendon are scarce. However, some authors assume that SEL image findings of common flexor tendon pathologies may be similar to common extensor tendon pathologies. Being a structure carrying similarities to other tendon structures as common extensor tendon and Achilles tendon, it should show the same pathology; hence, SEL findings are expected to be similar [49].

Klauser et al. [50] conducted a study on 25 common flexor tendons in 16 cadavers. They evaluated B-mode US and SEL in comparison to histologic examination. Tendons were classified into three grades according to their elastograms (hard, soft, and softest). They showed that the addition of SEL to B-mode US provided statistically significant improvement in correlation with histologic results compared with the use of B-mode US alone ($P < .02$). A finding that may express the value of using SEL in common flexor tendon evaluation and the value of adding it to B-mode US of these tendons in daily routine.

## 5.5 Patellar tendinopathy

Patellar tendinopathy is a common cause of knee pain among athletes, with a reported high prevalence rate about 30%–45% in jumping sports [51,52]. It is characterized by abnormal collagen, neovascularization, cellular infiltrates, and mucoid degeneration with pain and tenderness at the lower patellar pole representing the main symptoms at presentation [53,54].

Although B-mode US is helpful in diagnosing tendon pathologies in general, which may be impacted on patellar tendinopathy, lack of information about biomechanical characteristics of tissues may represent a limitation. In a study by Porta et al. [55], they examined 22 patellar tendons in 16 healthy subjects. Two sonographers of different experiences did compression SEL examinations independently after dividing the tendon into proximal, middle, and distal portions. The tendons turned out to be hard structures by qualitative evaluation of elastograms, which was shown by the predominance of green color. Further evaluation of SR showed values of (mean ± standard deviation) $1.47 \pm 0.64$, $4.38 \pm 1.36$, and $3.32 \pm 1.2$ for the proximal, middle, and distal thirds, respectively. In both qualitative and semiquantitative approaches of data analysis, good interobserver agreement was found.

In another study by Akkaya et al. [56], they postoperatively evaluated patellar tendons in patients with anterior cruciate ligament (ACL) reconstruction using autograft bone-tendon-bone. Patellar tendons of the operated knees were evaluated morphologically by B-mode US (length and thickness) and biomechanically by calculating SR using compression SEL. On the operated side, patellar tendons were found to be shorter and thicker with lower SR values than those on the other side, which may indicate increased stiffness on the operated side.

Prior studies showed controversy regarding the data about tissue elasticity in patients with patellar tendinopathy; one study showed increased elasticity compared with the control, whereas the others encountered no differences [57–59].

Zhang et al. [60] evaluated 20 healthy male athletes and 13 male athletes with unilateral patellar tendinopathy. They used both B-mode US and SWE to evaluate the morphologic and elastic properties of tendons. The pathologic tendons were significantly larger and showed higher shear elastic modulus than that of the normal tendons of patients and also than that of the dominant side of the healthy athletes.

Another study applied SWE in the evaluation of healthy patellar tendons in three different age groups [61]. They found a significant decrease in shear modulus and SWV values, with a significant increased side-to-side difference in the oldest of the three examined groups. This study may reflect the ability of SWE to depict subclinical changes in aging tendons before being depicted by B-mode US.

## 5.6 Quadriceps tendinopathy

There are only a few publications on SEL of the quadriceps tendon [62,63]. Evaluation of quadriceps tendons in chronic hemodialysis patients by B-mode US and compression SEL showed that quadriceps tendons are thinner and have lower elasticity scores than controls. This study was conducted on 53 patients with chronic renal failure under hemodialysis program and 25 healthy individuals as controls. Tendons in patients showed lower thickness values (4.9 mm) than the tendon thickness in healthy volunteers (5.4 mm), based on the proposed classification of elastograms into three scores: very stiff, stiff, and intermediate stiff tissues. Significant differences were found in the mean elasticity scores between both knees. The mean score of elasticity in the right knee was 3.14 in patients versus 3.79 in healthy subjects, and the mean score of elasticity in the left knee was 3.33 in patients versus 3.69 in healthy subjects ($P = .025$ and .018, respectively), which indicates softening of the tendons in patients [64].

The effect of changing the technical settings of SWE was evaluated in a previous study on both patellar and quadriceps tendons. Variations of the elastic modulus of tendon and muscle were found by altering transducer's pressure and ROI size [65].

## 5.7 Rotator cuff tendinopathy

Disorders of the rotator cuff or associated tissues are the most common shoulder problems, with progressive increase in rotator cuff disease in people over 40 years of age and the prevalence may reach up to 50% by the age of 60 years [66–69].

Diagnosis of tendinopathy cannot be achieved in some cases by B-mode US only because of the echogenic similarity between the diseased tendon and the surrounding tissues [70]. In such cases, SEL may have a role. In a study by Silvestri et al. [71], B-mode US was not able to differentiate between degenerated and enlarged supraspinatus tendon and subacromial bursitis; however, compression SEL enabled this differentiation based on tissue elasticity characteristics, as the bursa appeared soft against the hard tendon structure.

In another study, by Seo et al. 2015 [72], compression SEL was valuable in detecting intratendinous and peritendinous supraspinatus tissue pathology, with excellent interobserver reliability and excellent correlation between the MRI and B-mode US findings. A study of the long head of biceps tendon by the same group showed feasibility of SEL in detecting intratendinous and peritendinous alterations of the long head of biceps tendon, as it has excellent accuracy and excellent correlation with B-mode US findings. In 34 patients with clinical symptoms of biceps tendinosis, which was confirmed by B-mode US, and 98 patients without biceps tendinosis, compression SEL was performed in both transverse and longitudinal scans. The mean sensitivity, specificity, and accuracy were 69.4%, 95.6%, and

89.3% for transverse scans versus 94.4%, 92.1%, and 92.7% for longitudinal scans, respectively. For both approaches, good correlation with US findings ($P < .001$) and an almost perfect interobserver agreement were found [70,73].

Lee et al. 2016 [70] studied the feasibility of real-time compression SEL in diagnosing rotator cuff tendinopathy. In 39 patients with chronic pain the authors graded MRI findings of tendinopathy ranging from normal (0) to marked tendinosis (3) and correlated these grades with SRs. They evaluated SRs by two different methods: fat-to-tendon (Fat/T) and gel pad-to-tendon (Pad/T). Both the calculated SRs showed positive correlation with MRI grade. Based on these findings, they proposed that SEL might be a useful diagnostic tool for supraspinatus tendinopathy.

Multiple factors are believed to affect the outcome of rotator cuff tendon repair, of which the supraspinatus muscle and tendon are very important and crucial [74–77]. Until now, there are no definite diagnostic tools that can measure the alterations in tissue mechanical properties that may result from histopathologic changes of chronic tendinopathy. However, in the study by Tudisco et al. [78] they evaluated the feasibility of compression SEL both qualitatively and semiquantitatively in assessing small supraspinatus tendon tears and its relation to clinical findings. A strong positive correlation was found between strain index and the Constant-Murley score and the American Shoulder and Elbow Surgeons Shoulder Score. SEL may carry a great importance for the future in daily routine evaluation of rotator cuff tendon tears by providing important information to the surgeons that may help in operative decision-making.

## 5.8 Finger tendon and trigger fingers

Stenosing tenosynovitis represents a common hand problem presented mainly by snapping of the finger and pain over the A1 pulley. The thumb is the most commonly affected digit followed by the middle, index, and little fingers (in this order) [79]. The reported prevalence is about 3% in the normal population that may reach up to 17% in diabetic patients [80]. Thickness measurement of A1 pulley by high-resolution US showed a significant difference between trigger fingers and normal fingers previously. By applying SEL in diagnosing such pathology, a significant difference was found in both thickness and SR calculations between patients with trigger fingers and normal subjects, which was 4.2 versus 2.4 as determined by Miyamoto et al. [81], who showed significant reduction in this reported high value after steroid injection.

## 5.9 Joints and ligaments

### 5.9.1 Transverse carpal ligament

Progressive stiffness of transverse carpal ligament may constrain the carpal tunnel leading to median nerve (MN) compression. Carpal tunnel syndrome (CTS) is a compression neuropathy of the MN at the wrist level. So far, it is considered the most common upper limb neuropathy with an estimated prevalence rate of about

50 cases per 1000 subjects per year [82]. High-resolution US has been utilized efficiently in the diagnosis with multiple different parameters being evaluated: the utmost important of them being the delta and ratio. Both parameters incorporated the measurement of the cross-sectional area of the MN at the wrist (carpal tunnel) and proximally at the level of the distal third of pronator quadratus muscle. Delta represents the difference between the distal and proximal measurements and ratio represents the distal/proximal measurement. However, recent studies have been published evaluating the postinjection effect of corticosteroids in patients with CTS, as reflected in color elastograms before and after injection. In a previous study, Miyamoto et al. [14] evaluated both hands in 20 healthy volunteers and 22 hands in 20 patients with CTS by compression SEL. They estimated semiquantitatively the stiffness of intracarpal tunnel contents as the standardized acoustic coupler (AC)-to-intracarpal tunnel contents surrounding MN SR before and after 6 weeks of corticosteroid injection. A significant difference was detected between patients and normal subjects (12.6 in patients vs. 8.2 in normal subjects) and both before and after injection in the same patient (12.6 vs. 8.5). Another similar study by Klauser et al. [83] compared the perineural area in carpal tunnel between patients and healthy volunteers and in patients before and immediately after corticosteroid injection. They studied both hands in 15 healthy volunteers and 72 hands from 70 patients with symptomatic CTS. The obtained elastograms were classified according to a proposed grading system into 4 grades: grade 1 as red, indicating the softest structures; grade 2 as yellow, indicating soft structures; grade 3 as green, indicating hard structures; and grade 4 as blue, indicating the hardest structures. The perineural area in patients with CTS was grade 3 (hard), whereas it was grade 1 (softest) in healthy subjects, with a significant difference encountered between both groups ($P < .0001$). This area changed immediately after injection from grade 3 to grade 1 with also a significant difference encountered ($P < .0001$), which may indicate immediate distribution of the injectant in the perineural tissue with subsequent softening (Fig. 9.6).

### 5.9.2 Coracohumeral ligament

Coracohumeral ligament was also examined by SEL. It originates from the lateral aspect of the base of the coracoid process and inserts into the rotator interval or the supraspinatus tendon [84]. A tightened coracohumeral ligament may restrict external rotation in patients with adhesive capsulitis as suggested by Neer et al. [85]. By examining the shoulders of 30 healthy volunteers and 20 patients clinically presumed to have adhesive capsulitis, significant findings were detected regarding two main findings. The authors first found significant differences regarding tendon thickness studied by B-mode US between pathologic shoulders and normal shoulders ($P < .001$). Then significant differences in stiffness between pathologic and normal coracohumeral ligaments were registered by SWE and significantly higher values were reported for pathologic coracohumeral ligaments than for normal ligaments ($P = .004$) [86].

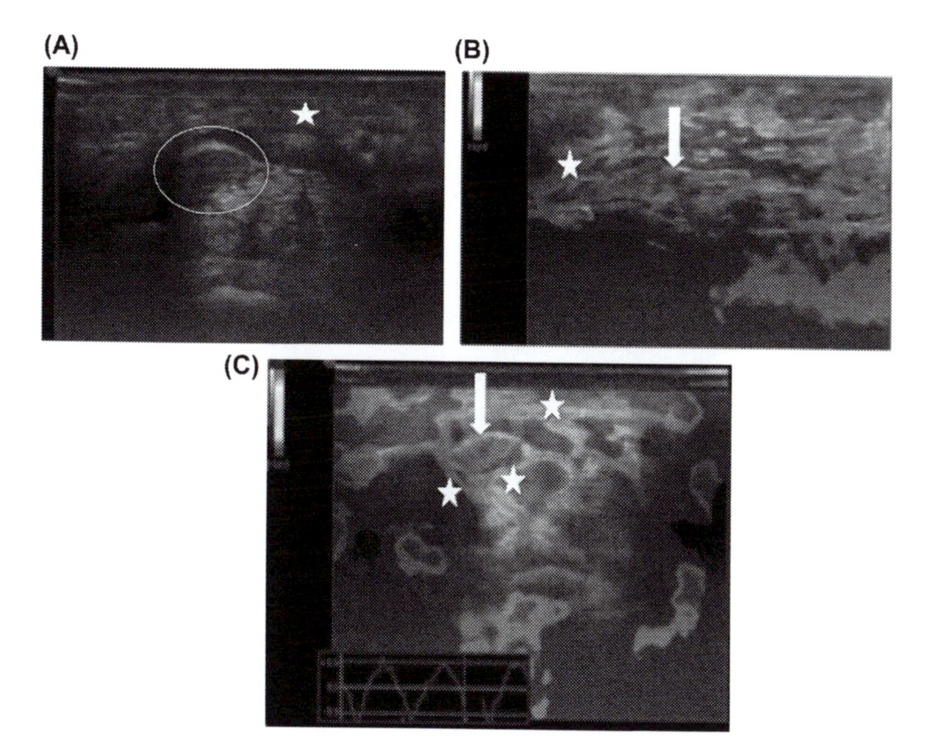


**FIGURE 9.6**

Example of a patient with carpal tunnel syndrome after ultrasonography-guided corticosteroid injection. Axial image of median nerve at the level of carpal tunnel: (A) Axial B-mode ultrasonography showing the median nerve in axial cut section with mild diffuse thickening and is relatively hypoechoic (*circle*) in comparison to the surrounding flexor tendons with a thin rim of hypoechoic fluid seen between it and the flexor retinaculum representing the injected corticosteroid (*star*). (B) The corresponding compression sonoelastographic image showing diffuse blue (dark gray in print version) and green (dark gray in print version) coloration of carpal tunnel contents, with the nerve showing diffuse blue (light gray in print version) coloration (*arrow*) and the injected corticosteroid appeared as a rim of red (black in print version) color palmar to it (*star*). (C) Compression sonoelastographic image obtained at the same level shortly after allowing distribution of the injected material in carpal tunnel shows the contents being mixed of green (dark gray in print version) and blue (light gray in print version) coloration with more abundant green (dark gray in print version) than that in the previous sonoelastography image (B). The nerve shows green (dark gray in print version) color (*arrow*) as well, with the injected corticosteroid appearing red (black in print version) around the median nerve and distributed within the carpal tunnel rather than seen palmar to the nerve (*stars*).

### 5.9.3 Anterior cruciate ligament

ACL is the responsible structure for mechanical stabilization of the tibiofemoral joint. ACL inserts directly into bone with insertion consisting of a linear transition from ligament to fibrocartilage to bone [87]. Based on this different tissue composition, ACL interface is expected to vary in cellular, chemical, and mechanical properties [88].

Recent studies conducted on bovine models have evaluated strain distribution at the ACL and its bone insertion using SEL algorithm intended for in situ monitoring of the mechanical properties of the ACL graft during healing. SEL analysis showed complex strains with compressive and tensile components at tibial insertions in knee joints subjected to tension [89].

## 6. Muscles

Muscles are considered suitable for SEL examination with a growing number of research studies and paucity of clinical publications. Different applications of SEL in muscle pathology evaluation have been published, but it is still insufficient for the adequate evaluation of the technique and for obtaining reproducible data.

Color elastograms of normal muscles in resting position usually show diffuse hardness or intermediate stiffness presented as a heterogeneous mosaic appearance (blue or green and yellow) with small peripheral areas of increased or reduced stiffness [15,90].

In one of those trials, Niitsu et al. [91] evaluated the effect of exercise on the hardness of flexor muscles of the elbow. They used a qualitative approach supported by semiquantitative analysis of the data using SR measurement. In six healthy male volunteers, they examined biceps brachii muscles with compression SEL and measured SR before, immediately after, and 1–4 days after exercise with the opposite arms being used as controls. Significant muscle hardening followed by recovery after exercise could be detected by SEL ($P < .01$). They registered increased hardness immediately after exercise and peak hardness after 2 days followed by a decrease until day 4.

Some congenital muscular disorders have already been evaluated by SEL, including muscle dystrophies due to a wide range of genetic disorders that may be characterized by muscle weakness and contractures.

Drakonaki and Allen [90] reported a case of Bethlem myopathy, which was diagnosed clinically by electromyography and confirmed by histopathologic examination, showing abnormal areas in muscle tissues of the thigh. They reported abnormal hyperechoic and hypersignal areas by US and MRI, respectively, detected peripherally at both vastus lateralis and the long head of biceps femoris muscles and centrally in the rectus femoris muscle. The selected regions showed corresponding stiff areas in color elastograms (blue) compared with the residual normal regions of involved muscles (green).

In case of cerebral palsy spasticity, US has gained much acceptance as a noninvasive modality that helps in the diagnosis and guidance of treatment. It was previously proposed that normal relaxed muscles are presented as soft structures in (green-yellow-red) elastograms, but contracted and degenerated muscles as hard structures (blue) in elastograms [15]. In a study by Vasilescu et al. [92], the authors showed better results regarding patient response to image-guided Botox therapy by using compression SEL as an adjunct tool to B-mode US, which facilitated and helped better in the determination of target sites for injection.

Congenital muscular torticollis (CMT) is a common congenital disorder in infants diagnosed mainly by US; however, it is subjected to some limitations that may interfere with the diagnosis [93–96]. According to Lee et al. [97], compression SEL may play a role in diagnosing CMT. It turned out to show stiffer elastograms. They proposed the usefulness of SEL as a subordinate tool in evaluating CMT in case of inconclusive B-mode US.

Idiopathic inflammatory myopathy is a group of muscle diseases characterized by changes in the muscular structures in addition to laboratory tests. Some pathologies, such as polymyositis, dermatomyositis, and inclusion body myositis, may be classified under this disease. In a study by Botar-Jid et al. [15] the authors evaluated the color composition of elastograms by specific software and compared the obtained values to laboratory test parameters. A proportional concordance was found between the average values of color parameters and serum creatine kinase and lactic dehydrogenase levels, so they suggested that SEL could be an important tool in the management of patients with myositis.

Trauma may lead to muscle rupture and hemorrhagic area formation, which may appear in SEL as a homogeneous very soft area (red) in color elastograms [16].

SEL can also detect muscle injuries as irregular areas of altered tissue elasticity, especially of the perilesional tissue. This may show an importance in lesion evolution, where lesions with favorable evolution show soft elastograms, while fibrosis may show blue elastograms (especially of the perilesional area) [16].

Determination of the degree of fatty infiltration and degeneration of the rotator cuff muscles in cases of tendon tears is considered crucial preoperative information for surgeons, as it is an important prognostic factor for postoperative anatomic and functional results [3,98,99].

The ability of SEL to assess fatty degeneration of the supraspinatus muscle was tested and compared to both B-mode US and MRI in a study conducted by Seo et al. [73]. The authors prospectively investigated 101 shoulders of 98 patients by MRI, B-mode US, and SEL. By comparing compression SEL and B-mode US to standard MRI, better results were detected by SEL than by B-mode US. They reported a sensitivity, specificity, and accuracy of 95.6%, 87.5%, and 91.1% for SEL versus 93.3%, 73.2%, and 82.2% for B-mode US, respectively. Both MRI and B-mode US grading had also positive correlation with SEL ($r = 0.855$, $P = <.001$ and $r = 0.793$, $P = <.001$ respectively).

A further study by Itoigawa et al. [20] evaluated the feasibility of SWE in assessing the supraspinatus muscle stiffness in six shoulders of three healthy volunteers along the long axis of the muscle fibers to get information about muscle extensibility, which is an important factor for planning of supraspinatus tendon repair. They divided the tendon into four quadrants (anterior deep, anterior superficial, posterior deep, and posterior superficial) and obtained the stiffness values for each.

The same approach was done by Rosskopf et al. [100], who evaluated SWE of the supraspinatus muscle in both control subjects and patients. They prospectively examined 22 healthy volunteers and 44 patients by SWE and both shoulders were examined in healthy volunteers by two independent examiners. The authors calculated for what they proposed mean total shear wave velocity (MTSWV). They compared SWV with tendon integrity, tendon retraction, fatty muscle infiltration, and muscle atrophy (tangent sign) on MRI. MTSWV was significantly higher in healthy volunteers than in patients ($P < .001$), with a significant difference between MTSWV and the degree of retraction ($P = .047$). Tangent sign describes the supraspinatus muscle volume atrophy, which is considered atrophic if it does not cross a line connecting the upper border of the scapular spine and the upper border of the coracoid process in the sagittal plane. MTSWV is significantly decreased in patients with positive tangent sign ($P = .015$). They could achieve a specific value of SWV for the normal supraspinatus tendons and showed its relation to tendon retraction and muscle volume loss.

Quantitative elastographic maps of different muscles have been evaluated during contraction and relaxation. The reported values were significantly higher at contraction versus relaxation for gastrocnemius, soleus, and tibialis anterior muscles. These results may indicate the feasibility of using this technique in the evaluation of neuromuscular disorders [101].

In a further study conducted on 13 healthy control subjects and 14 patients with Duchenne muscular dystrophy (DMD), Lacourpaille et al. [102] evaluated the stiffness of six different skeletal muscles at both contraction and relaxation status (medial head of gastrocnemius, tibialis anterior, vastus lateralis, biceps brachii, triceps brachii, and abductor digiti minimi). They reported significantly higher values in patients with DMD than in control subjects in all examined muscles except for the abductor digiti minimi.

## 7. Nerves

In one of the most common clinical pathologic conditions affecting the MN, i.e., CTS, MN was evaluated by compression SEL. MN elasticity was evaluated semiquantitatively and compared between both healthy subjects and patients with symptomatic CTS. Miyamoto et al. [14] examined both hands in 22 healthy subjects and 43 hands in 31 patients with CTS by compression SEL and calculated the AC/MN ratio. The mean AC/MN ratio was 6.9 in patients with CTS, which was significantly higher than the ratio in healthy subjects (4.1, $P < .001$). A cutoff value of 4.3 showed sensitivity of 82% and specificity of 68%. They concluded that the combination AC/MN ratio and cross-sectional area measurement by B-mode US improved the diagnosis of CTS, as significant improvement in diagnostic accuracy was achieved by SEL.

SWE has been also utilized in a study evaluating MN elasticity. By evaluating MN stiffness at the carpal tunnel, the mean stiffness value was found to be higher both in patients with CTS than in normal subjects and in the high-grade CTS group than in the mild- and moderate-grade CTS groups [103].

Peripheral nerve evaluation by SWE was also feasible in sciatic nerve and brachial plexus [104]. Andrade et al. [105] evaluated SWV of the sciatic nerve at 90-degree knee flexion and 180-degree extension with five times the ankle dorsiflexion. SWVs significantly increased during ankle dorsiflexion with the knee in maximum extension, but there was no change in the knee flexion.

## 8. Plantar fascia

Plantar fasciitis is the most common cause of heel pain. Diagnosis is usually achieved clinically; however, imaging may aid in the diagnosis, to exclude other possible causes, and in treatment guidance [106–109].

Plantar fascia is similar to tendons subjected to age-dependent changes, with reported elastic modulus changes (Fig. 9.7). Evaluation of the color histograms

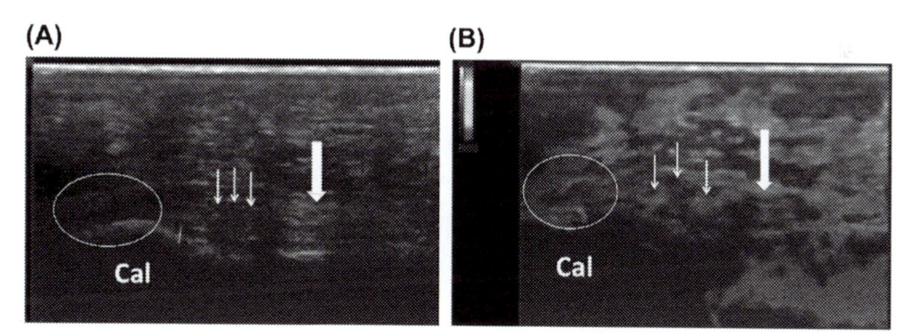


**FIGURE 9.7**

Example of pathologic plantar fascia. (A) Longitudinal B-mode ultrasonographic image of plantar fascia showing diffuse thickening with interruption of the parallel echotexture and multiple small hypoechoic areas (*small arrows*), with the largest (*circle*) seen at the site of insertion to the calcaneus (Cal), but exactly differentiating it from anisotropy is not possible. More distally the plantar fascia shows diffuse homogeneous hyperechogenicity (*large arrow*). (B) The corresponding sonoelastographic image of the same patient shows mild diffuse thickening of the fascia with heterogeneous green (dark gray in print version) and blue (light gray in print version) coloration (*small arrows*) and a small red (black in print version) irregular area (*circle*) at the site of insertion to calcaneus (Cal) corresponding to the previously described hypoechoic area in (A) denoting its softness, confirming it as a pathologic area. More distally the plantar fascia shows diffuse homogeneous blue (light gray in print version) coloration (*large arrow*).

obtained by compression SEL from 40 healthy subjects and 13 patients with plantar fasciitis enabled discrimination based on age difference and pathologic status. By dividing healthy subjects based on age difference, a predominant significance of blue (denoting soft) was detected in the older age group (over 50 years) than in the younger age group (18–50 years, $P = .002$). A predominantly significant red color (denoting hard) was also noted in histograms of older healthy subjects than in the fasciitis group ($P < .001$), which indicates softening of the plantar fascia with age and fasciitis [110].

Quantitative assessment of the stiffness of plantar fascia by SWE was also evaluated by Zhang et al. [111] in both healthy volunteers and patients with plantar fasciitis, showing a significant difference in elasticity moduli between elderly and younger volunteers and between patients and elderly volunteers ($P < .001$).

The diagnostic performance of SEL as an investigative tool for diagnosing plantar fasciitis has been the topic of several publications [112,113]. The diagnostic power of compression SEL was even better than that of B-mode US in the early diagnosis of plantar fasciitis [114]. In a case report by Wu et al. [115], the authors reported changes in compression SEL of a 30-year-old woman with recurrent heel pain. They showed multiple green and yellow areas in the symptomatic plantar fascia (denoting softening) and red areas in the normal asymptomatic fascia; however, B-mode US showed normal structure with normal echogenicity and thickness.

Recently, compression SEL was shown to be able to monitor treatment response after collagen injection. Plantar fascia hardening reported as increased SR 3 months after collagen injection was detected; however, no significant changes could be achieved by using B-mode US [116].

## 9. Tumor and tumorlike masses

Differences between the stiffness of malignant and benign tissues has been described in the literature. Malignant tissues have been assumed to have higher stiffness than benign tissue lesions. Compression SEL has shown a wide range of acceptance in the differentiation between benign and malignant lesions in different tissues and organs such as the liver, breast, thyroid, and lymph nodes [3,5,117–120].

Lalitha et al. [35] proposed that malignant soft tissue tumors are stiffer than benign tumors and show predominance of blue color (denoting hard) in elastograms. According to their experience, lipomas showed a variation in color elastograms from red to blue; however, vascular malformations were red to green and neurogenic tumors were green with no blue coloration detected. Ganglion cysts showed a predominance of green to red. Although cystic areas should appear red (very soft), they may sometimes show all three colors, which may be attributed to artifacts. Another characteristic pattern of cysts was also previously encountered, the so-called "bull's-eye" artifact, which may appear as a dark area with a central bright spot [121]. Benign schwannoma of the common fibular nerve was described by compression SEL as a stiff structure that was harder than the surrounding muscles and had a high elasticity index [122].

Magarelli et al. [123] analyzed the usefulness of compression SEL in evaluating superficial soft tissue lesions in a prospective study conducted on 32 patients.

SR was used for the evaluation of soft tissue tumors in the literature [124]. The reported SR for malignant tumors was 1.94, which was significantly higher than that for benign tumors (1.35, $P = .043$). SRs were determined for different superficial soft tissue lesions in another study. The median calculated values for lipoma, ganglia, and epidermal inclusion cysts were 0.83, 2.78, and 0.17, respectively, showing a statistically significant difference between lipoma and ganglia and between epidermal inclusion cysts and pilomatricomas, as well as between ganglia and epidermal inclusion cysts and matricomas ($P < .05$) [125].

To our knowledge, there is only one publication in the literature evaluating the role of SWE in evaluating soft tissue masses. Both qualitative and quantitative analyses of color elastograms showed no significant statistical association with malignancy and SWE has shown no clear additional information to B-mode US [126], this was in contradiction to what is expected.

## 10. Future perspectives

Fusion imaging by superimposing imaging datasets from computed tomography or MRI to SEL may represent an interesting development in the future [127].

Three-dimensional SEL may also represent an exciting development in the future. Three-dimensional elastograms can be reconstructed from two-dimensional elastograms [128–132]. Better-quality strain images may be achieved by using this technique, which may be of impact on better evaluation of mass lesions either in follow-up after chemotherapy or in preoperative and intervention planning.

## 11. Limitations and conditions of good practice

### 11.1 Strain elastography/sonoelastography

It is an operator-dependent sensitive imaging modality. Technique optimization is mandatory for better results. Obtaining artifact-free closed loops of successive compression-relaxation cycles is the main challenge. To obtain such loops of good diagnostic quality, suitable compression has to be adopted during the examination. Both high and very low pressures are not recommended, as they may lead to fake elastograms with false-positive or false-negative results. A strain indicator marker has been developed by manufacturers to overcome this problem. It helped the operator to adjust his/her manual compression according to an automatic scale recommendation.

It is also worthy to note that tissues are not examined separately, but in conjunction with surrounding tissues, which may affect their elasticity. Tissues near or closely related to hard structures are liable to be affected by their hardness. The hard shell of tissues as bones may interfere with and limit internal tissue strain readings of closely related soft tissues, which is termed the eggshell effect [133,134].

The ultrasound transducer has to be held, as much as possible, perpendicular to the skin surface with an optimum distance between the probe and target structure being $\geq$1.2 mm; shorter distances $<$1.2 mm may affect the results. A good solution to overcome this problem is to use a gel pad, when examining superficial structures. However, gel pads have to be kept outside the boundaries of ROIs during measurement, as their inclusion within ROIs may influence the results.

Care should be taken to put the patient in a comfortable position and to avoid any movement during the examination, which may affect the obtained elastograms and hence the obtained results.

Evaluating the whole loops of compression-relaxation cycles is more preferable than just static images, which may harbor inaccurate data. Optimum loops have to be composed of at least three successive compression-relaxation cycles. Avoidance of the analysis of elastograms at the beginning and the end of each cycle is necessary, as it does not provide correct elastograms as a result of incorrect elasticity calculations.

Furthermore, borders of elastograms should not be included in the evaluation. In longitudinal scans, borders of elastograms are liable to inhomogeneous pressure application and tissue shifting, which may contaminate the obtained elastograms.

As elastogram is obtained from the calculation of image pairs before and after compression application, it can be affected by the size of the sampling window. SEL window has to be standardized and kept constant in all examinations, especially during follow-up examinations. Its depth is advisable to be at least three times the examined structure and its width to be three-quarters the screen.

The lack of quantification of elastograms obtained by compression SEL has resulted in the implementation of multiple different approaches for semiquantitative analysis, including SR calculations, scoring of elastograms, and histogram analysis of colored elastograms. This wide diversity of data analysis and evaluation may interfere with analysis and repetition of results as well as with comparison of results between different studies, even if they use the same technique.

## 11.2 Shear wave elastography

It has overcome the elasticity quantification problem with its ability to obtain color elastograms and accurate quantitative data simultaneously, as the manual compression application in case of compression SEL was substituted by the ultrasound waves in case of SWE. However, there is still the lack of transferability of SWE values across machines and vendors, which may hamper its generalized application in the clinical routine. Researchers and radiologists should be totally aware of the differences in measurement values between the different devices upon use, as a certain value in one device may be represented differently in another device.

Some other artifacts are to be kept in mind during examination, including red (soft) lines around calcification, mosaic color that may appear during cyst evaluation, and artifacts occurring in lesions adjacent to large vessels, particularly when pulsatile [135].

## 12. Conclusion

SEL represents the most important achievement in US following Doppler imaging. It has enabled the evaluation of a tissue parameter, the tissue stiffness, that was not previously measurable by imaging modalities. Tissue stiffness as measured with SEL can be used to evaluate multiple tissues and organ systems in the body, many of which gained wide acceptance for the characterization of tissue pathology and disease processes. Both compressive SEL and SWE have been successfully applied in the MSK system, with SWE providing the additional advantage of quantitative analysis.

## References


[1] J. Ophir, et al., Elastography: a quantitative method for imaging the elasticity of biological tissues, Ultrason. Imaging 13 (2) (1991) 111–134.

[2] A. Pesavento, et al., A time-efficient and accurate strain estimation concept for ultrasonic elastography using iterative phase zero estimation, IEEE Trans. Ultrason. Ferroelectr. Freq. Control 46 (5) (1999) 1057–1067.

[3] A. Guibal, et al., Evaluation of shearwave elastography for the characterisation of focal liver lesions on ultrasound, Eur. Radiol. 23 (4) (2013) 1138–1149.

[4] A.E. Bohte, et al., Non-invasive evaluation of liver fibrosis: a comparison of ultrasound-based transient elastography and MR elastography in patients with viral hepatitis B and C, Eur. Radiol. 24 (3) (2014) 638–648.

[5] M. Ghajarzadeh, F. Sodagari, M. Shakiba, Diagnostic accuracy of sonoelastography in detecting malignant thyroid nodules: a systematic review and meta-analysis, Am. J. Roentgenol. 202 (4) (2014) W379–W389.

[6] A. Evans, et al., Invasive breast cancer: relationship between shear-wave elastographic findings and histologic prognostic factors, Radiology 263 (3) (2012) 673–677.

[7] Y. Zhang, et al., Differentiation of prostate cancer from benign lesions using strain index of transrectal real-time tissue elastography, Eur. J. Radiol. 81 (5) (2012) 857–862.

[8] A.S. Klauser, et al., Sonoelastography: musculoskeletal applications, Radiology 272 (3) (2014) 622–633.

[9] T. De Zordo, et al., Real-time sonoelastography: findings in patients with symptomatic Achilles tendons and comparison to healthy volunteers, Ultraschall der Med. 31 (4) (2010) 394–400.

[10] T. De Zordo, et al., Real-time sonoelastography findings in healthy Achilles tendons, Am. J. Roentgenol. 193 (2) (2009) W134–W138.

[11] T. De Zordo, et al., Real-time sonoelastography of lateral epicondylitis: comparison of findings between patients and healthy volunteers, Am. J. Roentgenol. 193 (1) (2009) 180–185.
[12] A.S. Klauser, R. Faschingbauer, W.R. Jaschke, Is sonoelastography of value in assessing tendons? Semin. Musculoskelet. Radiol. 14 (3) (2010) 323–333.
[13] A.S. Klauser, et al., Achilles tendon assessed with sonoelastography: histologic agreement, Radiology 267 (3) (2013) 837–842.
[14] H. Miyamoto, et al., Carpal tunnel syndrome: diagnosis by means of median nerve elasticity–improved diagnostic accuracy of US with sonoelastography, Radiology 270 (2) (2014) 481–486.
[15] C. Botar-Jid, et al., The contribution of ultrasonography and sonoelastography in assessment of myositis, Med. Ultrason. 12 (2) (2010) 120–126.
[16] C. Botar Jid, et al., Musculoskeletal sonoelastography. Pictorial essay, Med. Ultrason. 14 (3) (2012) 239–245.
[17] S. Aubry, et al., Viscoelasticity in Achilles tendonopathy: quantitative assessment by using real-time shear-wave elastography, Radiology 274 (3) (2015) 821–829.
[18] X.M. Chen, et al., Shear wave elastographic characterization of normal and torn achilles tendons: a pilot study, J. Ultrasound Med. 32 (3) (2013) 449–455.
[19] D. Cosgrove, et al., EFSUMB guidelines and recommendations on the clinical use of ultrasound elastography. Part 2: clinical applications, Ultraschall der Med. 34 (3) (2013) 238–253.
[20] Y. Itoigawa, et al., Feasibility assessment of shear wave elastography to rotator cuff muscle, Clin. Anat. 28 (2) (2015) 213–218.
[21] J. Bercoff, M. Tanter, M. Fink, Supersonic shear imaging: a new technique for soft tissue elasticity mapping, IEEE Trans. Ultrason. Ferroelectr. Freq. Control 51 (4) (2004) 396–409.
[22] L. Sandrin, et al., Transient elastography: a new noninvasive method for assessment of hepatic fibrosis, Ultrasound Med. Biol. 29 (12) (2003) 1705–1713.
[23] J. Bercoff, et al., In vivo breast tumor detection using transient elastography, Ultrasound Med. Biol. 29 (10) (2003) 1387–1396.
[24] J.H. Weinreb, et al., Tendon structure, disease, and imaging, Muscles Ligaments Tendons J. 4 (1) (2014) 66–73.
[25] L.M. Sconfienza, et al., Clinical indications for musculoskeletal ultrasound updated in 2017 by European Society of Musculoskeletal Radiology (ESSR) consensus, Eur. Radiol. 28 (12) (2018) 5338–5351.
[26] A.A. Schepsis, H. Jones, A.L. Haas, Achilles tendon disorders in athletes, Am. J. Sports Med. 30 (2) (2002) 287–305.
[27] N. Maffulli, P. Sharma, K.L. Luscombe, Achilles tendinopathy: aetiology and management, J. R. Soc. Med. 97 (10) (2004) 472–476.
[28] C.N. van Dijk, et al., Terminology for Achilles tendon related disorders, Knee Surg. Sport. Traumatol. Arthrosc. 19 (5) (2011) 835–841.
[29] M. Paavola, et al., Achilles tendinopathy, J. Bone Jt. Surg. Am. 84-A (11) (2002) 2062–2076.
[30] G. Riley, The pathogenesis of tendinopathy. A molecular perspective, Rheumatology 43 (2) (2004) 131–142.
[31] M. Astrom, A. Rausing, Chronic Achilles tendinopathy. A survey of surgical and histopathologic findings, Clin. Orthop. Relat. Res. (316) (1995) 151–164.

[32] F. Kainberger, et al., Imaging of tendons—adaptation, degeneration, rupture, Eur. J. Radiol. 25 (3) (1997) 209–222.
[33] P. Kannus, L. Jozsa, Histopathological changes preceding spontaneous rupture of a tendon. A controlled study of 891 patients, J. Bone Jt. Surg. Am. 73 (10) (1991) 1507–1525.
[34] E.E. Drakonaki, G.M. Allen, D.J. Wilson, Real-time ultrasound elastography of the normal Achilles tendon: reproducibility and pattern description, Clin. Radiol. 64 (12) (2009) 1196–1202.
[35] P. Lalitha, M. Reddy, K.J. Reddy, Musculoskeletal applications of elastography: a pictorial essay of our initial experience, Korean J. Radiol. 12 (3) (2011) 365–375.
[36] S. Tan, et al., Real-time sonoelastography of the Achilles tendon: pattern description in healthy subjects and patients with surgically repaired complete ruptures, Skelet. Radiol. 41 (9) (2012) 1067–1072.
[37] R.J. DeWall, et al., Spatial variations in Achilles tendon shear wave speed, J. Biomech. 47 (11) (2014) 2685–2692.
[38] Z. Ruan, et al., Elasticity of healthy Achilles tendon decreases with the increase of age as determined by acoustic radiation force impulse imaging, Int. J. Clin. Exp. Med. 8 (1) (2015) 1043–1050.
[39] L.C. Slane, et al., Quantitative ultrasound mapping of regional variations in shear wave speeds of the aging achilles tendon, Eur. Radiol. 27 (2) (2016) 474–482.
[40] W.L. Siu, et al., Sonographic evaluation of the effect of long-term exercise on achilles tendon stiffness using shear wave elastography, J. Sci. Med. Sport 19 (11) (2016) 883–887.
[41] R. Shiri, et al., Prevalence and determinants of lateral and medial epicondylitis: a population study, Am. J. Epidemiol. 164 (11) (2006) 1065–1074.
[42] F. Faro, J.M. Wolf, Lateral epicondylitis: review and current concepts, J. Hand Surg. Am. 32 (8) (2007) 1271–1279.
[43] R. Tosti, J. Jennings, J.M. Sewards, Lateral epicondylitis of the elbow, Am. J. Med. 126 (4) (2013) 357 e1–6.
[44] D. Levin, et al., Lateral epicondylitis of the elbow: US findings, Radiology 237 (1) (2005) 230–234.
[45] K.S. Ahn, et al., Ultrasound elastography of lateral epicondylosis: clinical feasibility of quantitative elastographic measurements, Am. J. Roentgenol. 202 (5) (2014) 1094–1099.
[46] V.C. Dones 3rd, et al., The diagnostic validity of musculoskeletal ultrasound in lateral epicondylalgia: a systematic review, BMC Med. Imaging 14 (2014) 10.
[47] A.S. Klauser, et al., Extensor tendinopathy of the elbow assessed with sonoelastography: histologic correlation, Eur. Radiol. 27 (8) (2017) 3460–3466.
[48] F. Kocyigit, et al., Association of real-time sonoelastography findings with clinical parameters in lateral epicondylitis, Rheumatol. Int. 36 (1) (2016) 91–100.
[49] M. Lasecki, et al., The snapping elbow syndrome as a reason for chronic elbow neuralgia in a tennis player – MR, US and sonoelastography evaluation, Pol. J. Radiol. 79 (2014) 467–471.
[50] A.S. Klauser, et al., Sonoelastography of the common flexor tendon of the elbow with histologic agreement: a cadaveric study, Radiology 283 (2) (2017) 486–491.
[51] E. Witvrouw, et al., Intrinsic risk factors for the development of patellar tendinitis in an athletic population. A two-year prospective study, Am. J. Sports Med. 29 (2) (2001) 190–195.

[52] O.B. Lian, L. Engebretsen, R. Bahr, Prevalence of jumper's knee among elite athletes from different sports: a cross-sectional study, Am. J. Sports Med. 33 (4) (2005) 561–567.
[53] K.M. Khan, et al., Patellar tendinopathy: some aspects of basic science and clinical management, Br. J. Sports Med. 32 (4) (1998) 346–355.
[54] M.E. Blazina, et al., Jumper's knee, Orthop. Clin. N. Am. 4 (3) (1973) 665–678.
[55] F. Porta, et al., Ultrasound elastography is a reproducible and feasible tool for the evaluation of the patellar tendon in healthy subjects, Int. J. Rheum. Dis. 17 (7) (2014) 762–766.
[56] S. Akkaya, et al., Real-time elastography of patellar tendon in patients with auto-graft bone-tendon-bone anterior cruciate ligament reconstruction, Arch. Orthop. Trauma. Surg. 136 (6) (2016) 837–842.
[57] C. Helland, et al., Mechanical properties of the patellar tendon in elite volleyball players with and without patellar tendinopathy, Br. J. Sports Med. 47 (13) (2013) 862–868.
[58] M. Kongsgaard, et al., Fibril morphology and tendon mechanical properties in patellar tendinopathy: effects of heavy slow resistance training, Am. J. Sports Med. 38 (4) (2010) 749–756.
[59] C. Couppe, et al., Differences in tendon properties in elite badminton players with or without patellar tendinopathy, Scand. J. Med. Sci. Sport. 23 (2) (2013) e89–95.
[60] Z.J. Zhang, et al., Changes in morphological and elastic properties of patellar tendon in athletes with unilateral patellar tendinopathy and their relationships with pain and functional disability, PLoS One 9 (10) (2014) e108337.
[61] M.Y. Hsiao, et al., Reduced patellar tendon elasticity with aging: in vivo assessment by shear wave elastography, Ultrasound Med. Biol. 41 (11) (2015) 2899–2905.
[62] S.G. Abram, A.D. Sharma, C. Arvind, Atraumatic quadriceps tendon tear associated with calcific tendonitis, BMJ Case Rep. 2012 (2012).
[63] L. Ventura-Rios, et al., Tendon involvement in patients with gout: an ultrasound study of prevalence, Clin. Rheumatol. 35 (8) (2016) 2039–2044.
[64] M.A. Teber, et al., Real-time sonoelastography of the quadriceps tendon in patients undergoing chronic hemodialysis, J. Ultrasound Med. 34 (4) (2015) 671–677.
[65] B.C. Kot, et al., Elastic modulus of muscle and tendon with shear wave ultrasound elastography: variations with different technical settings, PLoS One 7 (8) (2012) e44348.
[66] M.D. Chard, et al., Shoulder disorders in the elderly: a community survey, Arthritis Rheum. 34 (6) (1991) 766–769.
[67] P. Vecchio, et al., Shoulder pain in a community-based rheumatology clinic, Br. J. Rheumatol. 34 (5) (1995) 440–442.
[68] J.P. Iannotti, Full-thickness rotator cuff tears: factors affecting surgical outcome, J. Am. Acad. Orthop. Surg. 2 (2) (1994) 87–95.
[69] C. Milgrom, et al., Rotator-cuff changes in asymptomatic adults. The effect of age, hand dominance and gender, J. Bone Jt. Surg. Br. 77 (2) (1995) 296–298.
[70] S.U. Lee, et al., Real-time sonoelastography in the diagnosis of rotator cuff tendinopathy, J. Shoulder Elb. Surg. 25 (5) (2016) 723–729.
[71] E. Silvestri, et al., Sonoelastography can help in the localization of soft tissue damage in polymyalgia rheumatica (PMR), Clin. Exp. Rheumatol. 25 (5) (2007) 796.
[72] J.B. Seo, J.S. Yoo, J.W. Ryu, Sonoelastography findings of supraspinatus tendon in rotator cuff tendinopathy without tear: comparison with magnetic resonance images and conventional ultrasonography, J. Ultrasound 18 (2) (2015) 143–149.

[73] J.B. Seo, J.S. Yoo, J.W. Ryu, Sonoelastography findings of biceps tendinitis and tendinosis, J. Ultrasound 17 (4) (2014) 271–277.
[74] O. Boughebri, et al., Small supraspinatus tears repaired by arthroscopy: are clinical results influenced by the integrity of the cuff after two years? Functional and anatomic results of forty-six consecutive cases, J. Shoulder Elb. Surg. 21 (5) (2012) 699–706.
[75] C. Charousset, et al., Arthroscopic repair of full-thickness rotator cuff tears: is there tendon healing in patients aged 65 years or older? Arthroscopy 26 (3) (2010) 302–309.
[76] C. Chillemi, et al., Rotator cuff re-tear or non-healing: histopathological aspects and predictive factors, Knee Surg. Sport. Traumatol. Arthrosc. 19 (9) (2011) 1588–1596.
[77] J.L. Ellera Gomes, et al., Conventional rotator cuff repair complemented by the aid of mononuclear autologous stem cells, Knee Surg. Sport. Traumatol. Arthrosc. 20 (2) (2012) 373–377.
[78] C. Tudisco, et al., Tendon quality in small unilateral supraspinatus tendon tears. Real-time sonoelastography correlates with clinical findings, Knee Surg. Sport. Traumatol. Arthrosc. 23 (2) (2015) 393–398.
[79] A. Sungpet, C. Suphachatwong, V. Kawinwonggowit, Trigger digit and BMI, J. Med. Assoc. Thail. 82 (10) (1999) 1025–1027.
[80] L. Strom, Trigger finger in diabetes, J. Med. Soc. N. J. 74 (11) (1977) 951–954.
[81] H. Miyamoto, et al., Stiffness of the first annular pulley in normal and trigger fingers, J. Hand Surg. Am. 36 (9) (2011) 1486–1491.
[82] M.W. Keith, et al., American academy of orthopaedic surgeons clinical practice guideline on diagnosis of carpal tunnel syndrome, J. Bone Jt. Surg. Am. 91 (10) (2009) 2478–2479.
[83] A.S. Klauser, et al., Sonoelastographic findings of carpal tunnel injection, Ultraschall der Med. 36 (6) (2015) 618–622.
[84] H.F. Yang, et al., An anatomic and histologic study of the coracohumeral ligament, J. Shoulder Elb. Surg. 18 (2) (2009) 305–310.
[85] C.S. Neer 2nd, et al., The anatomy and potential effects of contracture of the coracohumeral ligament, Clin. Orthop. Relat. Res. (280) (1992) 182–185.
[86] C.H. Wu, W.S. Chen, T.G. Wang, Elasticity of the coracohumeral ligament in patients with adhesive capsulitis of the shoulder, Radiology 278 (2) (2016) 458–464.
[87] S.L. Woo, J.A. Buckwalter, AAOS/NIH/ORS workshop. Injury and repair of the musculoskeletal soft tissues. Savannah, Georgia, June 18–20, 1987, J. Orthop. Res. 6 (6) (1988) 907–931.
[88] X. Wei, K. Messner, The postnatal development of the insertions of the medial collateral ligament in the rat knee, Anat. Embryol. 193 (1) (1996) 53–59.
[89] J.P. Spalazzi, et al., Elastographic imaging of strain distribution in the anterior cruciate ligament and at the ligament-bone insertions, J. Orthop. Res. 24 (10) (2006) 2001–2010.
[90] E.E. Drakonaki, G.M. Allen, Magnetic resonance imaging, ultrasound and real-time ultrasound elastography of the thigh muscles in congenital muscle dystrophy, Skelet. Radiol. 39 (4) (2010) 391–396.
[91] M. Niitsu, et al., Muscle hardness measurement by using ultrasound elastography: a feasibility study, Acta Radiol. 52 (1) (2011) 99–105.
[92] D. Vasilescu, et al., Sonoelastography contribution in cerebral palsy spasticity treatment assessment, preliminary report: a systematic review of the literature apropos of seven patients, Med. Ultrason. 12 (4) (2010) 306–310.

[93] J.K. Bredenkamp, et al., Congenital muscular torticollis. A spectrum of disease, Arch. Otolaryngol. Head Neck Surg. 116 (2) (1990) 212–216.
[94] B.C. Ho, E.H. Lee, K. Singh, Epidemiology, presentation and management of congenital muscular torticollis, Singap. Med. J. 40 (11) (1999) 675–679.
[95] H.J. Park, et al., Assessment of follow-up sonography and clinical improvement among infants with congenital muscular torticollis, Am. J. Neuroradiol. 34 (4) (2013) 890–894.
[96] I. Dudkiewicz, A. Ganel, A. Blankstein, Congenital muscular torticollis in infants: ultrasound-assisted diagnosis and evaluation, J. Pediatr. Orthop. 25 (6) (2005) 812–814.
[97] S.Y. Lee, et al., Value of adding sonoelastography to conventional ultrasound in patients with congenital muscular torticollis, Pediatr. Radiol. 43 (12) (2013) 1566–1572.
[98] D. Goutallier, et al., Impact of fatty degeneration of the suparspinatus and infraspinatus msucles on the prognosis of surgical repair of the rotator cuff, Rev. Chir. Orthop. Reparatrice. Appar. Mot. 85 (7) (1999) 668–676.
[99] D. Goutallier, et al., Influence of cuff muscle fatty degeneration on anatomic and functional outcomes after simple suture of full-thickness tears, J. Shoulder Elb. Surg. 12 (6) (2003) 550–554.
[100] A.B. Rosskopf, et al., Quantitative shear-wave US elastography of the supraspinatus muscle: reliability of the method and relation to tendon integrity and muscle quality, Radiology 278 (2) (2016) 465–474.
[101] M. Shinohara, et al., Real-time visualization of muscle stiffness distribution with ultrasound shear wave imaging during muscle contraction, Muscle Nerve 42 (3) (2010) 438–441.
[102] L. Lacourpaille, et al., Non-invasive assessment of muscle stiffness in patients with Duchenne muscular dystrophy, Muscle Nerve 51 (2) (2015) 284–286.
[103] F. Kantarci, et al., Median nerve stiffness measurement by shear wave elastography: a potential sonographic method in the diagnosis of carpal tunnel syndrome, Eur. Radiol. 24 (2) (2014) 434–440.
[104] M.L. Palmeri, et al., On the feasibility of imaging peripheral nerves using acoustic radiation force impulse imaging, Ultrason. Imaging 31 (3) (2009) 172–182.
[105] R.J. Andrade, et al., Non-invasive assessment of sciatic nerve stiffness during human ankle motion using ultrasound shear wave elastography, J. Biomech. 49 (3) (2016) 326–331.
[106] M. DeMaio, et al., Plantar fasciitis, Orthopedics 16 (10) (1993) 1153–1163.
[107] E.G. McNally, S. Shetty, Plantar fascia: imaging diagnosis and guided treatment, Semin. Musculoskelet. Radiol. 14 (3) (2010) 334–343.
[108] D.J. Theodorou, et al., Plantar fasciitis and fascial rupture: MR imaging findings in 26 patients supplemented with anatomic data in cadavers, RadioGraphics 20 (2000) S181–S197. Spec No.
[109] A.S. Klauser, et al., Clinical indications for musculoskeletal ultrasound: a delphi-based consensus paper of the European society of musculoskeletal radiology, Eur. Radiol. 22 (5) (2012) 1140–1148.
[110] C.H. Wu, et al., Sonoelastography of the plantar fascia, Radiology 259 (2) (2011) 502–507.
[111] L. Zhang, et al., Assessment of plantar fasciitis using shear wave elastography, Nan Fang Yi Ke Da Xue Xue Bao 34 (2) (2014) 206–209.

[112] J. Rios-Diaz, et al., Sonoelastography of plantar fascia: reproducibility and pattern description in healthy subjects and symptomatic subjects, Ultrasound Med. Biol. 41 (10) (2015) 2605–2613.
[113] L.M. Sconfienza, et al., Real-time sonoelastography of the plantar fascia: comparison between patients with plantar fasciitis and healthy control subjects, Radiology 267 (1) (2013) 195–200.
[114] S.Y. Lee, et al., Ultrasound elastography in the early diagnosis of plantar fasciitis, Clin. Imaging 38 (5) (2014) 715–718.
[115] C.H. Wu, et al., Can sonoelastography detect plantar fasciitis earlier than traditional B-mode ultrasonography? Am. J. Phys. Med. Rehabil. 91 (2) (2012) 185.
[116] M. Kim, et al., Sonoelastography in the evaluation of plantar fasciitis treatment: 3-month follow-up after collagen injection, Ultrasound Q. 32 (4) (2016) 327–332.
[117] A. Lyshchik, et al., Thyroid gland tumor diagnosis at US elastography, Radiology 237 (1) (2005) 202–211.
[118] S. Wojcinski, et al., Acoustic radiation force impulse imaging with virtual touch tissue quantification: measurements of normal breast tissue and dependence on the degree of pre-compression, Ultrasound Med. Biol. 39 (12) (2013) 2226–2232.
[119] L.N. Nazarian, Science to practice: can sonoelastography enable reliable differentiation between benign and metastatic cervical lymph nodes? Radiology 243 (1) (2007) 1–2.
[120] F. Aigner, et al., Real-time sonoelastography for the evaluation of testicular lesions, Radiology 263 (2) (2012) 584–589.
[121] R.G. Barr, et al., WFUMB guidelines and recommendations for clinical use of ultrasound elastography: Part 2: breast, Ultrasound Med. Biol. 41 (5) (2015) 1148–1160.
[122] V. Cantisani, et al., Elastographic and contrast-enhanced ultrasound features of a benign schwannoma of the common fibular nerve, J. Ultrasound 16 (3) (2013) 135–138.
[123] N. Magarelli, et al., Sonoelastography for qualitative and quantitative evaluation of superficial soft tissue lesions: a feasibility study, Eur. Radiol. 24 (3) (2014) 566–573.
[124] I. Riishede, et al., Strain elastography for prediction of malignancy in soft tissue tumours–preliminary results, Ultraschall der Med. 36 (4) (2015) 369–374.
[125] Y.H. Lee, H.T. Song, J.S. Suh, Use of strain ratio in evaluating superficial soft tissue tumors on ultrasonic elastography, J. Med. Ultrason. 41 (3) (2014) 319–323.
[126] B. Pass, et al., Do quantitative and qualitative shear wave elastography have a role in evaluating musculoskeletal soft tissue masses? Eur. Radiol. 27 (2) (2016) 723–731.
[127] A.S. Klauser, et al., Fusion of real-time US with CT images to guide sacroiliac joint injection in vitro and in vivo, Radiology 256 (2) (2010) 547–553.
[128] B.F. Kennedy, et al., In vivo three-dimensional optical coherence elastography, Opt. Express 19 (7) (2011) 6623–6634.
[129] V. Egorov, S. Ayrapetyan, A.P. Sarvazyan, Prostate mechanical imaging: 3-D image composition and feature calculations, IEEE Trans. Med. Imaging 25 (10) (2006) 1329–1340.
[130] A. Sayed, et al., Nonlinear characterization of breast cancer using multi-compression 3D ultrasound elastography in vivo, Ultrasonics 53 (5) (2013) 979–991.
[131] J.J. Ou, et al., Evaluation of 3D modality-independent elastography for breast imaging: a simulation study, Phys. Med. Biol. 53 (1) (2008) 147–163.
[132] L.S. Taylor, et al., Prostate cancer: three-dimensional sonoelastography for in vitro detection, Radiology 237 (3) (2005) 981–985.

[133] H. Ponnekanti, et al., Fundamental mechanical limitations on the visualization of elasticity contrast in elastography, Ultrasound Med. Biol. 21 (4) (1995) 533–543.
[134] J. Ophir, et al., Elastography: imaging the elastic properties of soft tissues with ultrasound, J. Med. Ultrason. 29 (4) (2002) 155.
[135] L. Paluch, et al., Use of ultrasound elastography in the assessment of the musculoskeletal system, Pol. J. Radiol. 81 (2016) 240–246.

# Index

'*Note*: Page numbers followed by "f" indicate figures and "t" indicate tables.'


**A**

A fib. *See* Atrial fibrillation (A fib)
Abdominal aortic aneurysm (AAA), 89, 94
AC. *See* Acoustic coupler (AC)
Achilles tendon, 200–202, 201f
ACL. *See* Anterior cruciate ligament (ACL)
Acoustic coupler (AC), 207–208
Acoustic radiation force impulse imaging (ARFI imaging), 2, 23, 60–61, 111, 114–115, 143–145, 157–158
  excitation by acoustic radiation force, 57
ACT sequence. *See* Automated composite technique sequence (ACT sequence)
Activation maps, 79–83
Acute myocardial infarction, 67
AF. *See* Atrial fibrillation (A fib)
AFL. *See* Atrial flutter (AFL)
Age-standardized rate (ASR), 47
Alcohol abuse, 109
Alcoholic liver disease, 109
Alcoholic steatohepatitis (ASH), 110
Anechoic hematoma, 35–36
Angioplasty, 67
Anisotropy, 10–11
Anterior cruciate ligament (ACL), 205, 210
Area under receiver operating characteristic curve (AUROC curve), 113, 115, 171
ARFI imaging. *See* Acoustic radiation force impulse imaging (ARFI imaging)
Arrhythmogenic tissue, 138
Artifacts, 42–43
  bang artifact, 43
  BGR artifact, 43
  Bull's-eye artifact, 42–43
  sliding artifact, 43
  in thyroid elastography, 172–174
  worm artifact, 43
ASH. *See* Alcoholic steatohepatitis (ASH)
ASR. *See* Age-standardized rate (ASR)
Atrial arrhythmias, 76
  characterization in canines in vivo, 85
  clinical diagnosis, 76
  treatment, 76–77
Atrial fibrillation (A fib), 76
Atrial flutter (AFL), 76
Atrial tachycardia mechanisms, 77–78
AUROC curve. *See* Area under receiver operating characteristic curve (AUROC curve)
Autoimmune hepatitis, 109
Automated composite technique sequence (ACT sequence), 79–83

**B**

B-mode image, 23, 57
Back pain, 139
Bang artifact, 43
Barankin bound (BB), 84–85
Batts-Ludwig system, 14
BB. *See* Barankin bound (BB)
Beamformed signals, 71–72
Benign lesions, 29–36. *See also* Malignant lesions
  cysts, 29–30
  fat necrosis, 34
  fibroadenomas, 30
  fibrocystic change and related pathology, 30
  hematoma, 35–36
  mastitis, 34
  papillary lesions, 32–34
Benign prostatic hyperplasia (BPH), 57
Bethlem myopathy, 210
BGR artifact. *See* Blue, green, and red artifact (BGR artifact)
BI-RADS. *See* Breast Imaging Reporting and Data System (BI-RADS)
Bicubic Hermite interpolation, 92
Blue, green, and red artifact (BGR artifact), 29–30, 43
Body mass index (BMI), 113–114
Bone tumor ablation, 138
BPH. *See* Benign prostatic hyperplasia (BPH)
Bramwell-Hill equation, 92
Breast cancer, 137
Breast cancers, 6, 28
Breast elastography, 21. *See also* Cardiovascular elastography
  artifacts and limitations, 42–43
  diseases and applications, 29–41
  opportunities, 41–42
  principles/techniques, 21–29
    guidelines, 29
    SE, 21–27
    SWE, 27–29

Breast Imaging Reporting and Data System (BI-RADS), 26
Bull's-eye artifact, 29–30, 42–43, 214–215

## C

CAD. *See* Coronary artery disease (CAD)
Calcifications, 173–174
CAP. *See* Controlled attenuation parameter (CAP)
Cardiac ablation, 138–139, 145–147
Cardiac arrhythmias, 76
Cardiac imaging, 67–87. *See also* Vascular imaging
  electromechanical wave imaging, 76–87
  myocardial elastography, 67–76
Cardiac mapping, 76–78
Cardiovascular elastography. *See also* Breast elastography
  cardiac imaging, 67–87
  vascular imaging, 88–95
Carotid endarterectomy (CEA), 88
Carotid plaques, 95
Carpal tunnel syndrome (CTS), 207–208
CAT. *See* Chronic autoimmune thyroiditis (CAT)
CCI. *See* Cervical consistency index (CCI)
CDI. *See* Color-Doppler Imaging (CDI)
CDUS. *See* Color Doppler ultrasonography (CDUS)
CEA. *See* Carotid endarterectomy (CEA)
Central gland (CG), 56–57
Central papillomas, 32–33
Cerebral palsy spasticity, 211
Cervical consistency index (CCI), 184
Cervix, 181–188
  SE, 182–185
  SWEI, 185–188
CG. *See* Central gland (CG)
Cholestatic liver disease, 109
Chronic autoimmune thyroiditis (CAT), 171
Chronic hematoma, 35–36
Chronic hepatopathies, 113, 117, 123
Chronic liver disease (CVD), 109–112
Chronic viral hepatitis, 109
"Claim confirmation" stage, 15
Clinical elasticity estimation and imaging, 1
  clinical applications of strain and SWE, 6–8
  elastography, 2–6, 2f
    applications, 9–15
  future directions, 15–16
CMT. *See* Congenital muscular torticollis (CMT)
Collagen synthesis, 67
Color Doppler ultrasonography (CDUS), 9, 51, 203–204
Color elastograms, 210
Color image, 157–158
Color-based scoring system, 158–160
Color-Doppler Imaging (CDI), 60
Colorectal cancer, 137
Columbia University Medical Center (CUMC), 74–76
Compliance mapping, 92
Compression, 7
  elastography—Strain elastography (SE)
  SEL, 206–207, 211, 214
Computational models, 70–71
Computed tomography (CT), 76, 144, 157
Computed tomography angiography (CTA), 73–74
  validation of myocardial elastography against, 73–74
Congenital muscular disorders, 210
Congenital muscular torticollis (CMT), 211
Controlled attenuation parameter (CAP), 114
Coracohumeral ligament, 208
Coronary artery disease (CAD), 67, 73–74
Cosmetic surgery, 139
Cramér-Rao Lower Bound (CRLB), 84–85
Cross-correlation, 68–70
CT. *See* Computed tomography (CT)
CTA. *See* Computed tomography angiography (CTA)
CTS. *See* Carpal tunnel syndrome (CTS)
CUMC. *See* Columbia University Medical Center (CUMC)
CVD. *See* Chronic liver disease (CVD)
Cysts, 29–30

## D

DCIS. *See* Ductal carcinoma in situ (DCIS)
Delaunay interpolation, 79–83
Depth dependence, 13–14
Digital rectal examination (DRE), 47
DMD. *See* Duchenne muscular dystrophy (DMD)
Doppler myocardial imaging (DMI), 68
Doppler ultrasound, 3
DRE. *See* Digital rectal examination (DRE)
Duchenne muscular dystrophy (DMD), 212
Ductal carcinoma in situ (DCIS), 23, 36–37
Ductal epithelial cells, 32
Ductography, 32–33
Dynamic deformation exerted by external mechanical vibrators, 52–57
  MRE, 55–57
  sonoelastography, 52–54
  VE, 54–55

Dynamic elastography, 2
Dynamic programming algorithm, 91–92
Dysfunctional placenta, 188

## E

E/B ratio. *See* Elastographic to B-mode length ratio (E/B ratio)
ECG recordings. *See* Electrocardiographic recordings (ECG recordings)
ECGI. *See* Electrocardiographic imaging (ECGI)
Echocardiography, 67–68
ECM. *See* Extracellular matrix (ECM)
EDE. *See* Electrode displacement elastography (EDE)
EFSUMB. *See* European Federation of Societies for Ultrasound in Medicine and Biology (EFSUMB)
EI. *See* Elasticity index (EI)
Elastic modulus, 48
Elasticity
  contrast index, 160
  imaging, 181, 192
  moduli, 53–54
Elasticity index (EI), 170, 182–183, 188
Elastogram, 123–124, 157–158, 204–205
Elastographic to B-mode length ratio (E/B ratio), 23–26
Elastographic/elastography, 2–6, 2f
  applications, 9–15
  imaging, 1, 48
  methods, 61, 110
    for thermal therapy monitoring, 143–144
  principles for thermal therapy monitoring, 140–141
ElastPQ technique, 117–118, 117f–118f
Electrical isochrones, 85
Electrocardiographic imaging (ECGI), 78–79
Electrocardiographic recordings (ECG recordings), 76
Electrode displacement elastography (EDE), 143, 145
Electromechanical wave imaging (EWI), 76–87
  atrial arrhythmias
    characterization in canines in vivo, 85
    clinical diagnosis, 76
    treatment, 76–77
  cardiac arrhythmias, 76
  in normal human subjects and with arrhythmias, 85–87
  sequences, 79–85
  treatment guidance capability, 78–79
Electromechanical wave propagation, 85–87
Elevated temperature-induced bioeffects, 136
European Federation of Societies for Ultrasound in Medicine and Biology (EFSUMB), 14, 29
EWI. *See* Electromechanical wave imaging (EWI)
Examination technique, 111–112
Excitation by acoustic radiation force, 57–60
  ARFI imaging, 57
  SWE, 58–60
Extracellular matrix (ECM), 181

## F

Facet joint arthropathy, 139
Fat necrosis, 34
Fat-to-tendon method (Fat/T method), 207
Fat/T method. *See* Fat-to-tendon method (Fat/T method)
Fetal growth restriction (FGR), 188
Fibroadenomas, 30
Fibrocystic change and related pathology, 30
FibroScan® nonimaging elastography device, 7–8, 14
Field of view (FOV), 22
Fine-needle aspiration (FNA), 157
Finger tendon, 207
FNA. *See* Fine-needle aspiration (FNA)
Focused ultrasound (FUS), 135, 141
FOV. *See* Field of view (FOV)
Free-floating debris, 35–36
Functional brain surgery, 139
FUS. *See* Focused ultrasound (FUS)
Fusion imaging, 215

## G

GD. *See* Graves disease (GD)
Gel pad-to-tendon method (Pad/T method), 207
Generalized linear pressure-area relationship, 92
Gold standard, 110
Graves disease (GD), 171
Gynecomastia, 41

## H

Hanning apodization, 70–71
HBV. *See* Hepatitis B virus (HBV)
HCV. *See* Hepatitis C virus (HCV)
Heart valve surgery, 67
Hematoma, 35–36
Hemochromatosis, 109
Hepatitis B virus (HBV), 109
Hepatitis C virus (HCV), 109
High frame rate ultrasound imaging, 68
High-quality strain elastograms, 4

Histogram analysis of color elastograms, 158–160
Hitachi, pSWE from, 118–119
Homogeneous material, 6
Hook's law, 3–4, 48
Hypertension, 139

**I**

IDC. *See* Invasive ductal cancer (IDC)
Idiopathic inflammatory myopathy, 211
ILC. *See* Invasive lobular cancer (ILC)
Image texture analysis, 158–160
Image-guided Botox therapy, 211
Infarct model, 73
Inflammatory enthesitis, 200
Insufficient compression, 9–10
Interquartile range (IQR), 111–114
Invasive ductal cancer (IDC), 6, 23, 37
Invasive ductal carcinoma. *See* Invasive ductal cancer (IDC)
Invasive lobular cancer (ILC), 37–38
Invasive RF probes, 139
IQR. *See* Interquartile range (IQR)
Ishak score, 110
Isotropic material, 6

**J**

Japan Society of Ultrasonics in Medicine (JSUM), 14
Joints and ligaments
  ACL, 210
  coracohumeral ligament, 208
  transverse carpal ligament, 207–208
Juvenile papillomatosis, 32

**K**

Kelvin-Voigt fractional derivative viscoelastic model, 48
Knodell score/histology activity index, 110

**L**

LAD coronary artery. *See* Left anterior descending coronary artery (LAD coronary artery)
Lateral epicondylitis, 202–204, 204f
Lateral motion, 10
LB. *See* Liver biopsy (LB)
Left anterior descending coronary artery (LAD coronary artery), 71–72
LFI. *See* Liver fibrosis index (LFI)
LHM imaging. *See* Local harmonic motion imaging (LHM imaging)
Liver
  cancer, 137
  disease progression, 110
  elastography, 111–112
Liver biopsy (LB), 110
Liver fibrosis index (LFI), 123–124
Liver stiffness (LS), 110
Liver stiffness measurements (LSMs), 111
Local harmonic motion imaging (LHM imaging), 144–145
LS. *See* Liver stiffness (LS)
LSMs. *See* Liver stiffness measurements (LSMs)
Lymphoma, 38–40

**M**

Magnetic resonance (MR), 56
Magnetic resonance elastography (MRE), 55–57, 110–111
Magnetic resonance imaging (MRI), 50–51, 135, 200
  MRI-thermometry-monitored FUS treatments, 139–140
  thermometry, 146
Malignant lesions, 36–40. *See also* Benign lesions
  DCIS, 36–37
  IDC, 37
  ILC, 37–38
  lymphoma, 38–40
  mucinous cancer, 37
Malignant soft tissue tumors, 214–215
Mammogram, 32–33
Manual compression method, 23
Mastitis, 34
Mean total shear wave velocity (MTSWV), 212
Medial epicondylitis, 204–205
Median nerve (MN), 207–208
  elasticity, 212–213
MER. *See* Motion-estimation rate (MER)
Metastatic diseases of breast, 40–41
METAVIR score, 110
Metavir system, 14
MI. *See* Myocardial infarction (MI)
Minimally invasive thermal ablations, 146
MK equation. *See* Moens-Korteweg equation (MK equation)
MN. *See* Median nerve (MN)
MNSR. *See* Muscle to nodule strain ratio (MNSR)
Moens-Korteweg equation (MK equation), 89
Motion estimation, 68
Motion-estimation rate (MER), 83–84, 93–94

Motion-matching methodology, 79–83
Motion-sampling rate (MSR), 83–84
mpMRI–guided biopsies. *See* Multiparametric MRI–guided biopsies (mpMRI–guided biopsies)
MR. *See* Magnetic resonance (MR)
MRE. *See* Magnetic resonance elastography (MRE)
MRI. *See* Magnetic resonance imaging (MRI)
MSK elastography. *See* Musculoskeletal elastography (MSK elastography)
MSR. *See* Motion-sampling rate (MSR)
MTSWV. *See* Mean total shear wave velocity (MTSWV)
Mucinous cancer, 37
Multiparametric MRI–guided biopsies (mpMRI–guided biopsies), 51–52
Multiple papillomas, 32
Multiple probes, 114–115
Multiple sonication procedures, 143–144
Muscle to nodule strain ratio (MNSR), 160
Muscles, 210–212
Musculoskeletal elastography (MSK elastography), 197
  compression elastography, 198
  future perspectives, 215
  limitations and conditions of good practice
    strain elastography/sonoelastography, 215–216
    SWE, 216–217
  muscles, 210–212
  nerves, 212–213
  plantar fascia, 213–214
  sonoelastography applications, 200–210
  SWE, 199
  transient elastography, 199
  tumor and tumorlike masses, 214–215
Myocardial elastography, 67–76
  computational models, 70–71
  imaging deformation of myocardium, 68
    high frame rate ultrasound imaging, 68
    motion estimation, 68
  importance in clinic, 74–76
  mechanical deformation of normal and ischemic or infracted myocardium, 67
  myocardial ischemia and infarction detection in canines in vivo, 71–73
    infarct model, 73
    ischemic model, 71–72
  phantoms, 71
  validation against CTA, 73–74
Myocardial infarction (MI), 67

## N

NAFLD. *See* Nonalcoholic fatty liver disease (NAFLD)
Negative predictive value (NPV), 51–52, 158
Neighboring sectors, 79–83
Nerves, 212–213
Non-contact methods, 78–79
Nonalcoholic fatty liver disease (NAFLD), 109–110, 113
Noninvasive FUS ablation methods, 143
NPV. *See* Negative predictive value (NPV)
Nuclear perfusion, 67

## O

Open-architecture system, 91–92
Optical imaging techniques, 78–79
Optimal strain estimation, 84–85

## P

Pacemakers, 67
Pad/T method. *See* Gel pad-to-tendon method (Pad/T method)
Papillary lesions, 32–34
Papillomas of breast, 32
Parallel beamforming, 68
  PWI system using, 91–92
Parallel beamforming myocardial elastography (PBME), 70–71
Parenchyma to nodule strain ratio (PNSR), 160
Patellar tendinopathy, 205–206
2-Pattern scoring system, 158
PBME. *See* Parallel beamforming myocardial elastography (PBME)
PCa. *See* Prostate cancer (PCa)
PCa antigen 3 (PCA3), 52
  biomarker detection, 60
PCA3. *See* PCa antigen 3 (PCA3)
PDUS. *See* Power Doppler ultrasonography (PDUS)
Phantoms, 71
Phased array simulation model, 70–71
Piecewise PWI (pPWI), 93–94
Placenta, 181, 188–192
  elasticity, 191
  SE, 188
  shear wave methods, 189–192
Plantar fascia, 213–214
Plantar fasciitis, 213
PNSR. *See* Parenchyma to nodule strain ratio (PNSR)
5-Point color scale, 25
4-Point grayscale scoring system, 158–160

Point shear wave elastography (pSWE), 28, 111, 114–119, 166, 169
ElastPQ technique, 117–118, 117f–118f
from Hitachi, 118–119
VTQ, 111, 115–116, 116f
Poisson's ratio, 7
Polyvinyl alcohol (PVA), 92–93
Positive predictive value (PPV), 158
Power Doppler ultrasonography (PDUS), 51
PPV. *See* Positive predictive value (PPV)
pPWI. *See* Piecewise PWI (pPWI)
Precompression, 9, 11–13, 22
Pregnancy, elastography applications in
cervix, 181–188
placenta, 188–192
tissues, 181
Pressure-flow curves, 71–72
Probe-based thermal therapy methods, 143
Prostate cancer (PCa), 47
current status and future trends, 60–61
dynamic deformation exerted by external mechanical vibrators, 52–57
excitation by acoustic radiation force, 57–60
experimental ranges of elastic moduli, 49t
static deformation by compression, 49–52
Prostate cancer, 138, 145
Prostate-specific antigen (PSA), 47
PSA. *See* Prostate-specific antigen (PSA)
pSWE. *See* Point shear wave elastography (pSWE)
PubMed, 112
Pulse wave imaging (PWI), 89, 90f. *See also* Electromechanical wave imaging (EWI)
in aortic aneurysms and carotid plaques, 94–95
mechanical testing, 94
performance assessment in experimental phantoms, 92–94
system using parallel beamforming, 91–92
3D, 92
Pulse-wave induced displacement, 89
Pulse-wave velocity (PWV), 89, 95
automated PWV estimation, 91–92
mapping, 91
Pulsed elastography. *See* Transient elastography
PVA. *See* Polyvinyl alcohol (PVA)
PWI. *See* Pulse wave imaging (PWI)
PWV. *See* Pulse-wave velocity (PWV)

## Q

QIBA. *See* Quantitative Imaging Biomarker Alliance (QIBA)
QM. *See* Quality measure (QM)
Quadriceps tendinopathy, 206
Qualitative sonoelasticity map, 52–53
Quality Map, 42
Quality measure (QM), 21, 41–42
Quantitative ARFI, 58
Quantitative elastographic maps, 212
Quantitative Imaging Biomarker Alliance (QIBA), 14
Quantitative imaging elastographic techniques, 48
Quantitative ultrasound (QUS), 61
Quasi-static elastography, 2, 143

## R

Radiation-force-based methods, 143–144
Radical prostatectomy (RP), 51
Radio frequency (RF), 68, 135
ablation, 76
Radiological Society of North America (RSNA), 14
Real-time elastography. *See* Real-time tissue elastography (RTE)
Real-time SEL-based cross-sectional techniques, 197
Real-time tissue elastography (RTE), 111, 172
Real-time ultrasound, 115
Receiver operating characteristic curve (ROC curve), 26, 54–55
Region of interest (ROI), 25–26, 160, 164, 169, 186, 198
Renal denervation, 147
RF. *See* Radio frequency (RF)
ROC curve. *See* Receiver operating characteristic curve (ROC curve)
ROI. *See* Region of interest (ROI)
Rotator cuff tendinopathy, 206–207
RP. *See* Radical prostatectomy (RP)
RSNA. *See* Radiological Society of North America (RSNA)
RTE. *See* Real-time Tissue Elastography (RTE)

## S

SAD. *See* Sum-of-absolute-difference (SAD)
Scar formation, 67
SE. *See* Strain elastography (SE)
Sector-based sequence, 83–84
SEL. *See* Sonoelastography (SEL)
Shear wave (SW), 199
Shear wave absolute vibro-elastography (SWAVE), 190–191
Shear wave elasticity imaging (SWEI), 58, 182, 185–188, 191

Shear wave elastography (SWE), 1, 5, 7–8, 21, 27–29, 58–59, 111, 143–144, 166–172, 197, 199, 199f, 203f, 216–217. *See also* Strain elastography (SE)
clinical applications, 6–8
examination technique, 169
excitation by acoustic radiation force, 58–60
imaging of thyroid, 166
interobserver and intraobserver variabilities, 168
interpretation, 28, 170–172
practical advice and tips, 169–170
review of literature, 28–29, 167
techniques, 27–28
Shear wave imaging (SWI), 145
Shear wave methods, 189–192
Shear wave speed (SWS), 3, 5, 21, 53–54, 110, 166, 185, 189f, 190
Shear wave velocity (SWV), 171, 199
Signal-to-noise-ratio of displacement ($SNR_d$), 93–94
Single-heartbeat EWI, 84–85
Sliding artifact, 43
Society of Radiologists in Ultrasound (SRU), 14
Solitary papillomas, 32
Sonication, 145
Sonoelasticity imaging. *See* Sonoelastography (SEL)
Sonoelastography (SEL), 2, 52–54, 60, 197, 202–203, 211, 215–216
applications in musculoskeletal system, 200–210
Achilles tendon, 200–202, 201f
finger tendon and trigger fingers, 207
joints and ligaments, 207–210
lateral epicondylitis, 202–204, 204f
medial epicondylitis, 204–205
patellar tendinopathy, 205–206
quadriceps tendinopathy, 206
rotator cuff tendinopathy, 206–207
tendon disorders, 200
Sonography, 32–33
Speckle tracking, 68
Spline interpolation, 79–83
Spontaneous preterm birth (sPTB), 181–182
SR. *See* Strain ratio (SR)
SRI. *See* Strain rate imaging (SRI)
SRU. *See* Society of Radiologists in Ultrasound (SRU)
SSI. *See* Supersonic shear imaging (SSI)
Static deformation by compression, 49–52
SE, 49–52
Static elastography, 2
Static pressures, 94
Stenosing tenosynovitis, 207
Stiff. *See* Dysfunctional placenta
Stiffness, 21, 92
Strain, 3–4
histogram, 157–160
imaging, 197
rate, 3
Strain estimation techniques, 68
Strain elastography (SE), 3, 6–7, 9, 21–27, 49–52, 111, 123–124, 123f, 157–166, 182–185, 197–198, 215–216. *See also* Shear wave elastography (SWE)
cervix, 182–185
clinical applications, 6–8
E/B ratio, 23–25
examination technique, 160–164
specificity, sensitivity, NPV, and PPV in nodules, 162t–163t
interobserver and intraobserver variabilities, 164
interpretation, 23
placenta, 188
4-point scoring system for grayscale-based scoring, 160t
5-point color scale, 25
practical advice, tips, and limitations, 164–166
review of literature, 26–27
strain histograms, 158–160
strain ratio, 25–26, 160
techniques, 21–23
Tsukuba 5-point scoring system, 158t
Strain rate imaging (SRI), 8, 68
Strain ratio (SR), 25–26, 160, 198
Stress, 3–4
Stroke, 88
Sum-of-absolute-difference (SAD), 68
Supersonic shear imaging (SSI), 42, 58–59, 119, 143–144
Supersonic shear wave imaging. *See* Supersonic shear imaging (SSI)
Supraspinatus muscle volume atrophy, 212
Surgical scars, recurrence at, 41
SW. *See* Shear wave (SW)
SWAVE. *See* Shear wave absolute vibro-elastography (SWAVE)
SWE. *See* Shear wave elastography (SWE)
SWEI. *See* Shear wave elasticity imaging (SWEI)
SWI. *See* Shear wave imaging (SWI)
SWS. *See* Shear wave speed (SWS)
SWV. *See* Shear wave velocity (SWV)

## T

TE. *See* Transient elastography (TE)
Technical confirmation stage, 15
Temporally-unequispaced acquisition sequences (TUAS), 79, 83–84
Tendinitis, 200
Tendon disorders, 200
Tennis elbow. *See* Lateral epicondylitis
Thermal ablation, 135
Thermal therapy monitoring
  clinical use of thermal exposures, 137–139
    back pain, 139
    cardiac ablation, 138–139
    cosmetic surgery, 139
    functional brain surgery, 139
    hypertension, 139
    tumor ablations, 137–138
  diseases and applications, 144–146
    cardiac ablation, 145–146
    prostate, 145
    tumor treatments, 144–145
  elastographic methods for, 143–144
  elastography principles for, 140–141
  future opportunities, 146–147
  need for exposure monitoring, 139–140
  thermal effects on tissues, 136
Thermometry, 135
Three-dimensional (3D)
  elastograms, 215
  PWI, 92
  segmentation, 54–55
  ultrasound, 68
  wave field of displacement, 56–57
Thyroid
  nodules, 157
  US, 157
Thyroid elastography, 157. *See also* Musculoskeletal elastography (MSK elastography)
  artifacts in, 172–174
  SE, 157–166
  SWE, 166–172
Time delay, 79–83
Time-points, 79–83
Time-shift techniques, 68
Time-varying radiation force, 144
Tissue
  Doppler imaging, 8
  elasticity, 197
  heterogeneity, 10–11
  movement, 11–13
  stiffness, 135, 157–158
Toshiba, 2D-SWE from, 121
"Traffic light" display, 42
Transfer functions, 54–55
Transient elastography (TE), 3, 111–114, 190, 199
  values, 111
Transperineal prostate MRE, 56–57
Transrectal ultrasonography (TRUS), 47, 50–51, 61
Transverse carpal ligament, 207–208
Trauma, 211
Trigger fingers, 207
TRUS. *See* Transrectal ultrasonography (TRUS)
Tsukuba scoring system, 158
TUAS. *See* Temporally-unequispaced acquisition sequences (TUAS)
Tumor
  ablations, 137–138
    bone, 138
    breast, 137
    liver, 137
    malignancies, 138
    prostate, 138
  treatments, 144–145
  and tumorlike masses, 214–215
Two-dimensional (2D)
  matrix array, 92
  speckle tracking, 8
Two-dimensional shear wave elastography (2D-SWE), 28, 111, 119–121, 166
  SSI, 119, 120f
  from Toshiba, 121
  2D-SWE.GE, 119–120, 121f

## U

Ultrasonic correlation-based methods, 166
Ultrasonography (US), 51, 157, 197
Ultrasound, 135
  cardiac ablation catheter, 146, 146f
  elastography, 135, 146, 157
  system displays, 189
  transducer, 216
  ultrasound-based elastographic methods, 110–111
Ultrasound-based liver elastography, 111
  comparative studies, 122–123
  CVD, 109–112
  pSWE, 111, 114–119
  strain elastography, 123–124, 123f
  TE, 111–114
  2D-SWE, 111, 119–121
US. *See* Ultrasonography (US)

## V

Vascular imaging, 88–95. *See also* Cardiac imaging
  AAA, 89
  methods, 90–92
  PWI, 89
  PWV, 89
  stroke, 88
Vascular stiffening, 89
VE. *See* Vibroelastography (VE)
Ventricular arrhythmias, 76. *See also* Atrial arrhythmias
Ventricular bradycardia, 76
Ventricular fibrillation, 76
Ventricular tachycardia, 76
Vibroelastography (VE), 54–55
Virtual Touch imaging (VTI), 167
Virtual touch quantification (VTQ), 111, 115–116, 116f, 167
Voltage-sensitive dyes, 78–79
VTI. *See* Virtual Touch imaging (VTI)
VTQ. *See* Virtual touch quantification (VTQ)

## W

WFUMB. *See* World Federation for Ultrasound in Medicine and Biology (WFUMB)
Windowed voxel-to-voxel technique, 56–57
Wolff-Parkinson-White syndrome (WPW syndrome), 77–78
World Federation for Ultrasound in Medicine and Biology (WFUMB), 29
Worm artifact, 43
WPW syndrome. *See* Wolff-Parkinson-White syndrome (WPW syndrome)

## Y

Young's elasticity modulus, 53–54
Young's modulus, 5–6, 157–158, 197

## Z

Ziv-Zakai Lower Bound (ZZLB), 84–85

Printed in the United States
By Bookmasters